FUNDAMENTAL THEORY OF STRUCTURES

Second Edition

FUNDAMENTAL THEORY

Second Edition

Professor of Civil Engineering, Brigham Young University

Robert E. Krieger Pub. Co.

OF STRUCTURES

D. ALLAN FIRMAGE

Huntington
New York
1980

Original Edition 1963
Second Edition 1980

Printed and Published by
ROBERT E. KRIEGER PUBLISHING COMPANY, INC.
645 NEW YORK AVENUE
HUNTINGTON, NEW YORK 11743

Printed in the United States of America

Library of Congress Cataloging in Publication Data

Firmage, David Allen.
   Fundamental theory of structures.

   Includes index.
   1. Structures, Theory of.
[TA645.F57 1980]   624'.17   79-25213
ISBN 0-88275-443-2

PREFACE

This book is intended as a textbook in the first course in structural engineering. It is assumed that the student has completed courses in basic mechanics, that is, statics, dynamics, and mechanics of materials. The content of this book is developed from the principles of basic mechanics.

The primary purpose is to develop fundamental theories and procedures with a minimum of emphasis on specific applications. The analysis of very specialized structures has been avoided unless a useful principle can be illustrated. The intent is to develop a textbook as a teaching aid and not an encyclopedia of the subject.

The book covers primarily statically determinate structures. The analysis of statically indeterminate structures is treated only for continuous beams in Chapter 4 and approximate solutions for single-story frames in Chapter 7. Chapter 10 covers the deflection of structures, which serves as an introduction to a following course in the analysis of statically indeterminate structures.

In Chapter 3 is a discussion on the loading of common civil engineering-type structures. This material is included so that the student may have some background in the field of design loads. I have had the feeling that too many engineers blindly use design loads from some standard specification without any comprehension of the validity of such loading for his specific design problem. The determination of the proper loading is just as important as a correct analysis.

It will be observed in Chapters 5 and 8 that I have broken with tradition. I have termed the forces in the members of a truss or space frame as *force* and not stress. I believe this change is long overdue. All modern textbooks covering the mechanics of materials define stress as *force per unit area*. I have adhered to this correct definition of stress, thus avoiding confusion to the student and the continued propagation of an error in terminology.

v

The material can be covered in a normal semester of three lectures per week. More problems than will normally be assigned are given at the end of the chapters, thus giving the instructor some choice in the selection of problems.

Thanks are given to those students who offered suggestions while using this book in draft form. Special thanks are given to Mr. Richard H. Chiu for the checking of example problems, and to Professor Robert K. Thomas for his review of the manuscript. Acknowledgment is given to the several typists who worked at various times on the drafts of the manuscript, and also to Mrs. Leslie Morse, who did a major share of the typing.

The following acknowledgments are given for the chapter opening photographs: Chapter 1, Charles P. Cushing; 2, Cities Service; 3, 4, and 5, Bethlehem Steel; 6, Skidmore, Owings & Merrill; 7, J. P. Lohman, Inc.; 8, Benton and Bowles, Inc.; 9, John A. Roebling's Sons; 10, U. S. Air Force.

I am indebted to those organizations and individuals for permission to use photographs and other materials. Finally, I owe many thanks to those individuals, especially my family, who have given me encouragement in the preparation of this book.

D. ALLAN FIRMAGE

Provo, Utah
October 1962

Preface to Second Printing

In the second printing, errors have been corrected and other minor adjustments made. Two example problems have been added to Chapter 3 to help the student understand the use of wind loads and earthquake loads in the analysis of structures. Additional homework problems have been added to Chapters 5 and 8.

In this day of computers in Structural Engineering it is even more important that the student have an understanding of the analysis of basic structures. This is necessary before a person can prepare or use computer programs. Hand calculations will always be used on simple type of structures and are necessary in the training process. Therefore no attempt has been made to introduce any computer analysis in this book. Individual teachers may want to introduce computer programs for the solving of some of the problems.

Provo, Utah   D. ALLAN FIRMAGE
August 1971

PREFACE
Second Edition

The purpose of the enlarged Second Edition was to place under a single cover the basic fundamentals of structural theory and analysis. Chapters have been added covering the classical methods of structural analysis of statically indeterminate structures. In addition the basics of formulation of the matrix methods, stiffness and flexibility, have been included. From this introductory material work could be expanded to introduce more complex problems of larger magnitude wherein work on computer solutions could be utilized.

The basic concepts and operations of plastic analysis of steel structures are presented in Chapter 15. An understanding of this material will prepare the student to proceed into work in the design of steel building frames by plastic design method.

A short chapter is included in the use of SI units in structural analysis. The book is written in "English" units since a large majority of the structural engineering profession still uses the "English" units. The book could easily be used in SI units by simply converting the applied loads in pounds or kips to forces of Newtons and converting linear measurements. The last chapter explains this simple procedure.

Many thanks go to Susan Gardner for her careful and conscientious typing of the new chapters. Thanks also go to the many students who have used and reviewed the text material.

Provo Utah   D. ALLAN FIRMAGE
June 1979

CONTENTS

1

HISTORY OF STRUCTURAL ENGINEERING

1.1 Introduction

Man has been interested in structures for countless ages. His first need was structures for shelter and later public buildings for government and religious purposes. As he began to move about, bridges became a necessity. From ancient times certain men have been responsible for these structures. These men, who held a high position in their community, were undoubtedly above the average in intellect and leadership. How they obtained their training we are not sure; possibly through a form of apprenticeship. They were the structural engineers of the ancient world. They did not employ a knowledge of mechanics of materials, but their profession was based on a knowledge of "rules of thumb" with respect to proper proportion of structural members and construction methods.

The materials available in this dawn of civilization were just stone and wood. Because of their nature, the first structural forms were the post and beam and the arch. In tropical areas, where vines grew long and strong, crude suspension bridges took form.

During the period of the "Old Kingdom" in Egypt, and concurrently in Mesopotamia, great and inspiring structures were built. The greatness of an empire was measured by its buildings. The engineer of that period as he is today was continually meeting new challenges. He was not only called upon to build great buildings, but his talents were used by the men of war who went out to conquer or to repel the adversary.

1.2 Pre-Christian and Roman Period

The birth of science began with the emergence of the Greeks as the leaders of the known world. Previous to the Greeks the Egyptians had developed some mathematics such as plane geometry, but the Greek scholars such as Pythagoras (about 600 B.C.) and later Aristotle (384 to 322 B.C.) and Archimedes (287 to 212 B.C.) gave impetus to the knowledge of science. Archimedes, who is considered the founder of statics, introduced the term "center of gravity." He applied his knowledge to the invention of many machines, both for war and for peace. The ending of Archimedes' life by a Roman soldier seemed to signify the end of the great age of the Greek philosophers.

The leadership of engineering passed to the conquering Romans. Where the Greeks would probably be classed as scientists, the Romans were more nearly engineers. The Greeks were abstract thinkers; the Romans were practical builders. The Romans required great public buildings, roads, and aqueducts. The "greatness" of Rome and her engineers was realized throughout the empire in monumental bridges and aqueducts. The Roman engineer became master of the stone arch as testified by the Ponti di Augusto at Rimini, Italy, and the Pont du Gard at Nimes, France. The Roman engineer was fortunate in having many fine stone masons. Many of the ancient stone arches were built with dressed stones without mortar and exist in this condition to this day. The military engineers accompanying the conquering Roman armies were adept at pile bent bridges and timber arches. Caesar's bridge over the Rhine, built in about ten days, held the record for speed of erection until the Allied armies assaulted this famous waterway in World War II.

1.3 Medieval Period

This period is generally known as a period of the decline of civilization, specifically throughout Europe. No country or group of people predominated during this period. However, progress in science and engineering did not stop altogether. The great structural dome at the St. Sophia Church in Constantinople was constructed in 563 A.D. The Arabic system of numbers was developed about 600 A.D. The St. Benezet Bridge at Avignon, France, the Old London Bridge, and the Ponte Vecchio at Florence, Italy, are all credited to this era. Of all the achievements during this period the development of the Arabic numeral system is no doubt the most noteworthy. This new system was necessary to make possible future advances in science and engineering. The Greek and the Roman systems were not satisfactory for future scientific work.

1.4 The Renaissance and Post-Renaissance Period

Volumes could be written concerning the scientific and engineering developments during this period. Some of the great names are Da Vinci, Palladio, Galileo, Hooke, Newton, Bernoulli, Euler, and Coulomb. This period covering several hundred years (about 1450 to 1850) can be thought of as the beginning of scientific focus upon the unanswered questions of nature.

Da Vinci heralded this period with his ideas of flying machines, portable military bridges, and many other devices employing principles of mechanics. Many of his ideas were lost to man for several hundred years. To many he is remembered as a painter, but his talents went far into other realms.

This period can rightly be considered as the dawn of engineering mechanics (statics, dynamics, and resistance of materials). Galileo delved into problems of this nature when he considered the action of the material of a beam during bending. His results were erroneous, but he started men to thinking about these problems. Robert Hooke, who became interested in this same problem, is credited with the law relating to elasticity of materials which bears his name. During this same period of time the Bernoulli brothers, James and Johann, and Leonhard Euler were investigating the action of materials and structural shapes. The work of these men would not have been possible without the preceding work of the "father" of modern physics, Sir Isaac Newton. His laws of motion and force and his development of infinitesimal calculus preceded the work of Bernoulli and Euler.

In 1776 Coulomb published the first correct analysis for the fiber stresses in a beam of rectangular cross section subjected to a bending moment. He is probably best known for his theory of earth pressures. Coulomb was soon followed by others interested in the field of engineering mechanics. The student will be exposed to the work of such men as Navier, Claperyron, Saint-Venant, and others as he proceeds with his studies of structural engineering.

Another interesting development of this period of history was the truss. The Italian architect Andrea Palladio (1518–1580) is credited with the first use of trusses in structures. His material, of course, was wood, and his designs were based on "rules of thumb" or experimentation and not on rational analyses. Rational analyses of trusses did not come until about 300 years later when the American, Squire Whipple, published his treatise on "Bridge Building." It is interesting to note that one of the simplest problems of structural analysis, that is, analysis of simple trusses, was so long in developing, whereas more involved studies in theory of elasticity and buckling of bars preceded truss analysis by several years.

1.5   The Modern Period

This period can be dated from about 1850 to the present time—a little more than 100 years. It can be said that during this period structural engineering came of age. The structural engineer became a distinct personality. New methods of analysis and design developed along with new materials, making it possible to carry greater loads over longer spans.

Many men have been involved in the development and furthering of knowledge in structural engineering during this period. Those men who presented methods of structural analysis which bear their names are many.

Maxwell developed his method for statically indeterminate structures in the mid-nineteenth century as well as the graphical method for stress analysis of trusses. Culmann, a professor at Zurich, and Betti, the Italian, also presented work in graphical procedures. Mohr in Germany and Castigliano in Italy developed new approaches to the analysis of statically indeterminate structures.

A great contribution of an American during this period was by Professor Charles E. Greene of the University of Michigan who developed the moment-area method for determining slopes and deflections in beams. Another American professor, G. A. Maney of the University of Minnesota, published the important slope-deflection method for frame analysis in 1915.

No history of stress analysis could be written without mentioning the famous German professor and engineer Müller-Breslau. He wrote many papers on the theory of structures, developing further and emphasizing the previous works of Maxwell and Castigliano.

The most recently developed and one of the most practical methods of indeterminate structural analysis is the method of moment distribution contributed by Professor Hardy Cross of the United States. This method published in 1930 can be said to have revolutionized the analysis of continuous frames.

No new methods have been presented in recent years, but special adaptations and extensions of previously presented methods of analysis have been published from time to time. However, knowledge of materials and design and construction procedures have undergone many improvements and changes in the mid-twentieth century. The advent of welded steel structures and prestressed concrete structures has greatly advanced structural engineering technology. New developments are the ultimate strength method of concrete design and the plastic design of steel structures. The development and wide utilization of new structural forms have resulted in great economies of materials as well as esthetically pleasing structures. The thin shell and folded plate roof structures have reached a prominent place in the architecture and structural engineering of the present day. These new structural forms, new in the sense of wide usage, have required the modern structural engineer to employ more complex and detailed design procedures.

The great contributions to engineering in the nineteenth century are associated with the names of those who developed the classical methods of structural analysis, whereas the twentieth century contributions have been predominately in the field of design, research, and construction. Because of the great numbers of engineers who have made major contributions to the field of engineering in these lines, it is difficult to name names. The twentieth century has seen the development of the engineer administrator. His duties have been to plan and supervise the design and/or construction of mammoth engineering works. He has had to bring together a knowledge of engineering mechanics and materials along with a knowledge of business methods and administrative techniques to create a structure of utility to man. Many of these men have been extending the field of structural engineering into new forms, new processes, and new concepts. Their contributions stand side by side with those of the early developers of structural theories in furthering the progress of engineering.

The electronic computer has become a standard tool of the structural engineer. Structural engineering, particularly in the realm of statically indeterminate structures, has always been plagued with a quantity of laborious calculations. The electronic computer offers means of reducing this work as well as opening larger areas of investigation to the structural designer.

Active research along many avenues of structural engineering is taking place. The volume of technical writings has greatly increased. The advent of the atomic age has accelerated research in structural engineering as well as in many other fields. The engineer of today seems obsessed with a desire to improve upon the past. He who accepts what is known now as complete, or methods now used as final, will soon find himself passed by and left to contemplate the past.

The student should recognize that in a short time in the classroom he will glean the knowledge of centuries of the combined thinking of Bernoulli, Navier, Whipple, Maxwell, Castigliano, Müller-Breslau, Cross, and many others. However, his thoughts should not remain in the past, but he should train his vision toward those new ideas and methods to be developed in the future, and to which he can be a contributor.

References

1. Steinman, D. B. and S. R. Watson, *Bridges and their Builders*, Putnam, 1941.
2. Kinney, J. S., *Indeterminate Structural Analysis*, Addison-Wesley, 1957.
3. Girvin, H. F., *A Historical Appraisal of Mechanics*, International, 1948.
4. Timoshenko, S. P., *History of Strength of Materials*, McGraw-Hill, 1953.
5. Grinter, L. E., *Theory of Modern Steel Structures*, Vol. 1, Macmillan, 1949.
6. Finch, J. K., *The Story of Engineering*, Doubleday, 1960.
7. Oliver, J. W., *History of American Technology*, Ronald Press, 1959.
8. *Transactions of the American Society of Civil Engineers*, Centennial Transactions, Vol. CT, 1953.
9. Sandström, G.E., *Man the Builder*, McGraw-Hill, 1970.

2

REVIEW OF PRINCIPLES
OF STATICS

2.1 Equilibrium

All engineering structures whether they are a bridge, airplane, oil storage tank, or a rocket in space are acted on by forces. If the structure is static (not in motion), the resultant of all the applied forces is equal to zero. If the structure, such as an airplane or automobile, is moving at constant velocity, we can still say that it is in equilibrium and the resultant of all the applied forces is zero. If it is not moving at constant velocity, we can treat it as being in dynamic equilibrium. For dynamic equilibrium it is necessary to include not only the applied forces but also the inertia forces. The reader is referred to textbooks on *dynamics* for a more detailed explanation of dynamic equilibrium. Some structures which are normally thought of as static structures may be in dynamic equilibrium under certain types of loadings. Examples are buildings subjected to earthquakes, wind, or blast loads, or the dynamic load effect of moving vehicles on bridges. Whatever type of structure or whatever type of loading, it should be remembered that a structure can always be treated as being in equilibrium, that is, the sum of all the forces acting on the structure (summed up in any direction) is equal to zero. It should also be remembered that gravity forces and reaction forces (forces being applied by the supporting medium) should also be considered as applied forces.

In addition to the sum of the applied forces being equal to zero it is also a necessary condition for equilibrium that the moment taken about *any* axis (on or off the structure) of *all* the forces acting on the structure are equal to zero.

2.2 Two-Dimensional Force System

The foregoing concept of equilibrium can be stated in equation form. If we adopt a two coordinate axis system for two-dimensional force systems and a three coordinate axis system for three-dimensional force systems, we can sum up the applied forces in the direction of these axes. Considering a two-dimensional force system our two equations relating to forces are

$$\sum F_x = 0$$
$$\sum F_y = 0 \tag{2.1}$$

Consider a system of forces acting on a body. In the beginning some of the forces will be known and some will be unknown. By using the equations (2.1) it is possible to determine two of the unknown forces. If there are more than two unknown forces and the forces are not concurrent, it is necessary to use additional equations. One additional equation is determined from the concept of equilibrium of moments, which can be stated in equation form by

$$\sum M = 0 \tag{2.2}$$

Equation 2.2 is the third equation of statical equilibrium, and by using the three equations it is possible to determine three unknown forces acting on a body when the other forces are known. If it is possible to determine all the unknown forces acting on a structure by the use of these equations, the structure can be classified as *statically determinate*. When there are more unknown forces than applicable equations of statics, the structure is classified as *statically indeterminate*. For the analysis of statically indeterminate structures it is necessary to utilize other concepts in addition to those of static equilibrium. Structures that require additional equations beyond those of statical equilibrium can be analyzed by using only the equilibrium equations if certain assumptions are made with regard to distribution of forces and moments. However, these assumptions are usually only approximately true and thus the results are approximate. Chapter 7 covers some approximate methods used in the analysis of building frames. Later chapters cover the most used methods of analysis of *statically indeterminate* structures.

The student should not develop the idea that *only* three unknown forces can be determined by the three equations of statics for any complete structure. It is well to think of the idea that *only* three unknown forces can be determined by Eqs. 2.1 and 2.2 for any *rigid body* but that a complete structure may be composed of many rigid bodies. Examples are the three-hinged arch (Chapter 6) in which four unknown reactions are determined, or the cantilever bridge in which many unknown reactions can be determined by the three equations of statics.

Any structure can be considered as being composed of an infinite number of rigid bodies. If a structure is in equilibrium, any part of the structure can be considered as being in equilibrium.

Any structure can be considered as being composed of an infinite number of rigid elements. If a structure is in equilibrium, any part of the structure can be considered as being in equilibrium.

2.3 Free Body Diagrams

The selection of the size, shape, and material of a structural member requires the prior knowledge of the magnitude of the

(1) axial force
(2) shear force
(3) bending moment
(4) torsional moment

These four quantities must be known at all positions along the structural member. The science of structural analysis is the determination of the above four quantities for all types of structures. Most structures are composed of several elements fastened together. Complex structures can be best analyzed by "breaking down" the total structure into elements. The elements are called free bodies. The free bodies can be very small in size, being only a small portion of a structural member, or they might be an assemblage of several members. Developing a skill in the proper selection of a free body is one of the objectives of this book. The use of *free body diagrams* in structural analysis cannot be overemphasized.

When the equations of statics are applied, only the *external* forces are considered. However, it should be remembered that if a section is cut through a structure and the severed portion isolated from the rest of the structure, the *internal* forces acting at the cut section become *external* forces and are included in the computations for static equilibrium. An example is the beam shown in Fig. 2.1*a*. A section is cut through the beam at *a–a*; then in order to hold the left-hand portion in equilibrium it is necessary to change the *internal* forces and moments at section *a–a* to *external* forces and moments. The forces of equilibrium are shown in Fig. 2.1*b*.

The reader may recall from a course in mechanics of materials that in

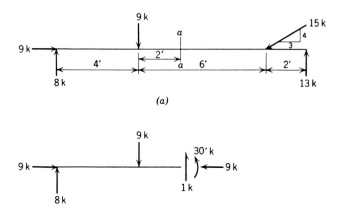

Fig. 2.1

order to determine the stresses at points in a structure it is first necessary to determine the internal forces and then, using these internal forces, the stresses can be determined. This method of determining internal forces in a structure by cutting sections through the structure and then converting the internal forces to external forces and determining their magnitude by employing the laws of statics is the standard basic procedure here.

2.4   Other Basic Concepts

It should be remembered from a study of statics that for purposes of structural analysis a force can be projected anywhere along its line of action, i.e., force vectors are sliding vectors. Very often time and work can be saved by considering a force acting at a particular point along its line of action. This will be readily apparent in the work in Chapter 5 dealing with forces in trusses.

Another important principle is the law first postulated by the French mathematician Varignon in the late seventeenth century. This law states that the moment of a force about any axis is equal to the sum of the moments of the components about that axis. For example, in Fig. 2.2, a force of 300 lb acts at a perpendicular distance of 12 ft from an axis passing through point A. The moment about A is then 3600 ft-lb. This force is then projected until it is acting on a point along its line of action (point B) and lying on a horizontal line through point A. The distance from point A to point B is then equal to $12/\sin 30° = 24.0$ ft. If the 300-lb force is broken into x and y components, these values are 260 and 150 lb respectively. The moment of the components about A is then $150 \times 24 = 3600$ ft-lbs.

2.5 Support Conditions

A structure may be supported by any of several types of supports (usually termed bearings) such as rollers, rockers, and pins. The number and direction of the reaction forces are dependent on the design of the bearing. To analyze correctly a structure we must, of course, know the final characteristics of the bearing. For example, if a planar structure were analyzed on the basis that one of the reactions could be only in one direction, the design of that bearing must conform to this condition. If a bearing were designed to permit movement in the x direction and some later physical condition such as rust and dirt prohibited this movement, the stress conditions in the structure could be measurably changed.

In the following discussions certain support symbols are used, which are explained by Fig. 2.3.

2.6 Graphical Methods

When a set of known coplanar forces are applied to a rigid body, it is possible to arrive at the solution of unknown forces acting on the body by graphical methods instead of algebraic analysis. The limitations are the same in both methods. Only two unknown force components can be found for a coplanar, concurrent force system, and only three unknown force components can be determined for a coplanar, nonconcurrent force system. The system of concurrent, coplanar forces acting on a rigid body can be formed into a force polygon by starting with one force and drawing a vector to represent this force. The length of the vector represents the magnitude of the force to some convenient scale. The direction of that vector is drawn in the same direction as the force acting on the rigid body. The second force vector is drawn similarly, with the tail of this vector at the tip of the first vector. All the known force vectors are drawn in this tip-to-tail fashion until all known force vectors have been plotted showing their true magnitude and direction. The two unknown force vectors then

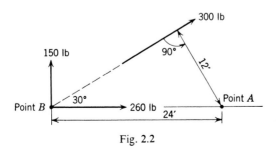

Fig. 2.2

Symbol	Displacement	Rotation	Number of Reaction Components
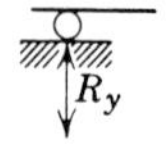 R_y	Lateral only	Possible	One-normal to plane of foundation*
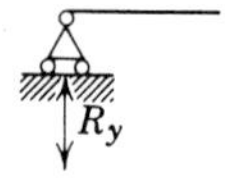 R_y	Lateral only	Possible	One-normal to plane of foundation*
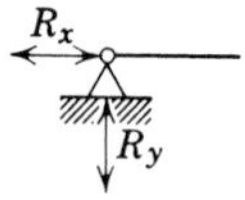 R_x R_y	None	Possible	Two
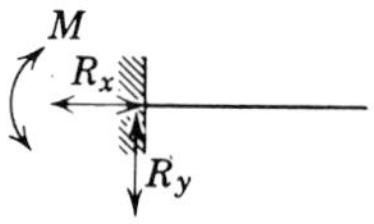 M R_x R_y	None	None	Three
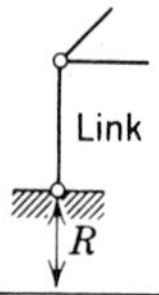 Link R	Perpendicular to link only	About ends of link	One-collinear with axis of link

*Note reactions can be two directional.

Fig. 2.3. Legend for support symbols.

have to close the polygon. The intersection of these two lines gives the length of the vectors, and by scaling the lengths the magnitudes can be determined. Figure 2.4a shows the known forces acting on a body. These forces are formed into a force vector polygon in Fig. 2.4b, and the length of vectors F_1 and F_2 gives the magnitude of the unknown forces when the directions of the unknown forces are known. It can also be seen that the two unknowns could be the magnitude and direction of one force. Direction can be an unknown as well as magnitude. To define a force it is necessary to give its direction and line of action as well as its magnitude.

It is not necessary to draw the known forces in the force vector polygon in any particular order. The student should make a proof of this by drawing the known vectors in a different order from Fig. 2.4b.

When solving for unknowns by graphical means in a nonconcurrent coplanar force system, two diagrams are necessary. One diagram is a

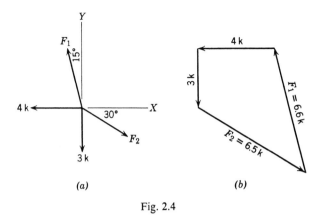

Fig. 2.4

layout of the forces drawn to a convenient scale on the diagram of the rigid body. The other is a layout of the forces as vectors to true direction and magnitude. Figure 2.5a and b shows the forces on a scaled diagram of the rigid body. Figure 2.5c shows the forces drawn as vectors to true direction and with their lengths proportional to the magnitude of the respective forces. The steps to determine the magnitude of the components of the reactions are as follows:

1. Letter the spaces between the forces and letter the beginning and end of each force vector in the force polygon with the corresponding letters. It is usual to use lower case on the body diagram and upper case on the force diagram. This notation is called Bow's notation.
2. Choose an arbitrary point (O) called the pole on either side of the line of forces and about midheight between the beginning and end of the force line (Fig. 2.5c).
3. From the beginning and end of each force in the force polygon draw lines to the pole (point O). These lines are called rays. Each pair of rays common to a force can be considered as components of that force; that is, rays BO and CO can be considered as components of force BC.
4. On the body diagram draw lines parallel to the rays between the forces. These lines are called strings. The first string is drawn from any point along the line of action of force ab, which is the reaction at the left end of the beam. However, it should be realized that this force ab is neither vertical nor horizontal. The only known point along its line of action is the point of contact between the beam and the left support. String bo *must* then start at this point.
5. At the point where string bo intersects the line of action of force bc draw the string co (parallel to ray CO) until it intersects the line of action of force cd. Since this string is common to forces BC and CD in the force diagram, it is drawn between the lines of action of forces bc and cd on the body diagram.
6. The subsequent strings (parallel to the rays in the force diagram) are drawn between lines of action of the forces in the body diagram until the line of

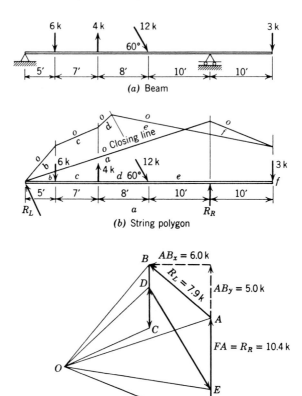

(a) Beam

(b) String polygon

(c) Force diagram

Fig. 2.5

action of the last force is reached. This last force is the right reaction. A line is next drawn from the starting point of the first string (*bo*) to the end point of the last drawn string (*fo*). This is the closing line, string (*ao*), of the string polygon.

7. Draw the last ray on the force diagram through *O* parallel to the closing string *ao*. This should be drawn until it intersects the force vector *FA* (the right reaction). As a result of the right support resting on rollers, the direction of the right reaction and thus vector *FA* is vertical. The left reaction is then represented by the vector extending from *A* to *B*. If desired, the *x* and *y* components of the left reaction can also be easily scaled.

The magnitudes of the left and right reactions from the illustrated problem are 7.9 kips and 10.4 kips respectively.

There are some types of structural problems in which it is necessary to draw the string polygon through two specific points on the structure. Particular uses are the three-hinged arch covered in Chapter 5 and the suspended cable in Chapter 9. In the problem of Fig. 2.5 it was only necessary to draw the string polygon through the point of the left support. A general case of the string polygon through two points is shown in Fig. 2.6. Part (a) shows the force system, with the two specific points of support labeled M and N. In part (c) the force diagram is shown and an arbitrary pole O_1 is selected. Rays to the beginning and end of each force vector are then drawn to the pole O_1. In part b the string polygon is drawn (shown by the full lines) starting at point M and ending at point (N') on a vertical line drawn through N. The closing line is then line o_1e. A line drawn parallel to this closing line from pole O_1 on the force diagram will intersect a vertical line, drawn from D, at point E. The vertical component of the reaction at N is then equal to the force vector DE. It is noted, however, that the string polygon did not pass through point N as required.

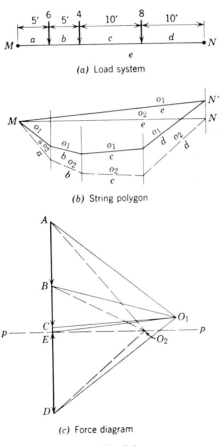

(a) Load system

(b) String polygon

(c) Force diagram

Fig. 2.6

If a particular pole location had been selected, the string polygon would have intersected point N instead of N'. However, simple reasoning indicates that the values of the reaction cannot change with a change in pole location. Therefore, DE must be the true value of the vertical reaction at N. The true closing line must then be parallel to the line connecting M and N and must pass through point E on the force diagram. This line is shown as line p–p on the force diagram. Since it is possible to have an infinite number of string polygons through two points, any pole along line p–p can be selected. The pole O_2 is selected and a new string polygon is then drawn as shown by the dashed lines in Fig. 2.6b. It is observed that this new string polygon passes through the points M and N. Any pole along line p–p will produce a string polygon through M and N.

2.7   Three-Dimensional Force Systems

A force in space can be resolved into three component forces parallel to three coordinate axes. By representing a force in space in Fig. 2.7 by a vector whose origin is at the origin of the coordinate axes, the components of the force parallel with each respective axis is

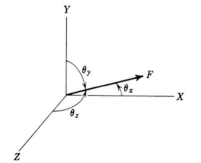

Fig. 2.7

$$F_x = F \cdot \cos \theta_x$$
$$F_y = F \cdot \cos \theta_y \qquad (2.3)$$
$$F_z = F \cdot \cos \theta_z$$

If there are several forces acting on a rigid body, the following equations must be satisfied in order for the body to remain in equilibrium.

$$\begin{array}{ll} \sum F_x = 0 & \sum M_x = 0 \\ \sum F_y = 0 & \sum M_y = 0 \quad (2.4) \\ \sum F_z = 0 & \sum M_z = 0 \end{array}$$

For a rigid body acted upon by nonconcurrent, nonplanar forces a maximum of six unknown forces can be determined by employing Eq. 2.4. In order to determine the moment of a force about an axis it is necessary to determine the perpendicular distance from the force vector to the axis. Usually it is easier when dealing with a coordinate axis system to resolve a force vector into components parallel to the coordinate axes. Then by Varignon's theorem the moment is equal to the sum of the moments of the components. It should be remembered that a force parallel to or intersecting an axis causes no moment about that axis.

Since a force can be represented by a vector, the relationship of Eq. 2.3 can be developed without resorting to the trigonometric functions. Referring to Fig. 2.8, the components of a force F in space can be developed to form a box of sides x, y, and z.

$$\cos \theta_x = \frac{x}{d}$$

$$\cos \theta_y = \frac{y}{d}$$

$$\cos \theta_z = \frac{z}{d}$$

$$F_x = F\frac{x}{d}$$

$$F_y = F\frac{y}{d}$$

$$F_z = F\frac{z}{d}$$

or

$$\frac{F_x}{x} = \frac{F_y}{y} = \frac{F_z}{z} = \frac{F}{d} \tag{2.5}$$

Equation 2.5 is very useful in the solution of three-dimensional force systems. For instance, if a vertical tower is guyed by three guy wires as shown in Fig. 2.9, the weight of the tower, the reaction at the base, and the forces in the three guy wires form a concurrent three-dimensional force system. The force in each wire can be considered as a vector whose length is equal to the length of the guy wire. The origin of each vector is con-

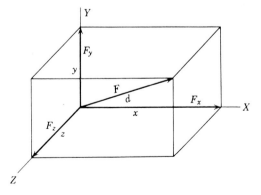

Fig. 2.8

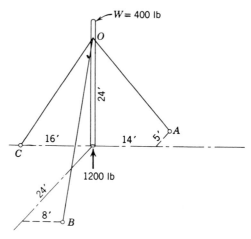

Fig. 2.9

sidered at the point of attachment to the tower. The solution of such a problem is best handled by tabulating the x, y, z, and d values for each component. Knowing the x, y, z, and d values for each vector, the x, y, and z components in terms of the force in each vector can be determined by Eq. 2.5. The summation of the component forces in the x, y, and z direction can be made, and the first three equations of Eq. 2.4 can be used to determine the force in each guy wire.

Table 2.1 is the solution to the problem. The bottom of the tower is considered as hinged, and the bottom of all guy wires is at the same elevation.

Table 2.1   Force in Guy Wires

Force	x	y	z	d	F_x	F_y	F_z
F_{OA}	+14	−24	−5	28.23	$+0.496F_{OA}$	$-0.850F_{OA}$	$-0.177F_{OA}$
F_{OB}	+8	−24	+24	34.86	$+0.229F_{OB}$	$-0.688F_{OB}$	$+0.688F_{OB}$
F_{OC}	−16	−24	0	28.84	$-0.555F_{OC}$	$-0.832F_{OC}$	0
External						+800	

$$\Sigma F_x = 0 = +0.496F_{OA} + 0.229F_{OB} - 0.555F_{OC}$$

$$\Sigma F_y = 0 = -0.850F_{OA} - 0.688F_{OB} - 0.832F_{OC} + 800$$

$$\Sigma F_z = 0 = -0.177F_{OA} + 0.688F_{OB}$$

$$F_{OB} = 0.257F_{OA}$$

$$+0.496F_{OA} + 0.229(0.257)F_{OA} = 0.555F_{OC}$$

$$F_{OC} = \frac{0.555}{0.555} F_{OA} = F_{OA}$$

$$F_{OA}(-0.850 - 0.688 \times 0.257 - 0.832) = -800$$

$$F_{OA} = \frac{800}{1.859} = 430 \text{ lb}$$

$$F_{OB} = 0.257 \times 430 = 111 \text{ lb}$$

$$F_{OC} = 430 \text{ lb}$$

References

1. Tuma, J. J., *Statics,* Quantum, 1974.
2. Langhaar, H. L., and A. P. Boresi, *Engineering Mechanics,* McGraw-Hill, 1959.
3. Beer, F. P., and E. R. B. Johnston, *Mechanics for Engineers,* McGraw-Hill, 1972.
4. Meriam, J. L., *Mechanics,* John Wiley, 1978.

Problems

1. Replace the system of forces acting on the frame in Fig. 2.10 by a resultant R at point A and a couple acting horizontally at C and D.
$\quad$ *Ans.* $R_x = +12$ lb, $R_y = -50$ lb, couple $\begin{cases} \rightarrow 92 \text{ lb} \\ \leftarrow 92 \text{ lb} \end{cases}$

2. Determine the reactions for the beam of Fig. 2.11(a) by numerical methods, (b) by graphical methods. $\quad$ *Ans.* $R_{ax} = 15$ kips $\leftarrow R_{ay} = 4$ kips $\uparrow$ $R_b = 12$ kips $\uparrow$

3. Determine the reactions at A and C for the frame of Fig. 2.12. W is equal to 10 kips.

4. Determine the reactions at the ends of the beam for the beam of Fig. 2.13.
$\quad$ *Ans.* $R_{ay} = 3.94$ kips, $R_{ax} = 2.14$ kips, $R_c = 2.23$ kips.

5. Determine the reactions for the beam of Fig. 2.14.

6. The beam of Fig. 2.15 is subjected to the loads shown. Determine the reaction at A and the force in the tie BC. $\quad$ *Ans.* $T_{BC} = 12.9$ kips.

7. Determine the force in the strut of the bulkhead wall of Fig. 2.16. Each strut supports a 10 ft length of wall.

8. If the weight of the crane as shown in Fig. 2.17 is 12,000 lb (not including counterweight W'), what is the required weight of the counterweight for a load of $W = 16,000$ lb at a safety factor against overturning of 1.5? What are the loads applied to each axle when $W = 16,000$ lb and W' is equal to the value determined?
$\quad$ *Ans.* Front axle $= 37$ kips. Rear axle $= 12$ kips.

9. A three-dimensional concurrent force system is shown in Fig. 2.18. Find the magnitude of the resultant force and the angles between the resultant and the coordinate axes. F_1 lies in the XZ plane and F_2 in the YZ plane.
$\quad$ *Ans.* $R = 683$ lb.

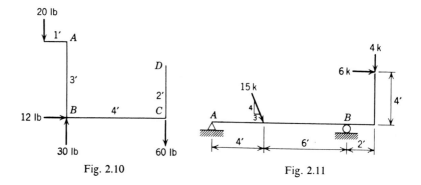

Fig. 2.10

Fig. 2.11

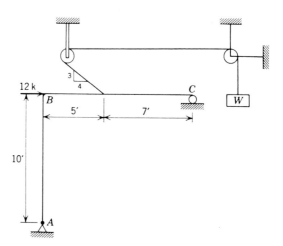

Fig. 2.12

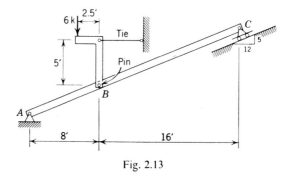

Fig. 2.13

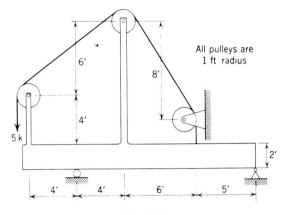

Fig. 2.14

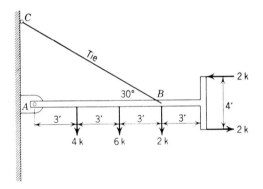

Fig. 2.15

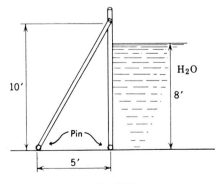

Fig. 2.16

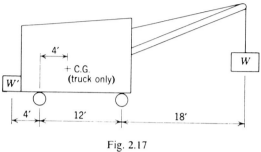

Fig. 2.17

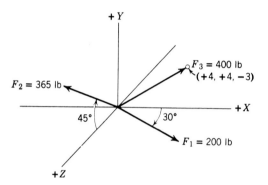

Fig. 2.18

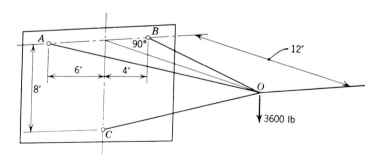

Fig. 2.19

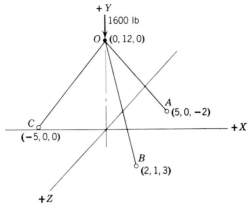

Fig. 2.20

10. Find the forces in the members *AO*, *BO*, and *CO* of Fig. 2.19. Point *O* is at the same elevation as *A* and *B*.

$$Ans.\ AO\ =\ +2414\ \text{lb},\ BO\ =\ +3416\ \text{lb},\ CO\ =\ -6480\ \text{lb}.$$

11. Determine the compression forces in the legs of the tripod of Fig. 2.20. The coordinates of the ends of the legs are as shown.

LOADS ON STRUCTURES

3.1 General

Nature of Loads. All structures on earth are subjected to loads. These loads in turn create forces on various parts of the structure, and these forces cause stresses and strains on the elements of the material of which a structure is composed. The magnitude of these stresses and strains determines whether the elements of the structural parts will hold together or will separate. Separation of the parts will, of course, mean failure of the structure. Failure as loosely used by most engineers does not always mean separation of the parts. Excessive strain is also usually classified as failure. It is apparent that in order to determine the worthiness of a structure it is necessary to know what loads the structure will be subjected to throughout its life.

Before a clear concept can be obtained with regard to loads it is necessary to know how loads can be defined. Loads are usually thought of as being defined only by a unit of force as related to some unit force of gravity such as pound, kip (1000 lb), gram, kilogram, etc. However, there are other qualities of a load that have to be defined before an evaluation of the effect of the load can be made. It is necessary to know where on the structure the load will be acting and in what direction. It is also necessary to know something about the distribution of the load, whether it is distributed over a small or large area. For most structures it is necessary to know the manner of application of the load. Is the load applied suddenly or gradually? If suddenly, what is the force-time relationship? Some structures will be subjected to loads that are only applied gradually, such as dead loads of buildings or bridges. However, other structures, such as airplanes, rockets, and automobiles, may be subjected to sudden loads. Loads that are applied gradually are classified as static loads, whereas those applied during a short period of time are called dynamic loads. Another characteristic of a load that should be known is how many times will the load be applied to the structure, and if applied more than once what is the time interval between loads. Repeated loads over a short time interval are classified as vibratory loads, whereas those over a long time interval cause no concern with respect to vibration; however, they may cause fatigue failure if repeated many times.

Static Loads.   As previously defined, a static load is a load that is applied to a structure gradually over a relatively long time interval. In this case all the parts of a structure are at any instant essentially in static equilibrium. For a static load condition, the magnitude of the load is independent of the elastic nature of the structure. Because of this, static loads are much more easily handled in structural design than are dynamic loads. Many dynamic loads are treated as equivalent static loads. Examples are wind loads, moving loads on bridges, and the usual treatment of earthquake loads. A static load carried by all structures is the dead weight of the structure.

Dynamic Loads.   A dynamic load is one that is applied over a short time interval or one that varies rapidly with time. Dynamic loads can be broken into three categories.

IMPACT LOAD.   This type of load is the result of a mass striking a body while traveling at a certain velocity, for example, a bullet striking a piece of armor plate or a truck wheel dropping into a depression in a bridge pavement. With impact the structure can be required to absorb a large amount of energy. The effect on the structure can be equivalent to several times the weight of the moving body applied as a static load.

VIBRATORY LOAD.   An example of this type of loading is oscillating machinery or a troop of soldiers marching in cadence across a bridge. Such loads can cause small or large effects, depending on the ratio of frequency of the applied load and the natural frequency of vibration of the structure.

IMPULSIVE LOAD.   This type of load is due to a rapid rise of the peak force. This load is produced by blasts, wind gusts, ocean waves, or a smooth wheel moving rapidly across the smooth floor of a bridge. It is

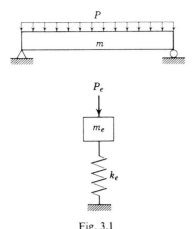

Fig. 3.1

quite often referred to as a sudden loading. The effect on a structure is never more than twice the effect of the peak force applied as a static load.

The first two types of dynamic loads mentioned are quite specialized in nature and therefore are not treated further. The impulsive type of dynamic load is more common, and a basic understanding of some aspects of this loading should be understood by the structural engineer. Only a brief introduction of the subject is considered in this book. Reference (1), however, gives a thorough coverage and also includes an extensive bibliography.

An exact solution of the effect on an actual structural system subjected to dynamic loads is not practical. It is necessary to idealize the arrangement of the weights and structural elements into a simple arrangement of masses and springs. For instance, Fig. 3.1 shows a simple beam idealized to a one degree of freedom system; a mass supported by a spring in which the position of the mass can always be described by the one variable y. In this equivalent system the deflection of the concentrated mass can represent the deflection at a particular point on the actual structure. The particular point is most often the midpoint of the span. For more complex structures the idealized system will contain several degrees of freedom. However, only a one degree of freedom system will be treated here.

The basic idea of an undamped elastic system with one degree of freedom is shown in Fig. 3.2. A rigid body of mass m is supported and guided by a horizontal unyielding base. An external force P is applied to the mass in a horizontal direction. This force acts against the mass, which in turn acts against a spring of constant stiffness. The reactive force R of the spring is the required information. The reactive force of the spring is equal to the displacement x multiplied by the spring coefficient k. It is

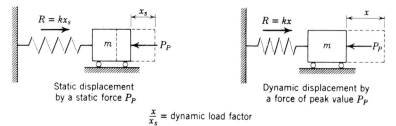

$\dfrac{x}{x_s}$ = dynamic load factor

Fig. 3.2. Single degree of freedom systems.

considered that the spring coefficient remains constant. This is the equivalent of saying that the material in a structure remains in the elastic region.

If the forces acting on the body at any instant are added, the dynamic equilibrium equation is as follows:

$$m\frac{d^2x}{dt^2} + kx = P(t)$$

where $P(t)$ denotes that the applied force P is a variable with respect to time. The solution of the preceding equation for particular values of $P(t)$ will give values of x as a function of time. Substituting various values of time into the solution of the differential equation will give a displacement time relationship. If the ratio of the dynamic displacement x at any time t to the static displacement x_s is termed the response, then a response time curve can be plotted. If the maximum stress in the spring is the desired result, then only the maximum response is required. This maximum response can be termed the *dynamic load factor*. If the dynamic load factor is known for any particular type of dynamic loading, the maximum stress in the spring can be determined by multiplying the stress found by applying the peak load as a static load by the dynamic load factor. It is therefore apparent that if a structural element can be idealized as a one degree of freedom system, the dynamic loading condition can be simplified to an equivalent static system providing the stresses in the structure remain elastic. *The equivalent static load system is a static load whose magnitude is equal to the peak value of the dynamic load multiplied by the dynamic load factor.*

The dynamic load factors for various load functions are shown in Fig. 3.3. In these plots the value T is the period of natural frequency of the one degree of freedom system.

The period of natural frequency of a one degree of freedom system can be determined from the equation

$$T = 2\pi\sqrt{\delta_{st}/g}$$

where δ_{st} is the static deflection, in displacement per unit load, of the spring, and g is the acceleration of gravity.

It is noted from Fig. 3.3 that the maximum dynamic load factor never exceeds 2.0 and under some loadings may even be less than one. For instance, for a rectangular type of pulse (Fig. 3.3*b*) the dynamic load factor will be less than one if the duration of the pulse is less than $\frac{1}{6}$ the natural period of frequency of the structure.

From these charts or others that have been developed[1] it is possible to

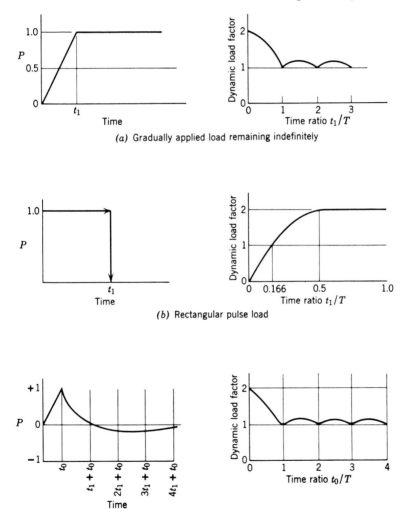

(a) Gradually applied load remaining indefinitely

(b) Rectangular pulse load

(c) Blast type load

Fig. 3.3. Dynamic load factors.

determine the effect of dynamic loads. It is seen that the effect of the dynamic load on the structure is not only a function of the peak value of the load but also the rate of application of the load and the elastic characteristics of the structure.

From the information presented it is seen that the treatment of dynamic loads is more involved than for static loads, but by applying principles of dynamics to the problem, workable methods can be developed for the design of structures subjected to dynamic loads.

3.2   Loads on Buildings

General.   A building may be subjected to a wide variety of loads during its lifetime. The magnitude and nature of some of the loads can be predicted quite accurately, whereas an evaluation of the other loads to any high degree of accuracy is not possible. The purpose of the building, its structural nature, its shape and size, as well as its location are all influencing factors with respect to the imposed loads.

The structural engineer usually uses a set of loadings that have been established by the engineering profession as design loads for his structures. These established design loads will probably be those adopted and published by a governmental agency as a Building Code. All the larger cities of the Unites States have an adopted building code.

The primary loads for which buildings are designed are the dead load, live load, and wind loads. In some localities seismic forces due to earthquakes should also be considered. These four loadings are discussed in detail in the following sections.

Dead Loads.   The dead loads are considered as those applied to the structure at time of construction and generally remaining on the structure throughout its lifetime. These loads would be the weight of the structure itself and any permanently attached equipment. The weight of the structural parts of the structure has to be estimated at the beginning of the structural design work and then revised as the design proceeds. Usually the final dead loads carried by the structure can be ascertained quite accurately. Table 3.1 gives the unit weights of common materials used in building construction.

For other materials see the *Steel Construction Manual* of the American Institute of Steel Construction.

Live Loads.   The live loads carried by a building are those of a temporary nature such as people, merchandise, snow, etc. They are usually static

Table 3.1 *Unit Weights of Common Building Materials*

	Pounds per cubic foot
Steel	490
Aluminum	165
Reinforced concrete	150
Lightweight concrete	90–120
Timber: Douglas fir	35
Timber: Southern yellow pine	40
Brick masonry	100–135
Stone	135–175
Five-ply felt and gravel (or slag) roofing	$6\frac{1}{2}$ lb per square foot

loads but might be dynamic loads such as cranes in industrial buildings, or automobiles within parking garages. The design live load for a building depends on its use. Live loads are generally applied as a uniform load over floor or roof areas. Live loads are to be considered as moving loads insofar as they can be applied to particular parts of a structure so as to produce the maximum stress. Table 3.2 gives the building code requirements for minimum live loads as published by the Uniform Building Code (2).

Snow Loads. Snow loads are the usual design live loads for roofs of buildings. The magnitude of snow load if not governed by a local building code should be based on the maximum expected snow accumulation. The accumulation depth will depend upon the slope of the roof. The steeper the slope, the smaller the accumulation. Snow weights vary from 8 lb per cubic foot for dry snow to 12 lb per cubic foot for wet snow. Table 3.3 gives snow load data for typical areas in the United States. The snow loads shown in this table are average. Values could be considerably higher in mountain regions or where roof slopes were such that drifting or piling could take place. Careful consideration should be given to possible heavy roof loads in regions of moderate and heavy snowfall.

Wind Loads. The wind forces on any structure result from a change of velocity of the air due to the obstruction of the structure, friction between the air and the structure, and possible periodic "aerodynamic" forces. The effect of friction is minor and the "aerodynamic" effect is only important for flexible structures such as suspension type structures, tall stacks, etc. On most structures such as buildings and bridges, the time rate of build-up of peak wind velocity is large with respect to the natural frequency of the structure. If this is so, the wind force can be treated as a static force.

The simplest approach to the determination of the pressure on a flat object normal to the flow of air is to consider that a particle of air has a

Table 3.2(a)   Uniform Live Loads

USE OR OCCUPANCY		
CATEGORY	DESCRIPTION	UNIFORM LOAD lbs/sq. ft.
Armories		150
Assembly areas and auditoriums and balconies therewith	Fixed seating areas	50
	Moveable seating and other areas	100
	Stage areas and enclosed platforms	125
Cornices, marquees and residential balconies		60
Exit facilities, public		100
Garages	General storage and/or repair	100
	Private pleasure car storage	50
Hospitals	Wards and rooms	40
Libraries	Reading rooms	60
	Stack rooms	125
Manufacturing	Light	75
	Heavy	125
Offices		50
Printing plants	Press rooms	150
	Composing and linotype rooms	100
Residential		40
Rest rooms		40
Reviewing stands, grandstands and bleachers		100
Schools	Classrooms	40
Sidewalks and driveways	Public access	250
Storage	Light	125
	Heavy	250
Stores	Retail	75
	Wholesale	100

The above loads have special methods of application as designated
in the Uniform Building Code.

velocity V just before striking the object and a velocity of zero just after striking the object. If this is so, the pressure from Bernoulli's theorem is $p = wV^2/2g$, where w is the unit weight of the air and g is the acceleration of gravity. If the density of the air is taken as 0.0765 lb per cubic

Table 3.2(b) Minimum Roof Live Loads[1]
(lbs./sq.ft.)

ROOF SLOPE	TRIBUTARY LOADED AREA IN SQUARE FEET FOR ANY STRUCTURAL MEMBER		
	0 to 200	201 to 600	over 600
Flat or rise less than 4 inches per foot Arch or dome with rise less than one-eighth of span	20	16	12
Rise 4 inches per foot to less than 12 inches per foot Arch or dome with rise one-eighth of span to less than three-eights of span	16	14	12
Rise 12 inches per foot and greater Arch or dome with rise three-eights of span or greater	12	12	12
Awnings except cloth covered	5	5	5
Greenhouses, lath houses and agricultural buildings	10	10	10

[1] Where snow loads occur, the roof structure shall be designed for such loads as determined by the Building Official.

foot corresponding to 15°C at 760 mm Hg and V is expressed in miles per hour, then the velocity pressure is

$$p = 0.002558V^2 \qquad (3.1)$$

The velocity pressure expression (Eq. 3.1) cannot be used to predict wind forces on structures. The velocity of the air after striking an object varies over the surface of the object. If a steady wind strikes a flat plate that is a great deal longer than the length dimension (parallel to the plate) of the wind, the wind will only be able to escape over or under the plate. If the plate is narrow, the moving air could also escape around the sides. This then indicates that the wind pressure on a flat surface depends on the dimensions of that surface. The equation for wind pressure on a flat-sided object can then be predicted by the equation

$$p = 0.002558C_DV^2 \qquad (3.2)$$

Table 3.3   Snow Load Data*

(Values in pounds per square foot of roof area)

Flat and Low-Rise	Medium-Rise	High-Rise	Typical Areas
40	30	15	Minnesota, New York, Maine
30	25	12	Colorado, Missouri, Wisconsin, Washington
25	20	12	Illinois, Oregon
20	15	10	Mississippi, Texas, Louisiana
15	12	10	California

*See Reference (3).

Low-rise: rise less than 4 in. per foot, or, if curved, rise less than $\frac{1}{8}$ of span length.

Medium-rise: rise more than 4 in. and less than 12 in. per foot. If curved, rise between $\frac{1}{8}$ and $\frac{3}{8}$ of span length.

High-rise: rise more than 12 in. per foot, or, if curved, rise greater than $\frac{3}{8}$ of the span length.

For flat plates normal to the wind the values of C_D, termed the drag coefficient, have been found to be the following:

$$L/h = 1.0 \quad 2.0 \quad 5.0 \quad 10.0 \quad \infty$$
$$C_D = 1.12 \quad 1.19 \quad 1.20 \quad 1.23 \quad 1.98$$

where L = length of plate and h = height of plate.

Part of the force as predicted by Eq. 3.2 is due to pressure on the windward side and the remainder is due to suction on the leeward side. Wind tunnel tests have been conducted on a number of different-shaped models to determine specific values of C_D. The value of C_D for an average building with plane surfaces normal to the direction of the wind is approximately 1.3, of which 0.8 is due to the pressure on the windward wall and 0.5 the suction on the leeward wall. For a building the coefficient 1.3 is spoken of as the shape factor. If a value of 1.3 is inserted into Eq. 3.2, the result is

$$p = 0.0033V^2 \qquad (3.3)$$

For a wind velocity of 100 mph, Eq. 3.3 yields $p = 33$ lb per square foot.

For the external effects on inclined or nonplanar surfaces, it is necessary to determine the proper shape factor. The design pressure would be the product of the velocity pressure (Eq. 3.1) and the shape factor. Many studies have been conducted to determine the proper value of wind force for roof structures. Reference (4) gives a very extensive bibliography on wind forces on buildings. In the 1930's the American Society of Civil Engineers (ASCE) had a committee (Sub-Committee 31) study the known information then available, and in 1936 the Fifth Progress Report[5] recommended the following values for gabled roofs:

Windward slope: for values of roof slope θ.
$\qquad \theta$ = zero to $20°$
$\qquad\qquad p' = -0.7p$ (minus means suction)
where p is the velocity pressure from Eq. 3.1.
$\qquad \theta = 20°$ to $30°$
$\qquad\qquad p' = (0.07\theta - 2.10)p$
$\qquad \theta = 30°$ to $59°$
$\qquad\qquad p' = (0.030\theta - 0.90)p$
$\qquad \theta = 60°$ to $90°$
$\qquad\qquad p' = 0.90p$
In the equations for p' use value of θ degrees.
Leeward slope: a suction value of $p' = -0.60p$ for all values of slope.

A series of wind tunnel tests were made in Switzerland and the results have been incorporated into the Swiss Building Code.[6] These data are probably the most extensive and reliable to date. The coefficients for gabled roofs with length to width ratio of the building ranging from $\frac{5}{2}$ to $\frac{2}{5}$; and the width to height ratio from $\frac{3}{2}$ to $\frac{2}{3}$ are as follows:

Windward roof slope
$\qquad \theta$ = zero to $20°$
$\qquad\qquad p' = -1.0p$
$\qquad \theta = 20°$ to $50°$
$\qquad\qquad p' = 0.01(5\theta - 200)p$
$\qquad \theta = 50°$ to $90°$
$\qquad\qquad p' = 0.01\theta p$
Leeward slope
$\qquad\qquad p' = -0.70p$ for all values of slope

A comparison of the 1936 ASCE Sub-Committee 31 report with the

Swiss Building Code is shown in Fig. 3.4. The coefficients for structures other than gabled roofs are given in References (4), (5), and (6). Also given are the internal pressure effects due to wind leakage into the building.

In 1940 the ASCE Sub-Committee 31 published a final report.[7] In this final report the design wind force recommendations differed somewhat in value as well as form of presentation from the 1936 progress report. In the final report the recommended wind loads were in the form of

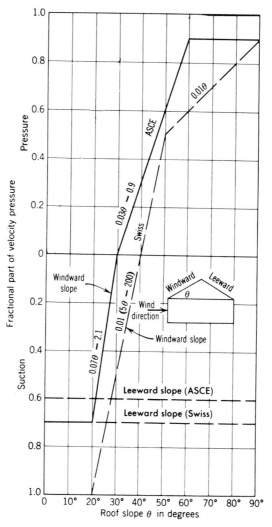

Fig. 3.4. Wind pressures and suctions on symmetrical gable roofs external coefficients. (From "Wind Forces on Structures: Forces on Enclosed Structures," T. W. Singell, Paper 1710, *Proceedings of ASCE Structural Division*, Vol. 84 ST 4, July 1958.)

pressures or suctions in pounds per square foot. The recommended pressure on vertical surfaces of normal size buildings was 20 lb per square foot. This pressure corresponds to a wind velocity of 77.8 mph in Eq. 3.3. For sloping roof surfaces the recommended external wind pressures are as given in Fig. 3.5a, and the internal effects are given in 3.5b.

In 1961 the Task Committee on Wind Forces of the Committee on Loads and Stresses of the Structural Division of the American Society of Civil Engineers published[8] a final report on Wind Forces on Structures. In this report the Committee presented the coefficients from the

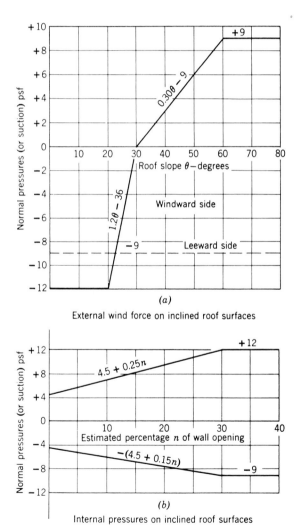

(a)

External wind force on inclined roof surfaces

(b)

Internal pressures on inclined roof surfaces

Fig. 3.5. ASCE Committee (1940) wind loading on sloping roofs.

1936 report (Fig. 3.4) but did not recommend the general wind loading of the 1940 final report (Fig. 3.5). The 1940 recommendation for external effect is the same as the 1936 ASCE Report but for the single wind velocity of approximately 80 mph. The 1961 Report gave values for various wind velocities. This 1961 Report contains many tables listing wind pressure coefficients, at different positions for various types of buildings, taken from the work of V. Ackeret of the Institutes of Aerodynamics of Zurich. These coefficients are contained in the Swiss Building Code. These tables are a very valuable addition to the information available on wind loads and are recommended at the present time above all other data.

The ASCE method also takes into account an additional wind force. The additional loading is the pressure or suction inside the building. When a building has windows or doors that may be open during a wind storm, an additional wind force on the roof structure is created. If the openings are on the windward side of the building but not on the opposite side, an internal pressure can build up within the building. If there are openings on the leeward side but not on the windward side of the building, the wind will create a suction on the interior surface of the roof. Unless either internal condition is specifically known and controlled, the roof should be designed for the internal condition which when combined with the external effects will require the larger sizes of members of the roof structure. In most cases this will result in two separate analyses. One with external wind combined with internal pressure, and the second case of external wind combined with internal suction. The following example shows how to combine these two effects.

EXAMPLE PROBLEM 3.1   Given the gable roof shown below, determine the net pressures on the roof surfaces using ASCE (1940) procedure. The value of n is estimated at 25 percent.

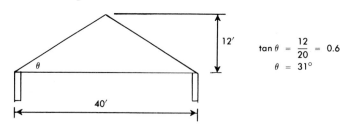

$$\tan \theta = \frac{12}{20} = 0.6$$
$$\theta = 31°$$

External force on windward side $= 0.3\theta - 9$
$$= 0.3(31.0) - 9.0 = 0.3 \text{ psf (pressure)}$$
External force on leeward side $= 9.0$ psf (suction)

Internal force:

Pressure $= 4.5 + 0.25(n) = 4.5 + 0.25(25) = 10.75$ psf

Suction $= -(4.5 + 0.15(n)) = -(4.5 + 0.15(25)) = -8.25$ psf

Net Pressures

(A)

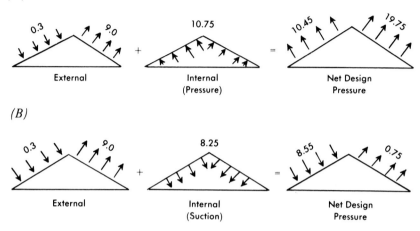

(B)

Note: Roof structure would be required to be designed for net pressures of both *(A)* and *(B)*.

Applications of this wind loading method to the design of roof structures are treated in Chapter 6.

In addition to selecting the proper coefficient in Eq. 3.2 the structural engineer has also to select the proper wind design velocity. Building codes sometimes specify the design wind velocity. If the engineer is not subject to a code, he should seek the latest information on probable maximum wind velocities for the region where his building will be erected. The best source for such information in the United States is the United States Weather Bureau. Figure 3.6 is a map of the United States which gives contour lines indicating the velocity of the fastest mile of wind that would be expected to occur in a fifty-year interval.[9]   The given reference also gives maps showing fastest mile of wind that may occur in two-year and one hundred-year intervals. These maximum velocities are to be considered as occurring at an elevation of 30 ft above the ground. Shielding by other buildings or trees will reduce the expected maximum velocities, but few engineers take into account any reduction in design wind velocity due to shielding. The uncertainty of any shielding is usually too great to rely on. It is known that the velocity of wind varies with height from ground. The values as given in the 1940 ASCE report were recommended for heights from zero to 300 ft. However, in the absence of more definite data, it is generally considered that the wind velocity for heights above 30 ft can be predicted by the following formula:

$$V_h = V_{30}\left(\frac{h}{30}\right)^{\frac{1}{7}}$$ (3.4)

where V_h is the wind velocity at any height h above the ground and V_{30} is the wind velocity at 30 ft above the ground.

Studies of wind forces on structures is continuing and engineers who are designing unconventional type structures in high wind velocity locations will want to research the literature on this subject. There have been several recent international conferences sponsored by the Wind Engineering Research Council and this organization has published extensive bibliographies on the subject of wind forces on structures.

The structural engineer is sometimes confronted with determining wind forces on circular elements such as standpipes, stacks, poles, and ropes. The Swiss data present information for such structures. The total force against such a structural element is given as

$$P = C_n q A$$ (3.5)

where C_n is the shape factor as given in Table 3.4; p is the velocity pressure from Eq. 3.1; and A is the area as seen in elevation of the structure.

As an example, if a smooth metal smokestack is 35 ft high and has a diameter of 5 ft, the total wind force for a wind velocity of 100 mph is

$$\frac{h}{d} = \frac{35}{5} = 7$$
$$C_n = 0.5$$
$$p = 0.002558(100)^2 = 25.58 \text{ psf}$$
$$A = 35 \times 5 = 175 \text{ sq ft}$$
$$P = 0.5 \times 25.58 \times 175 = 2,240 \text{ lb}$$

Such structural elements as listed in Table 3.4 may be subjected to vibration during wind storms. This vibration will take place in a direction perpendicular to the direction of air flow. The forcing vibration is due to the phenomenon called vortex shedding. When the frequency of the vortex shedding is equal to the natural frequency of the structure, large amplitudes of vibration will be induced into the structure and failure can result. An example of this action was the failure of the Tacoma Narrows Suspension Bridge. Failures of other types of struc-

Table 3.4*

Upright cylinder for $d\sqrt{q}>1.5$			
Slenderness h/d	25	7	1
Surface condition	C_n	C_n	C_n
Smooth surface Metal, timber, concrete	0.55	0.5	0.45
Rough surface Round ribs Rib ht. = 2%d	0.9	0.8	0.7
Very rough surface Sharp ribs Rib ht. = 8%d	1.2	1.0	0.8
Smooth rough surface Sharp edges	1.4	1.2	1.0
Long cylinder for $L/d>100$	$d\sqrt{q}$	< 1.5	> 1.5
Surface condition		C_n	C_n
Smooth wires, rods, pipes		1.2	0.5
Rough wires rods		1.2	0.7
Fine wire cables		1.2	0.9
Thick wire cables		1.3	1.1

*Data from Reference (6).

tures have resulted from similar action. This action for long slender members should be investigated as well as the total wind load.

Earthquake Forces.   In many areas throughout the world earthquakes of varying intensity have occurred. When an earthquake occurs, all structures within the active region are subjected to forces due to the acceleration of the supporting earth. These forces will be of a different nature than either the dead or live loads. Engineers and seismologists have been studying the effects of earthquakes on structures for a considerable period of time, but the problem is very complex. Research is difficult because it is not possible to produce in the laboratory a natural earthquake, and of course, it is not yet possible to know when and where such an act of nature will occur. Because of the frequency of earthquakes in California many buildings in this state have been instrumented with recording accelerometers that have produced response records of buildings subjected to earthquakes. Many investigations[10] have been conducted to determine the effects of earthquakes on structures.

Accelograms of several large earthquakes show that ground motions are very complex and that they vary widely with respect to individual earthquakes. Attempts have been made to reduce the ground motion of an earthquake to a few sinusoidal motions; however, in most cases this is far from the true condition. Horizontal ground accelerations of over 30% g have been recorded at distances of 30 to 40 miles from the epicenters of earthquakes. Since the problem of earthquake "forces" is a dynamic problem, it involves not only magnitudes of acceleration but also the frequency of the motion. A flexible building may be able to absorb a great deal more energy than a rigid building. When dealing with static loads the magnitude of the load is irrespective of the nature of the building. This, of course, it not true when dealing with dynamic loads. The dynamic loading will be a function of the nature of the structure itself. This makes for greater complexity.

The analog computer has been used[11] for analyzing structures subjected to complex ground motions. New procedures and methods can lead to more rational and correct approaches to the design of earthquake resistant structures. Earthquakes in Mexico, Japan and California have proven that structures can be designed with very little increase in cost to resist earthquakes of moderately severe intensity, and experience in locations where structures have been built with no provision to resist earthquake forces gives proof that the ignoring of such forces in the construction of buildings can lead to disaster.

Building codes in California have for many years required structural engineers to include earthquake loadings in their designs. The codes have been changed from time to time and also varied somewhat among cities. In

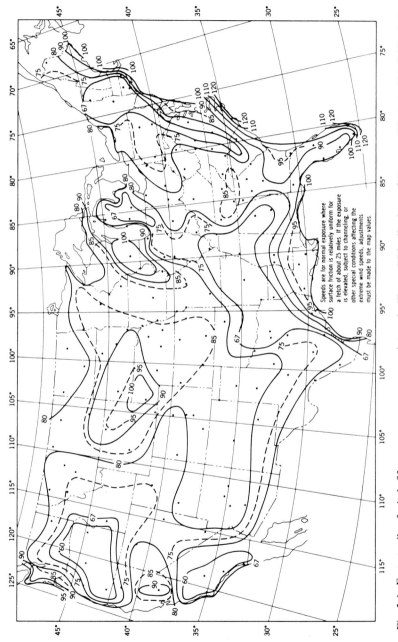

Fig. 3.6. Fastest mile of wind, fifty year mean recurrence interval. (From: "Distributions of Extreme Winds in U.S.," H. C. S. Thom, *Proceedings of the Structural Division of ASCE*, Paper 2433, April 1960.)

Speeds are for normal exposure where surface friction is relatively uniform for a fetch of about 25 miles. If the exposure is elevated, subject to channeling, or other special conditions affecting the extreme wind speeds, adjustments must be made to the map values general, however, they have approached the problem by requiring the structural engineer to design the structure for a static lateral load that is a percentage of the weight of the structure. The Structural Engineers Association of California has published[12] "Recommended Lateral Force Requirements" with the suggestion that such requirements be included in building codes for areas subjected to earthquake shocks.

The SEAOC publication states "The proper application of these lateral force requirements, both in design and construction, are intended to provide minimum standards toward making buildings and other structures earthquake resistive."

There is not room in this text to present all the aspects of the SEAOC Lateral Force Requirements. Designers of structures in earthquake zones should obtain the complete publication. The basic provisions only are included here.

Due to ground accelerations in the lateral direction, inertia forces are induced in the structure. The total lateral force V acting on a structure is determined by the equation

$$V = ZIKCSW \qquad (3.6)$$

Z is a factor dependent upon the possible intensity of earthquakes in the region. The Uniform Building Code divides the U.S.A. into four zones. The zone with the highest seismicity is assigned a $Z = 1.0$. The other three zones have Z's of $0.75, 0.375$ and 0.1875.

The value of I is dependent upon how essential the structure is for survival of the population during an earthquake emergency. I is 1.5 for most essential facilities such as hospitals, fire stations, etc. It is recommended that I not be less than 1.0 for any structure.

The value of K is dependent upon the type of construction. A rigid type of structure will have a high value of K while more flexible structures have lower values of K. The recommended values of K are given in Table 3.5.

W is the total dead load of the structure plus any possible snow load. In storage and warehouse occupancies a minimum of 25 percent of the live load should be included in the value of W.

The values of C and S are determined as follows but CS need not exceed 0.14. The value of C is determined from the formula

$$C = \frac{1}{15\sqrt{T}} \qquad (3.7)$$

but need not exceed 0.12.

The value of T is the period of natural frequency of the structure. This value can be determined from a dynamic analysis of the structure using the

*Table 3.5   Horizontal Force Factor "K" for Buildings
or Other Structures*

TYPE OR ARRANGEMENT OF RESISTING ELEMENTS	Value of K
All building framing systems except as hereinafter classified.	1.00
Building with a box system as defined in Section 1(B).	1.33
Buildings with a dual bracing system consisting of a ductile moment resisting space frame and shear walls or braced frames designed in accordance with the following criteria: 1. The frame and shear walls or braced frames shall resist the total lateral force in accordance with their relative rigidities considering the interaction of the shear walls and frames. 2. The shear walls or braced frames acting independently of the ductile moment resisting space frame shall resist the total required lateral force. 3. The ductile moment resisting space frame shall have the capacity to resist not less then 25 percent of the required lateral force.	0.80
Buildings with a ductile moment resisting space frame designed in accordance with the following criteria: The ductile moment resisting space frame shall have the capacity to resist the total required lateral force.	0.67
Elevated tanks plus full contents, on four or more cross-braced legs and not supported by a building.	2.5
Structures other than buildings.	2.0

structural properties and the deformational characteristics of the resulting elements of the structure. In the absence of a more exact period determination, the value of T for buildings may be determined by using the following formula

$$T = \frac{0.5\,h_n}{\sqrt{D}} \qquad (3.8)$$

where h_n is the height from the base to that level which is uppermost in the main portion of the structure, and D is the direction parallel to the applied forces.

For buildings that have a lateral force resisting system which consists of a moment resisting space frame capable of resisting 100 percent of the required lateral forces the value of T may be determined by the formula

$$T = 0.10 N \tag{3.9}$$

where N is the total number of stories above the base.

In order to use the simple equation (3.9) for T the structural system must not be prevented from resisting the lateral forces by more rigid elements.

The value of S is determined by the following formulas but shall not be less than 1.0:   for

$$\frac{T}{T_s} \leqslant 1.0 \, S = 1.0 + \frac{T}{T_s} - 0.5 \left(\frac{T}{T_s} \right)^2 \tag{3.10}$$

for

$$\frac{T}{T_s} > 1.0 \, S = 1.2 + 0.6 \left(\frac{T}{T_s} \right) - 0.3 \left(\frac{T}{T_s} \right)^2 \tag{3.11}$$

In determining the value of S in the above two formulas T shall not be less than 0.3 seconds. The term T_s in the above formulas in the characteristic site period. This would have to be established from a study of the site geology and the mechanics of the soil on which the building is founded. This value should not be taken as less than 0.5 seconds nor greater than 2.5 seconds.

When T_s cannot be properly established from substantiated geotechnical data then the value of S shall be 1.5.

Exception: Where T has been established by a properly substantiated analysis and exceeds 2.5 seconds, the value of S may be determined by assuming a value of 2.5 seconds for T_s.

After all the variables in equation (3.6) have been determined and the value of V calculated, the distribution of V to the various levels of the

A concentrated force F_1 acting in a lateral direction at the top level structure have to be determined from the formula

$$F_t = 0.07 \, TV \tag{3.12}$$

This value (F_t) need not exceed 0.25 V and may be considered as zero where T is 0.7 seconds or less.

The remaining lateral shear $(V - F_t)$ is distributed over all the levels of the building in accordance with the formula

$$F_x = \frac{(V - F_t) w_x h_x}{\sum\limits_{i=1}^{i=n} w_i h_i} \qquad i = \text{floor number} \tag{3.13}$$

At each level x the force F_x is applied over the area of the building in accordance with the mass distribution on that level.

EXAMPLE PROBLEM 3.2   A small office building is to be designed with the structural steel frame resisting the total lateral load. The dimensions of the frame are shown with the total dead load contributing to each floor (includes wts. of walls). Determine the lateral forces at each floor due to an earthquake in a region of maximum earthquake forces.

The sum of $F_1 + F_2 + F_3$ will equal the quantity V from the following equation

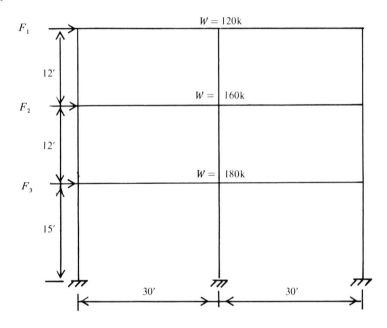

$$V = ZIKCSW$$

for this problem:

$Z = 1.0$, region of maximum seismicity;

$I = 1.0$, not an essential facility;

$K = 0.67$, ductile steel frame resists the total lateral load. (Table 3.5.)

$T = 0.10N = 0.10(3) = 0.30$ sec.

$$C = \frac{1}{15\sqrt{T}} = \frac{1}{15\sqrt{0.3}} = 0.122$$

T_s is established from properly substantiated geotechnical data. If this study showed

$$T_s = 0.8 \text{ sec};$$

then

$$T/T_s = 0.3/0.8 = 0.375 < 1.0$$

$$S = 1.0 + \frac{T}{T_s} - 0.5\left(\frac{T}{T_s}\right)^2$$

$$= 1.0 + 0.375 - 0.5(0.375)^2 = 1.30$$

$$CS = 0.122(1.30) = 0.159; \text{ use maximum of } 0.14$$

$$W = 120 + 160 + 180 = 460\,k$$

Since all variables in equation (3.6) have been determined

$$V = ZIKCSW = 1.0(1.0)(0.67)(0.14)(460) = 43\,k$$

$$F_t = 0.07\,TV = 0.07(0.3)(43) = 0.90\,k$$

$$\sum_{i=1}^{i=3} w_i h_i = (120)(39) + (160)(27) + (180)(15) = 11{,}700$$

$$F_1 = 0.9 + \frac{42.1(120)(39)}{11{,}700} = 17.8\,k$$

$$F_2 = \frac{42{,}1(160)(27)}{11{,}700} = 15.5\,k$$

$$F_3 = \frac{42.1(180)(15)}{11{,}700} = 9.7\,k$$

$$\Sigma = 43\,k$$

F_1 was increased slightly so that in rounding off to the nearest $0.1\,k$ the sum total would equal $43\,k$. In actual practice the lateral forces would most likely be rounded off to the nearest kip.

3.3   Loads on Highway Bridges

General.   The three general types of loads to which a highway bridge is subjected can be classified as: (1) loads due to the weight of the structure itself, (2) vehicular loads, and (3) loads due to nature such as wind, ice, temperature, earth, and water. The magnitude and distribution of some of these loads can be determined quite accurately, whereas others, at best, can only be approximated. A considerable amount of effort has been spent in trying to determine the true nature and magnitude of many of these loads. New information is appearing from time to time in the technical press with regard to bridge loadings. Much more research will be necessary for a close approach to the true nature of many of these loading conditions. The bridge engineer should be aware of the state of knowledge of highway bridge loadings and keep this in mind during the progress of his designs. He should also keep himself well informed on the latest results of investigations even though he may not immediately apply this information because of the natural lapse of time between publication of results of research and the incorporation of such results in the standard specifications.

The bridge designer usually does not have a completely free hand in choosing the loads for which he will design his bridge. In most cases this freedom of choice would neither be desirable nor advisable. In order to have a uniformity of design in our highway bridges and also to insure that all bridges will be adequately designed it is necessary to have a standard specification covering bridge loadings, materials, construction procedures, and major items of design procedure. The standard specification used by the states of the U.S.A. is the *Standard Specifications for Highway Bridges* as adopted by the American Association of State Highway and Transportation Officials (AASHTO). From time to time, for particular jobs, some states do deviate from the standard Specifications. However, generally speaking, all the states use the published AASHTO *Standard Specifications for Highway Bridges.*[13] All bridges on the national Interstate System are designed under these specifications. Here uniformity of design standards is very necessary. All bridges on the Interstate System have a load capacity of HS-20. In the discussion to follow on highway bridges when reference is made to the Specifications, it means the AASHTO *Standard Specifications for Highway Bridges.*

Dead Loads.   Dead loads are defined as the weights of the structure itself. These weights are usually considered as loads that are applied to the structure when it is first constructed and remain on the structure throughout its lifetime. If all the dimensions of the structure are known, the dead loads can usually be determined with a high degree of accuracy. However, the paradox is that the structure cannot reach a final design stage until the final dead loads are known, and the final dead loads cannot be determined until the final design is made. The solution to this problem is, of course, to assume dead load values, and then as the design progresses to revise the dead loads and the design accordingly until a sufficiently close convergence is reached. Experience will indicate how close a convergence is necessary. An experienced designer's first estimate will probably be so close to the computed load that a revision of the design will be unnecessary.

The inexperienced bridge designer should seek information that will aid him in making his first dead load estimate. However, he will have some degree of difficulty because there are so many factors that affect the weight of a highway bridge: length of span, width of roadway, live load, weight of flooring, and others. The procedure is first to design the floor slab and then the stringers and floor beams (see Fig. 5.15). A considerable percentage of the dead load of a highway bridge comes from the weight of the floor system. After the floor system is designed, the weight of the trusses or girders can be estimated and the computations for shear, moments, and forces due to the dead load can proceed. Studies of previous bridge designs will usually aid the designer in his assumptions of dead loads. Some texts have presented formulas for estimating the dead load of highway bridges, but because of the many factors involved and the vintage of some of these formulas their worth is doubtful.

Live Loads. Because of the wide range of vehicles on the highways today it would be a perplexing problem to decide for which actual vehicle or vehicles a bridge should be designed if actual vehicles were to be used. Because of this difficulty this approach is not used. A hypothetical vehicle is the design loading incorporated in the Specifications. This vehicle is not intended to conform to any actual vehicle, but it is a design standard for which bridges can be designed, and the designer can be sure that the bridge will safely carry existing vehicles which conform to the load limit laws enforced by the various states.

The standard AASHTO bridge loadings are divided into five classifications: H10-44, H-15-44, HS15-44, H20-44, and HS-20-44. These loadings consist of a standard truck of variable weight depending on the classification. This standard truck is shown in Fig. 3.7.

It should be noted that the HS loadings incorporate the standard truck from the H loading plus a semitrailer whose axle weight is the same as the rear axle of the truck. This semitrailer axle is placed at a distance varying from 14 to 30 ft. The spacing between these two axles to be used is that which produces maximum stresses. For simple-span structures this spacing will usually be 14 ft, whereas for continuous structures the critical spacing may be greater than 14 ft. The standard truck occupies a 10-ft width lane. This means that in multiple lane bridges the trucks are not spaced any closer than 10 ft center to center. The width of a rear tire for the H and H-S loadings is equal to 1 in. for each ton of total truck weight. For instance, the width of rear tire for the H20 or HS20 is 20 in.

An alternate loading, used in the design of structures on the Interstate Highway System, in lieu of the HS20 AASHTO truck is a pair of axles spaced at 4 ft and weighing 24 kips each. This loading will be critical only for short-beam spans. Comparing moments on simple beam spans it can be shown that the single 32-kip axle will produce the maximum moment for spans from zero to 11.5 ft. The 24-kip pair of axles will be the critical loading for spans from 11.5 to 37 ft, and the full vehicle will be critical for spans above 37 ft. The same relationship for maximum shear on simple beam spans can be determined. The division points for

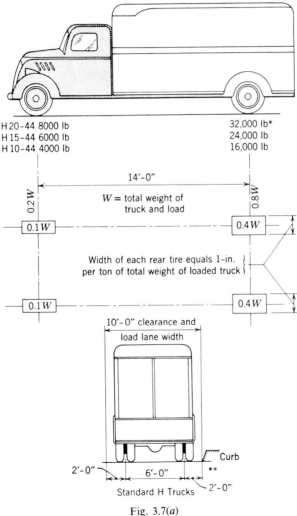

Fig. 3.7(a)

shear among the three types of loadings will be at 6 ft and 22 ft respectively.

An alternate loading to the truck in lieu of a train of vehicles is the lane loadings shown in Fig. 3.8. The Specifications state, the type of loading, whether lane loading or truck loading, to be used, and whether the spans be simple or continuous, shall be the loading which produces the maximum stress.*

An investigation of moments and shears on simple spans using the H-loadings shows that for the shorter spans the truck loading produces

* *Standard Specifications for Highway Bridges*, AASHTO, 1977, p. 21.

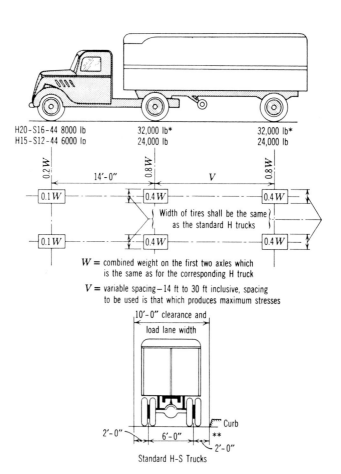

Fig. 3.7(*b*) Standard loadings of the AASHTO bridge specification.

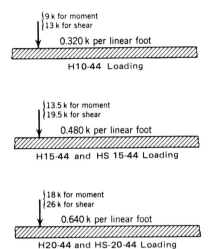

Fig. 3.8   H and HS lane loading (AASHTO).

the greater values, whereas for longer spans the lane loadings control. Figure 3.9 shows the relationship for moments of simple spans from 0 to 100 ft for the H20-44 loading. Appendix A of the AASHTO Specifications contains tables that indicate which loading is the most critical for various simple beam spans. These tables can also be used to an approximate extent

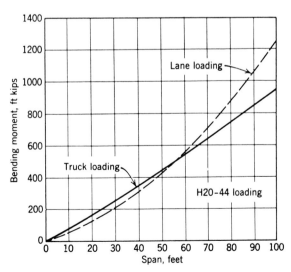

Fig. 3.9

to indicate which loading controls for truss spans. Where there is doubt, computations should be made for both loadings and the maximum value used.

Preliminary to the design of a highway bridge, it is necessary to decide for which one of the five load classes the bridge will be designed. The usual procedure is that the H10-44 is used only for lightly traveled roads carrying only very light truck traffic. Very few present-day bridges are designed for the H10-44 loading. The wisdom of designing bridges for such a light loading in view of our present-day traffic and vehicle weights is questionable. The H15 and HS15 loadings are used very often for designing bridges for county or state roads that are not part of the national highway system. One factor should be recognized in considering the H15 or HS15 loading. For long-span bridges the stresses caused by the dead load are likely to be greater in magnitude than those caused by the live load plus impact. As the span becomes greater the saving in cost of the structure by using the H15 instead of the H20 loading becomes smaller. An example of which the writer is familiar is a bridge that was designed for a HS15 in which the total cost of the bridge would have been increased approximately 5 to 10% for an increase in loading to HS20. A factor of considerable importance in this example was that the size of a number of the compression members due to minimum L/r requirements would have been adequate for the forces under the HS20 loading. In view of the uncertainty of future loadings and the increased utility of a structure designed for the heavier loading, this example shows false economy.

Investigations have been conducted in the past,[14], [15], [16] and the ASCE has a working committee studying highway bridge loadings. Future specifications may contain changes in specified loadings as the results of future investigations and studies are made known.

It is not unusual to have a truck of unusual or unlawful weight or size requesting passage over a section of highway from the law enforcement branches of our state highway departments. This happens when an unusual piece of military or industrial equipment has to be moved from one place to another. When such a request is made, the state engineering department has to check the effect of such a vehicle on the bridges en route. It is not necessary to do a complete stress analysis of every bridge that the unusual loading is to use, which might be a formidable task. It is, however, only necessary to compare the subject loading against the standard design loading. The comparison of wheel and axle loads, shears, and bending moments for the range of span lengths to be encountered en route can readily indicate if the bridges have adequate strength for the unusual vehicle.

Impact.   Live loads applied to bridges are in nature moving loads. However, in structural analysis it is usually necessary to treat all loads as stationary. It is a well-proven fact that the stresses produced by a vehicle moving across a bridge will be different from the stresses produced if the vehicle were in a stationary condition. This increment of stress due to a moving load instead of a stationary one is usually termed the *dynamic effect*. In bridge design this dynamic effect is called *impact*. The problem of determining the dynamic effect from an analytical basis is very complex, and no practical methods for use by the bridge designer have been developed. However, the advent of the bonded electrical strain gage and associated electronic recording equipment has made research investigations practicable in this area. The American Railway Engineering Association has had a vigorous research program aimed at the determination of more correct values of impact to use in design of railway bridges.[17] The AREA Bridge Design Specifications have been revised in the light of these investigations. Some investigations in impact for highway bridges have been conducted[18] and it is expected that the future will produce more investigations and more information as regards the dynamic effect of moving loads on highway bridges.

The dynamic effect of a vehicle moving across a bridge is due to two causes: the *sudden loading* of the structure and the *blow effect* caused by the wheels passing over rough surfaces of the roadway and the resulting bouncing of the load on the springs of the truck. This latter effect can be more correctly termed *impact*; however, this term which is loosely applied to bridge design specifications includes both of the causes just mentioned.

The maximum stress resulting from sudden loading is a function not only of the force (or weight) of the load but also the rate of time it takes the force to reach its peak value (see Fig. 3.3). This time function is not a direct time function but a time rate of loading versus period of natural frequency ratio. For beam bridges this time rate of loading for maximum moment will be shorter for shorter spans. It can then generally be said that the longer the span the less the effect of sudden loading on maximum bending stresses. However, this is not necessarily so for maximum shear in beam or truss spans. Another factor entering into the effect of sudden loading is the ratio of the mass of the load to the mass of the structure. The greater the ratio the greater the effect of the sudden loading. It can be stated generally that the longer the span the less the effect of the sudden loading. This is also generally true for the effect of direct impact. It should be remembered that the previous statements are general, and particular items of the structure may not conform exactly to these relationships. The reader is referred to the references at the end of the chapter[18,19] for a more complete treatment of the effect of dynamic loads.

The AASHTO Bridge Specifications make allowance for the dynamic effect by considering this effect as a fraction of the live load. This fraction is called impact and is determined by the following formula:

$$I = \frac{50}{L + 125}$$

where L is the length in feet of the portion of the span which is loaded to produce the maximum stress in the member. The maximum value of I is limited to 0.30. This maximum value of 0.30 is used for loaded lengths of 42 ft and less.

In order to obtain the moment, shear, force, or stress due to impact, the live load value is multiplied by the value of I obtained from the preceding equation. This value is then added to those quantities determined for dead load and live load.

At best, the present specifications for determining impact in highway bridges are a rough approximation, taking into effect only one factor— length of span—of the many factors that have an effect on the dynamic response of the structure. As impact load is only a minor portion of the total load on any bridge; any inaccuracy from the use of the present formula is small in magnitude. Nevertheless this should not deter engineers from seeking more correct methods.[20]

Wind Loads. It has been known to bridge engineers for a long period of time that the force exerted by a wind on a bridge structure could cause appreciable forces in the parts of the structure. This has been dramatically proven on several occasions by the collapse of bridges subjected to moderate and high winds. Engineers have for some time tried to arrive rationally at methods for predicting the effect of wind forces on bridges. The problem is complex because of the many variables such as the shape and size of the members of the structure, probable angles of attack of the wind, shielding effects of terrain, and velocity of the wind. The gust rather than the steady-state nature of wind further complicates the problem. Until recent years the design approach to wind effects on bridge structures was not much better than just a guess.

The AASHTO Specifications relating to wind loads were the result prin-cipally of past experience. Standard bridge specifications have treated wind loads as a static force, the magnitude of the force being a function only of the velocity of the wind and the surface area restricting the free movement of the air. The pressure was no doubt determined by inserting a probable maximum wind velocity into the equation $p = 0.0033V^2$. A 100 mph velocity gives a pressure of 33 lb per square foot on the exposed surface. The problem has been approached by several investigators by subjecting bridge models to wind tunnel tests.[21,22] The results of these tests have been published in the technical press, and the bridge specifications have been revised according to these investigations.

The present specifications require that the wind load on a bridge super-structure be considered as acting at right angles to the longitudinal axis of the bridge and be of similar nature as a moving load. The intensity of the load should be determined by considering a specified pressure acting on the exposed area of all members including railing and floor system as seen in elevation. The following pressures are specified for a wind velocity of 100 mph:

Trusses and arches	75 lb per square foot
Girders and beams	50 lb per square foot

The total force should not be less than 300 lb per linear foot in the plane of the loaded chord and 150 lb per linear foot in the plane of the unloaded chord for truss spans, and not less than 300 lb per linear foot for girder spans.

These wind loads are to be used without any live load on the structure and with an increase in the allowable working stress of 25%. When wind loads are to be combined with live loads, then only 30% of the mentioned wind load should be applied to the superstructure. However, in addition, a wind load on the live load should be included. This wind loading is 100 lb per linear foot applied at right angles to the longitudinal axis of the structure and 6 ft above the deck. This wind load is to be considered as applied to a moving live load. The combination of loading-dead load, live load, and wind load also utilizes a 25% increase in allowable working stress.

The designer has the option of designing for wind velocities other than 100 mph when he deems it is warranted. Then, the pressures for velocities should be determined by multiplying the already indicated pressures by the ratio of the square of the design wind velocity to the square of 100.

Miscellaneous. Several other forces of nature act on highway bridge structures such as thermal forces, earth pressure, ice and current pressure, and earthquake stresses. Since the scope of this textbook does not warrant their discussion, for more details the reader should consult the **AASHTO** *Standard Specifications for Highway Bridges.*

The bridge design engineer should be sure to include all possible forces the bridge is apt to be subjected to during its life. Combining the effects of many forces should be done on a rational and reasonable basis.

3.4 Railroad Bridges

Dead Load. Various formulas have been proposed in the past for estimating in advance of design the weights of railroad bridges.[23] However,
since changes in design specifications and live loadings will render these
formulas obsolete, they can only serve as a very rough guide. Reference
to actual past designs is the best source if readily available. As in
highway bridges the floor system is designed first and only the weight of
the main trusses or girders requires an estimate.

Live Load. The design live load for railway bridges has for many years
been essentially one devised by Theodore Cooper in 1894. This consists
of two locomotives followed by a string of cars. When Mr. Cooper, a
prominent bridge design engineer, first proposed his loading, the weight
of the driver axles were 40 kips and the uniform load representing the cars
was 4 kips per foot of track. As the size of locomotives increased with the
years, the design vehicle was increased in weight but the makeup and
spacing of axles remained the same. Today the American Railway
Engineering Association (AREA) Specification recommends a loading
with a 72-kip weight on each of the driver axles. This loading is classed as
a Cooper E72 loading as contrasted to Cooper's E40 original loading.
Railway bridge design offices have developed tables and charts giving
shears and moments for a wide range of simple spans loaded with the
various Cooper loadings. Figure 3.10 shows the Cooper E72 loading.
The table below the loading diagram is very useful for analyzing structures
loaded with Cooper E72 load.

Impact. The additional stress effect due to the live load being a moving
load and not a static load is evaluated for railroad bridges in the same
manner as for highway bridges. The impact effect is taken as a percentage
of the live load effect. The amount of impact in railway bridges has been
determined after an extensive investigation as previously noted in Section
3.3.

The effect of impact to be considered in the design of railway bridges is
given by the latest AREA Specification as follows:

1. To the axle loads specified (Cooper loading), there shall be added impact
forces, applied at the top of rail and distributed thence to the supporting members,
comprising:

   (a) The rolling effect. Vertical forces due to the rolling of the train from
   side to side, acting downward on one rail and upward on the other, the forces
   on each rail being equal to 10% of the axle loads.

   (b) The direct vertical effect. Downward forces, distributed equally to the
   two rails and acting normal to the top-of-rail plane, due, for steam locomotives,

to hammer blow, track irregularities, speed effect and car impact, and equaling the following percentage of the axle loads:

1. For beam spans, stringers, girders, floor beams, posts of deck truss spans carrying load from floor beam only, and floor beam hangers:

$$\text{For } L \text{ less than 100 ft} \qquad 60 - \frac{L^2}{500}$$

$$\text{For } L \text{ 100 ft or more} \qquad \frac{1800}{L - 40} + 10$$

2. $\qquad \text{For truss spans} \qquad \frac{4000}{L + 25} + 15$

or due, for rolling equipment without hammer blow (diesels, electric locomotives, tenders alone, etc.) to track irregularities, speed effect and car impact, and equaling the following percentage of axle loads:

$$\text{For } L \text{ less than 80 ft} \qquad 40 - \frac{3L^2}{1600}$$

$$\text{For } L \text{ 80 ft or more} \qquad \frac{600}{L - 30} + 16$$

$L =$ length, in feet, center to center of supports for stringers, transverse floorbeams without stringers, longitudinal girders and trusses (main members).

$L =$ length, in feet, of the longer adjacent supported stringer, longitudinal beam, girder, or truss for impact in floorbeams, floorbeam hangers, subdiagonals of trusses, transverse girders, supports for longitudinal and transverse girders and viaduct columns.

(c) For members receiving load from more than one track, the impact percentage shall be applied to the static live load on the number of tracks shown here:

Load received from:
 Two tracks:

For L less than 175 ft	Full impact on two tracks
For L from 175 ft to 225 ft	Full impact on one track and a percentage of full impact on the other as given by the formula, $450 - 2L$
For L greater than 225 ft	Full impact on one track and none on the other

 More than two tracks:

For all values of L	Full impact on any two tracks

Wind Loads. The wind forces on railway bridges are treated in a manner similar to that for highway bridges. The wind loads given by the 1959 AREA *Specifications for Steel Railway Bridges* are as follows:

The wind force shall be considered as a moving load acting in any horizontal direction. On the train it shall be taken at 300 lb per linear foot on the one track, applied 8 ft above the top of rail. On the bridge it shall be taken at 30 lb per square foot of the following surfaces:

1. For girder spans, $1\frac{1}{2}$ times the vertical projection of the span.
2. For truss spans, the vertical projection of the span plus any portion of the leeward trusses not shielded by the floor system.
3. For viaduct towers and bents, the vertical projections of all columns and tower bracing.

The wind force on girder spans and truss spans, however, shall not be taken at less than 200 lb per linear foot for the loaded chord or flange, and 150 lb per linear foot for the unloaded chord or flange.

If a wind force on the unloaded bridge of 50 lb per square foot of surface as just defined, combined with the dead load, produces greater stresses than those produced by the specified wind forces, combined with the stresses from dead load, live load, impact, and centrifugal force, the members wherein such greater stresses occur shall be designed with this in mind.

Miscellaneous.   There are also other loads to consider in the design of railway bridges. The forces are centrifugal force for bridges on curves, lateral forces due to nosing of the locomotive, and longitudal forces due to braking or traction. Temperature effects may also have to be taken into account for certain type structures.

Substructures will possibly also have to be designed for the usual earth pressures, stream flow, ice pressures, etc. The reader is referred to the AREA Specifications for details on these other loadings.

Problems

1. An office building has a floor system composed of girders spanning between columns spaced 24 ft. in one direction and 30 ft. in the other direction. Beams, spaced 7.5 ft. center to center, span between girders with the beams parallel to the 24 ft. direction. The floor consists of 4 in. of reinforced concrete, and the live load is in accordance with the Table 3.2(a). A typical interior bay of the building is shown in Fig. 3.11. Make a line sketch of a beam showing the magnitude of dead and live load per ft. of beam. Also do the same for a girder. Beam loads to girders are considered as concentrated loads.

2. If the same structural framing plan is used for the roof as for the main floor of the building of Problem 1, and the roof consists of 3 in. of lightweight concrete (110 pcf) and a 5 ply felt and gravel roof, what is the dead load per beam? Show the loading also for the girders.

3. What would be the snow load for the roof of Problem 2 if the building were to be erected in Chicago?

4. A symmetrical gable roof has a pitch (rise to span) of 1:4. Compare the external pressures due to wind on the windward and leeward roof surfaces by the 1936 ASCE and Swiss methods. Use wind velocity of 100 mph.

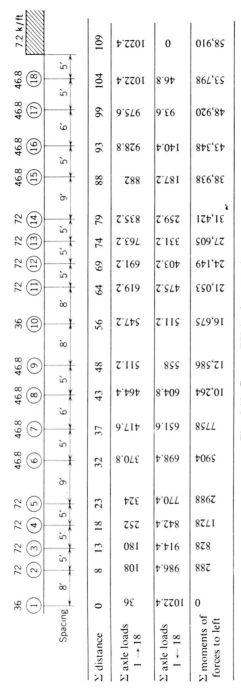

	1	2	3	4	5	6	7	8	9	10	11	12	13	14	15	16	17	18	7.2 k/ft
Axle load	36	72	72	72	72	46.8	46.8	46.8	46.8	36	72	72	72	72	46.8	46.8	46.8	46.8	
Spacing		8'	5'	5'	5'	9'	6'	5'	8'	8'	5'	5'	5'	5'	9'	5'	6'	5'	5'
Σ distance	0	8	13	18	23	32	37	43	48	56	64	69	74	79	88	93	99	104	109
Σ axle loads 1→18	36	108	180	252	324	370.8	417.6	464.4	511.2	547.2	619.2	691.2	763.2	835.2	882	928.8	975.6	1022.4	1022.4
Σ axle loads 1←18	1022.4	986.4	914.4	842.4	770.4	698.4	651.6	604.8	558	511.2	475.2	403.2	331.2	259.2	187.2	140.4	93.6	46.8	0
Σ moments of forces to left	0	288	828	1728	2988	5904	7758	10,264	12,586	16,675	21,053	24,149	27,605	31,421	38,938	43,348	48,920	53,798	58,910

Fig. 3.10. Cooper E72 load per track.

65

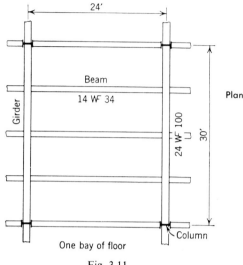

Fig. 3.11

5. Determine the external and internal pressures or suction on a symmetrical gable roof whose slope is 40 degrees and the percentage of openings is 20. Use 1940 ASCE loading.

6. A skyscraper is to be constructed with a total height of 500 ft. Show a plot of wind pressure versus height of building for the walls of this building using the basic formula of Eq. 3.3 and a wind velocity of 100 mph at 30 ft. above ground.

7. A six story office building ($I = 1.25$) is to be built in a region of maximum seismicity. The structural frame is as shown in Fig. 3.12 with the total dead loads per floor level per frame. The frame is to be designed as a moment resistant space frame capable of resisting 100% of the total lateral forces. A geotechnical study showed Ts to be 2.2 seconds. Determine the lateral load at each level of the frame due to an earthquake. Using the SEAOC Recommended Lateral Force Requirements.

8. Determine the lateral load at each level for the frame of Problem 7 but the frame is a braced frame (crossed diagonals in the middle bay at each level). The braced frame resists all of the lateral load.

9. The dimensions of a highway bridge are as shown in Fig. 3.13. The size of the interior stringers is W 24 X 76 (24 in. high and 76 lbs./ft.) and the exterior stringers W 21 X 68. The floor beams are W 36 X 170. The estimated weight of each truss (together with bracing) is 400 lbs. per linear foot. Draw a sketch showing magnitude and position of the dead loads on the stringers (uniformly distributed load), and floor beams, and the dead loads at each panel point of the truss. Take truss dead loads as acting at the lower chord panel points only.

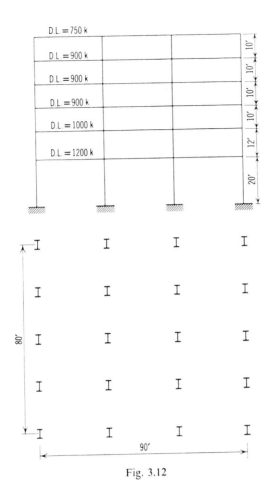

Fig. 3.12

10. A highway bridge consists of simple beam spans of 60-ft. length. What is the impact percentage to be used for determining the maximum moment.

Ans. 27%.

11. For the truss of Fig. 3.13 what is the maximum total lateral wind force on the structure in accordance with the AASHTO specification? Use 15 in. as the width of the chords and end post and 10 in. as average width of the web members. Take length of members as the distance between panel points and neglect areas of gusset plates (plates to which members are connected). Wind should also be considered as acting on stringers and curb.

12. A single-track railway bridge carrying steam locomotives consists of 80-ft. simple girder spans. What is the total impact percentage to be applied to each girder in accordance with AREA specification? Consider each girder supports one rail.

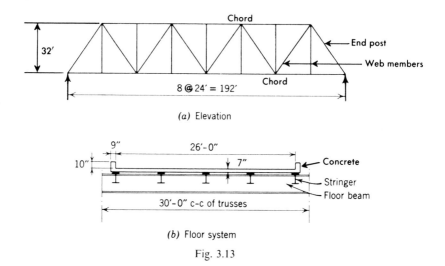

(a) Elevation

(b) Floor system

Fig. 3.13

13. A double-track railway bridge carrying diesel locomotives consists of a deck truss of 360-ft. simple span. What is the impact percentage to be applied to the main members of the truss in accordance with the AREA specification?

Ans. 17.8% on one track, none on the other.

References

1. Norris, C.H., R.J. Hansen, M.J. Holley, J.M. Biggs, S. Namyet, and J.K. Minami, *Structural Design for Dynamic Loads,* McGraw-Hill, 1959.
2. International Conference of Building Officials, *Uniform Building Code.*
3. Ketchum, V. and J.H. Cissel, "Snow Load on Buildings," *Engineering News Record,* July 7, 1949, p. 76.
4. Singell, Thomas W., "Wind Forces on Structures: Forces on Enclosed Structures," *Proc. ASCE, Structural Div.,* Paper 1710, July 1958.
5. "Fifth Progress Report ASCE Sub-Committee 31," *Proc. ASCE,* p. 398, 1936.
6. "Schweizerischer Ingenieur and Architekter Verein," *Technische Normin* No. 160, 1956.
7. "Wind Bracing in Steel Buildings," Final Report ASCE Sub-Committee No. 31, *Trans. ASCE,* Vol. 105, p. 1712, 1940.
8. Paper No. 3269, *Trans. ASCE,* Vol. 126, Part II, pp. 1124-1198, 1961.
9. Thom, H.C.S., "Distribution of Extreme Winds in U.S.," *Proc. ASCE Structural Div.,* Paper 2433, April 1960.
10. Blume, John A., "Structural Dynamics in Earthquake-Resistant Design," *Proc. ASCE Structural Div.,* Paper 1695, July 1958.
11. Bycroft, G.N., "White Noise Representation of Earthquakes," *Proc. ASCE Eng. Mechanics Div.,* Paper 2434, April 1960.
12. "Recommended Lateral Force Requirements of Seismology Committee, *Structural Engineers Association of California,* 1974.

13. American Association of State Highway and Transportation Officials, *Standard Specifications for Highway Bridges,* 1977.

14. "Live Loads for Long-Span Highway Bridges," *Proc. ASCE, Separate No. 198,* June 1953.

15. Asplund, S.O., "Probabilities of Traffic Loads on Bridges," *Proc. ASCE, Separate No. 585,* January 1955.

16. Texas Eng. Experiment Station, *Texas A. and M., Bulletin* 127, 131, 135.

17. Ruble, E.J., "Impact in Railroad Bridges," *Proc. ASCE Structural Div.,* Paper No. 735, July 1955.

18. Edgerton, R.C. and R.C. Beecroft, "Dynamic Stresses in Continuous Plate Girder Bridges," *Trans. ASCE,* Vol. 123, 1958.

19. Tung, T.P., L.E. Goodman, T.Y. Chen, and N.M. Newmark, "Highway Bridge Impact Problems," *Bulletin 124, Highway Research Board,* 1956.

20. Wen, R.K., "Dynamic Response of Beams Traversed by Two-Axle Loads," *Proc. ASCE, Eng. Mechanics Div.,* Vol. 86 EM 5 pp. 91-111, October 1960.

21. Biggs, John M., Namyet, Saul, and Adachi, Jiro, "Wind Loads on Girder Bridges," *Proc. ASCE, Separate No. 587,* January 1955.

22. Biggs, John M., "Wind Loads on Truss Bridges," *Proc. ASCE, Separate No. 201,* July 1953.

23. Merriman-Wiggin, *American Civil Engineers Handbook,* John Wiley and Sons, 5th edition, p. 1162.

ANALYSIS OF BEAMS

4.1 Introduction

The beam is probably the oldest of all structural forms and is the most widely used today. A beam is generally defined as a structural member that is principally subjected to transverse loads. In some structures beams are subjected to axial loads as well as transverse loads. However, if the major loads are transverse to the axis of the member, the structural element is termed a beam. A member whose major loading is parallel to its axis is classified as a compression member or tension member.

Beams can be categorized into two classes: (1) statically determinate beams and (2) statically indeterminate beams. Statically determinate beams are those in which all the unknown forces acting on the beam can be determined by the equations of statics (Eqs. 2.1 and 2.2). If the number of unknown components of reaction are more than the applicable number of equations of statics, then the beam is classified as statically indeterminate. Figure 4.1 shows three beams acted upon by known loads P_1 and P_2. Beam (a) is statically determinate. There are three unknown components of reaction: two vertical and one horizontal. Since there are three applicable equations of statical equilibrium, the unknown forces can be determined by these equilibrium equations. Beam (b) is statically indeterminate as there are four unknown reaction forces. Beam (c) is statically determinate although for the entire structure there are four unknown reaction forces. However, the structure is not just one rigid body but two

71

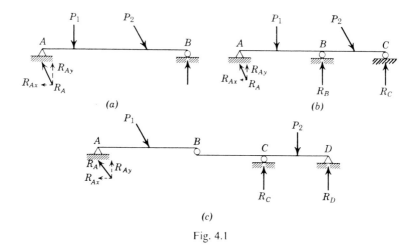

Fig. 4.1

rigid bodies, whereas beams (a) and (b) are single rigid bodies. Each rigid
body of beam (c) has the same number of unknown reaction forces as
applicable equations of equilibrium. The reactions at A and B for beam
AB is first determined. The reaction at B is then applied as a force at B
on the beam BCD (in opposite direction to the direction of the reaction).
The reactions at C and D are then found by the usual methods of statics.

4.2   Shears

In order to determine the necessary dimensions of a beam so that it can
safely carry the loads imposed on it, it is necessary to determine the shears
and bending moments that the imposed loads cause in the beam.

The shear at any point a on a beam is defined as the algebraic summation
either to the left or the right of point a of all the forces perpendicular to the
axis of the beam. In structural engineering computations it is usually
advantageous to plot a diagram showing the value of shear at all points
along the beam for a particular loading. This diagram is called a shear
diagram. To plot it, it is necessary to define positive and negative shear.
The usual sign convention is as follows:

+SHEAR.   If the beam were suddenly severed at a point and the loads
remained acting on the beam, a positive shear would cause the left-hand
portion of the beam to move up with respect to the right hand-portion,
that is, the resultant of the normal forces to the left of the point would be
upward.

—SHEAR. Negative shear would cause the left-hand portion of the beam to move down with respect to the right-hand portion; that is, the resultant of the normal forces to the left of the point would be downward.

Figure 4.2 shows a simply supported beam AE with concentrated loads acting at B and C and a uniform load between D and E. The first step is to determine all the external forces acting on the beam. The unknown external forces are the reactions at A and E. The reaction at E is determined by writing the equation for moment about point A and equating it to zero. There can be no moment at A when the end of the beam rests on a simple support. The reactions at A are then found by writing $\Sigma F_x = 0$ and $\Sigma F_y = 0$ for the beam as a whole. The next step in the process of determining the shear diagram is to proceed along the beam (usually from left to right) determining the algebraic sum of the forces normal to the axis of the beam that are to the left of the point under consideration. It is readily seen that the only force normal to the beam and to the left of any point between A and B is the y component of the reaction at A (13 kips). Therefore, the shear diagram will have a constant value of $+13$ kips between A and just to the left of B. The sum of the normal forces to the left of any point between B and C will be $13 - 11 = +2$ kips. This value of shear will remain constant up to a point just to the right of point C where the shear will suddenly change to $+2 - 8 = -6$ kips. Only the y component of the force at point C causes shear in the beam. Between D and E the value of shear will be constantly changing due to the uniform load. The shear diagram line always begins and ends with a value of zero at the end of the beam.

The beam of Fig. 4.2 is idealized to a certain extent. The beam itself is represented by a line that is the centroidal axis of the beam. For the purposes of the problem the reactions and the loads are considered to be applied to the beam at the centroidal axis.

(1) $\Sigma M_A = 0$ $R_E = \dfrac{11 \times 4 + (10/5) \times 4 \times 10 + 2 \times 4 \times 16}{18}$

$$= \frac{252}{18} = 14 \text{ kips}$$

(2) $\Sigma F_y = 0$ $R_{Ay} = 11 + 8 + 8 - 14 = 13 \text{ kips}$

(3) $\Sigma F_x = 0$ $R_{Ax} = \frac{3}{5} \times 10 = 6 \text{ kips}$

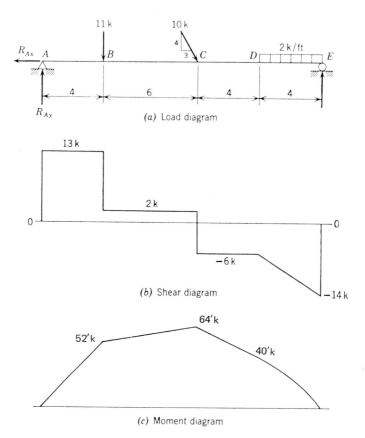

(a) Load diagram

(b) Shear diagram

(c) Moment diagram

Fig. 4.2

An important relationship between load and shear is shown in Fig. 4.3.

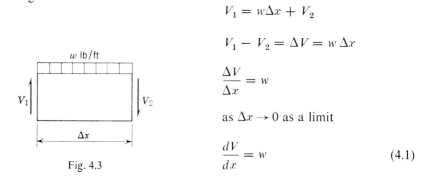

Fig. 4.3

$$V_1 = w\Delta x + V_2$$

$$V_1 - V_2 = \Delta V = w\,\Delta x$$

$$\frac{\Delta V}{\Delta x} = w$$

as $\Delta x \to 0$ as a limit

$$\frac{dV}{dx} = w \qquad (4.1)$$

This development shows that at a point the value of dV/dx, the slope of the shear diagram, is equal to the rate of loading at that point. It should also be noted that the change in shear between two points on a beam is equal to the total load between the two points.

4.3 Bending Moment

The bending moment at any point a in a beam is determined by taking the algebraic sum of the moments of the forces either to the left or the right of point a. The procedure for plotting a bending moment diagram is quite similar to the plotting of a shear diagram. First, all the external forces acting on a beam are determined. Second, the value of moment at specific points along the beam is computed. A sign convention must be established for bending moment. The standard convention is that moments causing compression on the top fiber and tension on the bottom fiber of the beam are positive. Moments producing the opposite effect are negative.

+ Moment   − Moment

The moment at point A and point E of the beam of Fig. 4.2 is zero by nature of the condition of the support. For all points on the beam between A and B, the bending moment is equal to R_{Ay} times the distance x from the origin at point A to the point in question. The equations for moment for points along the beam can be written.

From A to B $\qquad\qquad\qquad M = R_{Ay} \cdot x$

This is a first-degree equation resulting in a straight line for the bending moment diagram between A and B.

From B to C $\qquad\quad M = R_{Ay} \cdot x - 11(x - 4)$

where x is still the distance to R_{Ay}. This is also a first-degree equation. The slope dM/dx of the bending moment diagram, however, will be different between B and C from what it is between A and B.

From C to D

$$M = R_{Ay} \cdot x - 11(x - 4) - 10(\tfrac{4}{5})(x - 10)$$

From D to E

$$M = R_{Ay} \cdot x - 11(x - 4) - (\tfrac{40}{5})(x - 10) - 2(x - 14)\frac{x - 14}{2}$$

$$M = R_{Ay} \cdot x - x^2 + 9x - 72$$

This is a second-degree equation and results in a parabolic bending moment diagram between points D and E. For determining the bending moment between D and E it would have been easier to write the equation going from E to D with the value of $x = 0$ at point E. This equation would be

$$M = R_{Ey}(x) - 2x\frac{x}{2}$$

As previously mentioned the reaction at A and the load at C were considered to be acting at the centroidal axis of the beam. In actual beams this may not be so. If the reaction at A and the load at C were applied at the bottom and top fiber, respectively of the beam, small additional bending moments would be induced into the beam at these points.

4.4 Relationship between Shear and Moment

A segment of a beam is shown in Fig. 4.4 with the loads, shears, and moments indicated.

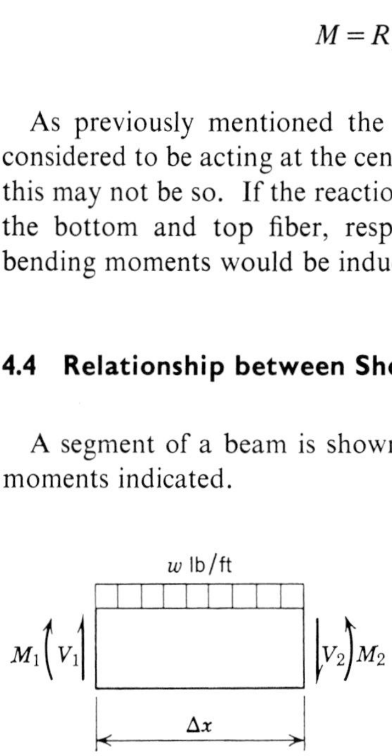

w lb/ft

Fig. 4.4

$$M_2 = M_1 + V_1 \Delta x - w\frac{\Delta x^2}{2}$$

$$M_2 - M_1 = \Delta M = V_1 \Delta x - w\frac{\Delta x^2}{2}$$

$$\frac{\Delta M}{\Delta x} = V_1 - \frac{w}{2}\Delta x$$

as $\quad \Delta x \to 0 \qquad \dfrac{dM}{dx} = V \qquad (4.2)$

This development shows that the value of dM/dx, the slope of the moment diagram, at a point is equal to the shear at that point. The change in moment ΔM between any two points is equal to the area of the shear diagram between the two points. The validity of these two statements can be verified by the shear and bending moment diagrams of Fig. 4.2.

An exception to the rule that the change in moment is equal to the area under the shear diagram takes place when an external moment is applied to the beam at some point. As an example in Fig. 4.5 a moment of 16 ft-kips is applied to the beam at C. The moment diagram changes suddenly from 38.4 ft-kips just to the left of C to 54.4 ft-kips just to the right of C.

The shape of the shear diagram is easily determined from an inspection of the load diagram, and the shape of the moment diagram can be quickly sketched from a study of the shear diagram. It should be noted that if $dM/dx = V$, a point of maximum moment (where $dM/dx = 0$) always

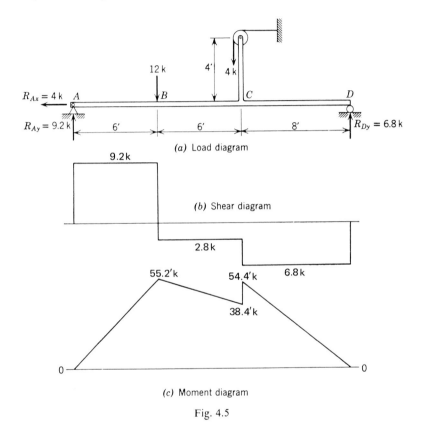

(a) Load diagram

(b) Shear diagram

(c) Moment diagram

Fig. 4.5

occurs where the value of shear is zero. The shear can be zero at several points along a beam. At each point of zero shear the slope of the moment diagram is zero. The moment diagram changes from a slope of one sign to a slope of the opposite sign.

In calculating the bending moment at any location on a beam that is loaded with several loads, it is less time consuming to determine the bending moment by remembering that the increment of bending moment between two points is equal to the area of the shear diagram between the two points. As an example of this refer again to the beam in Fig. 4.2. It is required to calculate the bending moment at a point midway between C and D. This moment can be found by the sum of the bending moment at $C(64\,ft\text{-}k)$ plus the shear at $C(-6\,k)$ multiplied by the distance of two feet.

$$M = 64 - 6 \times 2 = 52\,ft\text{-}k$$

This value of bending moment could also be determined by considering the segment of the beam to the right of the location under consideration, as a free body. The bending moment is then

$$M = 40 + 6 \times 2 = 52\,ft\text{-}k$$

The bending moment due to the incremental area of the shear diagram adds to the moment at D because the resultant of forces to the right of D is upward. This upward force to the right will cause compression in the top of the beam which is a positive bending moment.

4.5 Influence Diagrams

An influence diagram is a graph representing a *load function* as ordinate, and *position of load* on the beam as abscissa. The *load function* for a beam is usually the reaction, shear, or bending moment at a *particular point* on a beam. The difference between an influence diagram and a shear or moment diagram should be firmly understood. A shear or moment diagram is drawn for a condition in which all the loads are *fixed* in position on the beam. An influence diagram is drawn when the loading condition consists of loads moving across the structure. A shear diagram, for instance, represents the value of shear at all points on a beam with *one* particular loading position. An influence diagram for shear gives the value of shear at *one* particular point on the beam for a unit concentrated load moving across the beam.

Consider first an influence diagram for the left reaction of a beam simply supported at the ends as shown in Fig. 4.6a. As a unit load moves across the beam from left to right, the value of R_L is equal to $1 - x/L$. A plot of R_L as ordinate and x as abscissa is drawn in Fig. 4.6b. Suppose it is desirable to plot the influence diagram for shear at point A as shown in Fig. 4.6a. The shear at A with the unit load between the left reaction and A is $V_A = R_L - 1 = x/L$. This value of V_A is a straight line function of x and can be plotted for x from zero to A. With the unit load between point A and the right reaction, the value of $V_A = R_L = 1 - x/L$. The diagram can be completed by plotting values of x from a to L. Figure 4.6c shows the influence diagram for shear at point A. The influence diagram for moment at point A is drawn in Fig. 4.6d. It is observed that the influence diagram for moment consists of a triangle with its apex at the

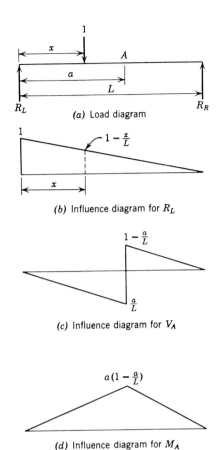

(a) Load diagram

(b) Influence diagram for R_L

(c) Influence diagram for V_A

(d) Influence diagram for M_A

Fig. 4.6

point on the beam under consideration. This is characteristic of all influence diagrams for moment at any point on a simply supported beam. It should be realized that there can be an infinite number of influence diagrams for shear or moment drawn for any one beam.

The influence diagram is always drawn for a unit load. This type of diagram is a handy tool for determining the load function for any magnitude of load or loads. The magnitude of shear at point A on the beam of Fig. 4.7a can easily be determined by using an influence diagram. The shear at A is, of course, numerically equal to the normal component of the reaction at A. The shear at a distance dx to the right of A for the loads P_1 and P_2 could be found from computing the reaction at A for the given positions of loads. If the value of V_A, however, is required for several

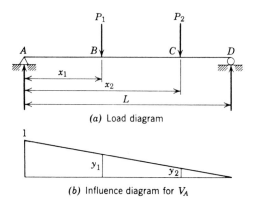

(a) Load diagram

(b) Influence diagram for V_A

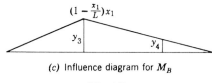

(c) Influence diagram for M_B

Fig. 4.7

positions of the load, it may be more expedient to determine the several values of V_A by use of the influence diagram. The influence diagram for V_A is shown in Fig. 4.7b. The value of V_A for loads P_1 and P_2 at B and C respectively will be: $V_A = P_1 y_1 + P_2 y_2$. The values of y_1 and y_2 can easily be determined from similarity of triangles.

The moment at B can be ascertained from an influence diagram for M_B by multiplying the intensity of each load by the respective ordinate of the influence diagram at the point of load.

$$M_B = P_1 y_3 + P_2 y_4$$

The influence diagram can be used for determining the shear or moment at a point on a beam for a uniform load of definite or indefinite length. For instance, consider the problem of finding the shear and moment at point B on the beam of Fig. 4.8a for a uniform load.

If the load over a dx length is treated as a concentrated load of $w\,dx$ intensity, then as shown by Fig. 4.8 the effect of the load over a dx length is $w(dx)y$, where y is equal to the average ordinate of the influence diagram in the dx length. The total effect of the entire uniform load over length L is

$$M = \int_0^L wy\,dx$$

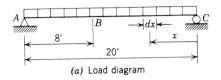

(a) Load diagram

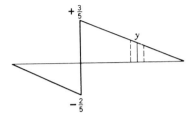

(b) Influence diagram for V_B

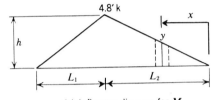

(c) Influence diagram for M_B

Fig. 4.8

$$y = \frac{h}{L_2} x$$

$$M = \int_0^L wy\,dx = \frac{wh}{L_1} \int_0^{L_1} x\,dx + \frac{wh}{L_2} \int_0^{L_2} x\,dx = w\,\frac{hL}{2}$$

The expression $w(hL/2)$ is the intensity of the uniform load multiplied by the area of the influence diagram. It can be stated that the magnitude of the load function for a uniform load is equal to the intensity of the load multiplied by the area of the influence diagram.

The shear and moment at point B on the beam of Fig. 4.8a for a uniform load of 2.6 kips per foot can be found by using this method. The uniform load is taken as variable in length.

$$\text{Maximum positive } V_B = \frac{3}{5} \times \frac{12}{2} \times 2.5 = 9 \text{ kips}$$

$$\text{Maximum negative } V_B = -\frac{2}{5} \times \frac{8}{2} \times 2.5 = -4 \text{ kip}$$

$$\text{Maximum } M_B = \frac{4.8}{2} \times 20 \times 2.5 = 120 \text{ ft.-kips}$$

AS AN EXAMPLE: The influence diagrams will be developed for (a) reaction at A; (b) shear just to the left of the reaction $B(V_{BL})$; (c) shear just to the right of the reaction $B(V_{BR})$; (d) bending moment midway between A and B for the beam shown below.

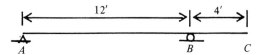

(a) Reaction at A can be determined by starting with a unit concentrated load at end C and moving it along the beam toward end A. With the unit load at C, $R_A - 4/12$. Therefore the ordinate to the influence diagram at C for value of R_A is $-1/3$. When the unit load is at B, R_A is zero and when the unit load is at A, R_A is $+1.0$. These two values are then plotted at the respective points B and A.

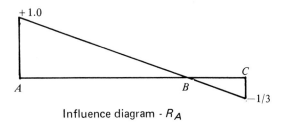

Influence diagram - R_A

(b) The shear just to the left of the reaction B can be determined by summing the normal forces either to the left or right of that point on beam. With the unit load between C and B the sum of the forces to the left of B will just be the reaction at A. Since this reaction is down, the influence diagram for V_{BL} will go from $-1/3$ at C to zero at B. When the unit load moves to the left of B, then the shear just to the left of B can most easily be determined by summing the normal forces to the right of a point just to the left of B. This of course will just be the value of R_B when the unit load is to the left of the point where the shear is being determined. The value of shear will then be negative since R_B is up, and thus the right hand portion of the beam moves up if the beam was suddenly severed at this location of shear. The value of V_{BL} is zero when the unit load is at A.

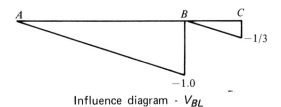

Influence diagram - V_{BL}

(c) The value of shear at a location just to the right of the reaction R_B is most easily calculated by summing the normal forces to the right of B. For a unit load between C and B, this value of shear is equal to -1.0 since the sum of normal forces to the right is just the unit load. When the unit load is to the left of the point where the shear is being determined there are no loads to the right of B and thus the shear is zero.

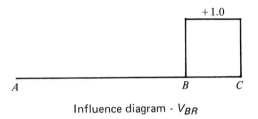

Influence diagram - V_{BR}

(d) To determine the bending moment midway between A and B the unit load will first be placed at point C. The bending moment at the selected location will then be

$$M = R_A(6) = -\frac{1}{3}(6) = -2\,ft\text{-}k$$

As the unit load moves toward B the value of R_A reduces linearly to zero. Thus the value of bending moment at the selected location reduces to zero and the influence diagram goes from -2 at C to zero at B. As the unit load moves to the left of B, but is still to the right of the midpoint, the reaction at A is upward and the bending moment is $+R_A(6)$. With the unit load between the midpoint and A, the value of bending moment is $R_B(6)$. It is thus seen that the influence diagram is a simple triangle between A and B—the same as for a beam without the cantilever overhang.

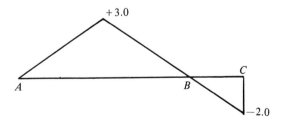

Influence diagram for bending moment 4.5 ft. from A

Figure 4.9a shows two cantilever beams with a center suspended beam. The influence diagrams for reaction at A, moment at B, shear at D, and moment at G are shown in Fig. 4.9b. Memorization of the shapes of the

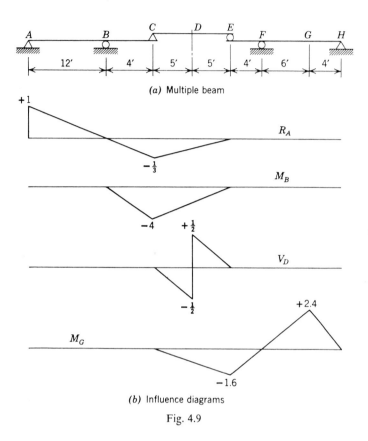

(a) Multiple beam

(b) Influence diagrams

Fig. 4.9

influence diagrams for particular beams should be avoided; instead a procedure should be developed whereby a unit load is moved across the beam and the shape of the influence diagram determined by deductive reasoning. The ordinates at the peak or valley points of the diagram should be computed. For the influence diagram for reaction at A, it is readily seen that with the unit load at A the reaction will be unity. As the unit load moves from A to B the value of the reaction will decrease linearly to zero at point B. The reaction at A will then become negative as the load moves from B to E. The magnitude of the reaction at A will increase as the load moves from B to C but will start to decrease as the load moves past C toward E. In the region C to E the value of the reaction at A is a direct function of the value of the reaction at C for the beam CE. As the load moves to the right from E, there will be no reaction at A as there can be no moment transferred through the pin at E.

A moment at B can be induced only if there is a force acting between

B and *C*. There can only be a force acting between *B* and *C* when the unit load is between *B* and *C* or between *C* and *E*, thus causing a reaction at *C*.

For a value of shear at point *D* there must be a load between *C* and *E*. Any loads on the spans *A* to *C* or *E* to *H* will cause no force at *C* or *E* and thus no moment or shear anywhere in the suspended span.

A moment at *G* can only be induced when there is a normal reactive force at *H*. There will be a reaction at *H* only for a unit load between *C* and *H*.

EXAMPLE PROBLEM 4.1. For a uniform dead load of 1.2 kips per foot, a uniform moving live load of 2 kips per foot and a moving concentrated load of 6 kips, the maximum values of R_A, M_B, V_D, and M_G are found from the influence diagrams of Fig. 4.9b as follows:

Dead load $R_A = 1.2(\frac{1}{2} \times 1 \times 12 - \frac{1}{2} \times \frac{1}{3} \times 14) = +4.4$ kips
Live load R_A (positive) $= 2 \times \frac{1}{2} \times 1 \times 12 + 6 \times 1 = 18$ kips
Live load R_A (negative) $= 2 \times \frac{1}{2} \times \frac{1}{3} \times 14 + 6 \times \frac{1}{3} = 6.7$ kips
Total maximum $R_A = +22.4$ or -2.3 kips

Dead load M_B (negative) $= 1.2 \times \frac{1}{2} \times 4 \times 14 = 33.6$ ft-kips
Live load M_B (negative) $= 2 \times \frac{1}{2} \times 4 \times 14 + 6 \times 4 = 80$ ft-kips
Total M_B (negative) $= 113.6$ ft-kips

Dead load $V_D = 0$
Live load $V_D = 2 \times \frac{1}{4} \times 5 + 6 \times \frac{1}{2} = 5.5$ kips (positive or negative)
Total $V_D = 8$ kips (positive or negative)

4.6 Moving Loads—General

Beams are sometimes subjected to moving loads. Such loads may consist of a series of concentrated loads of variable spacing and magnitude, a uniform load of variable length, or a combination of both. With one or two concentrated loads or a uniform load, the placement of the load for maximum effect on the beam is usually apparent from the shape of the influence diagram. However, when the beam is loaded with several concentrated loads, the position of the loads to cause maximum effect, hereafter termed *critical* position, is usually not obvious. For the more complicated loads the procedure for determining the critical position and likewise the maximum stresses can follow one of two methods. A cut-and-try procedure can be used, placing the load in several different positions and computing the shear or moment. This may be time consuming and tedious. The other method is to use some criteria that can be applied easily and that can readily establish the position of loading to cause the maximum effect on the beam.

In the work to follow, criteria will be developed for determining the critical position for maximum reactions, shears, and moments in simple end-supported beams.

4.7   Moving Loads—Maximum Reactions

The influence diagram for reaction for a beam simply supported at the ends consists of a triangle with the apex at the reaction under consideration. It is then obvious that with a single concentrated load the critical position would be with the load right over the reaction. A moving uniform load would be placed over the whole length of beam for maximum reaction. For a series of concentrated loads it is apparent that the critical position would be with one of the concentrated loads directly over the reaction. The question is: Which concentrated load should be placed at the reaction? Figure 4.10 shows a simple beam subjected to a series of concentrated loads, P_1 to P_5. In position 1 the first load P_1 is directly over the left reaction. R_{L1}, the value of the left reaction with the loads in position 1, is then

$$R_{L1} = \frac{G_1 x_1}{L}$$

where G_1 is the total weight of all the loads on the beam for this position of loading, and x_1 is the distance to the resultant G_1 from the right reaction.

In position 2 the loads have moved to the left until the second load P_2 is over the reaction and P_1 is off the span. Then

$$R_{L2} = \frac{G_2 x_2}{L}$$

where G_2 is the total weight of the loads on the beam in this position and x_2 is the distance as shown.

Solving for L from both preceding equations and setting them equal to each other, we obtain

$$\frac{G_1 x_1}{R_{L1}} = \frac{G_2 x_2}{R_{L2}} \quad \text{or} \quad \frac{R_{L2}}{R_{L1}} = \frac{G_2 x_2}{G_1 x_1} \tag{4.3}$$

If there is an increase in reaction from position 1 to position 2, then

$$\frac{R_{L2}}{R_{L1}} > 1 \quad \text{and} \quad G_2 x_2 > G_1 x_1$$

It is necessary only to determine whether $G_2 x_2$ is greater or less than $G_1 x_1$. It may be necessary to compare several products of Gx for complex loadings, but ordinary inspection will usually narrow the choice to two or three possibilities.

This procedure is general and may be simplified for specific loading conditions. This situation occurs when in moving the loads from position 1 to position 2 the leading load moves off the span and no additional loads move onto the beam (Fig. 4.11). Then the change in R_L can be written

$$\Delta R_L = R_{L2} - R_{L1}$$

The value of the left reaction for position 1 is

$$R_{L1} = \frac{G_2 x'_1}{L} + P_1$$

where G_2 is the total weight of all loads on the span exclusive of P_1, and x'_1 is the distance from the right reaction to the centroid of the loads comprising G_2. The value of the left reaction for position 2 is

$$R_{L2} = \frac{G_2(x'_1 + b_1)}{L}$$

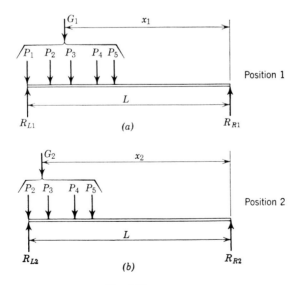

Fig. 4.10

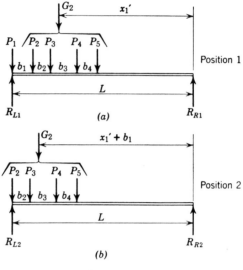

Fig. 4.11

Therefore,

$$\Delta R_L = \frac{G_2 b_1}{L} - P_1 \tag{4.4}$$

ΔR_L will increase in moving the loads from position 1 to position 2 if $G_2 b_1/L$ is greater than P_1. In this specific example of loading, time is saved in establishing the criterion because the location of the centroid of the loads on the beam does not have to be computed in the determination of the location of the critical position.

The procedure is to start with the first load and consecutively move the loading so that each load will be over the reaction. ΔR is calculated for each move. The process can stop as soon as ΔR becomes negative. The critical position for maximum reaction is the last position that resulted in a positive value of ΔR.

4.8   Moving Loads—Maximum Shears

The absolute maximum shear in a simply supported beam occurs when the reaction is maximum. The critical position for maximum absolute shear is then the same as the critical position for maximum reaction. For the maximum shear at any point other than just adjacent to the support, the following procedure establishes the criterion for critical position of a loading consisting of a series of concentrated loads. From the influence diagram it is obvious that the maximum shear will occur with one of the concentrated loads at the point in question (theoretically an infinitesimal distance to the right of the point).

As the positive area and the maximum positive ordinate of the influence diagram for shear at point A in Fig. 4.12 is greater in value than the corresponding negative values, it is apparent that the maximum shear at point A will occur with the majority of the loads to the right of point A instead of to the left of point A. However, it is not apparent by observation whether the shear would be greater, considering the loads are a vehicle, with the vehicle crossing the span from right to left or from left to right. Since the loads can usually move in either direction, two critical positions have to be established, the maximum shear determined for both cases, and the two results compared.

Considering the first position with P_1 at point A,

$$R_{L1} = \frac{Gx}{L} \quad \text{and} \quad V_{A1} = R_{L1} = \frac{Gx}{L}$$

Next consider the loads moved to the left so that P_2 is at A.

If no additional loads come on the beam and no loads pass off the beam in going from position 1 to position 2, then

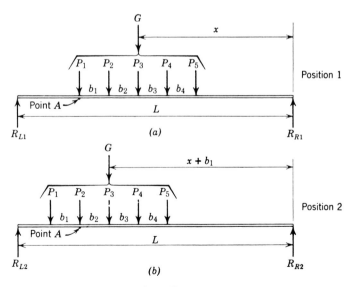

Fig. 4.12

$$R_{L2} = \frac{G(x + b_1)}{L} \quad \text{and} \quad V_{A2} = R_{L2} - P_1 = \frac{G(x + b_1)}{L} - P_1$$

The change in shear at point A in moving the loads from position 1 to position 2 is

$$\Delta V_{A(1-2)} = V_{A2} - V_{A1} = \frac{Gb_1}{L} - P_1$$

and likewise for additional movements of the load

$$\Delta V_{A(2-3)} = V_{A3} - V_{A2} = \frac{Gb_2}{L} - P_2$$

If there is an increase in shear from position 1 to position 2, $\Delta V_{A(1-2)}$ will be plus, and likewise for a reduction in shear $\Delta V_{A(1-2)}$ will be minus.

This development occurred when during the movement of the vehicle from one load at the point in question to the following load at the point no loads moved off the span and no additional loads moved onto the span. *When a load moves off the span (Fig. 4.13) or a load moves onto the span or when both events occur,* the procedure is similar, but the detail computations are as follows:

$$V_{A1} = R_{L1} = \frac{G_1 x_1}{L}$$

$$V_{A2} = R_{L2} = \frac{G_2 x_2}{L}$$

$$\Delta V_{A(1-2)} = \frac{G_2 x_2}{L} - \frac{G_1 x_1}{L} = \frac{1}{L}(G_2 x_2 - G_1 x_1)$$

Since the total of the loads G on the span changes and the position of the centroid changes in relationship to the loads, it is necessary to determine the product of Gx for each position of the loads.

The procedure in all cases for determining the critical position of the loads is as follows: compute $\Delta V_{A(1-2)}$ If the increment in shear is minus, position 1 is the critical position, and the shear can be computed for this position. This value of shear will be the maximum. If $\Delta V_{A(1-2)}$ is a plus value, position 2 is more critical than position 1. The next step is to determine $\Delta V_{A(2-3)}$. If the increment in shear is minus, position 2 is the critical position. If $\Delta V_{A(2-3)}$ is plus, then calculate the increment of shear with moving the load to the next position. The critical position is the last position for which the increment in shear is a positive value.

It is apparent that the determination of the critical position and the calculation of the value of shear for the critical position would be less time consuming than a cut-and-try procedure.

EXAMPLE PROBLEM 4.2.   Given a series of concentrated loads shown in

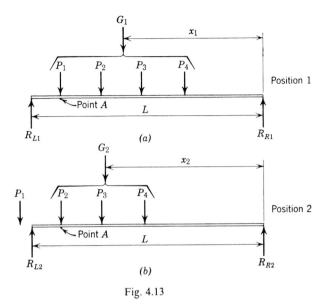

Fig. 4.13

Fig. 4.14 the maximum left and right reactions for a 32-ft simple beam span is determined.

As all the loads are on the span when load ① is at the left reaction, Eq. 4.4 is employed in determining the critical position. In moving the load to the left from a position of load ① at R_L to load ② at R_L

$$\Delta R_L = \frac{G_2 b_1}{L} - P_1 = \frac{38 \times 10}{32} - 4 = +$$

As ΔR_L is plus position 2 > position 1. In moving the load again to the left until load ③ is at R_L

$$\Delta R_L = \frac{26 \times 6}{32} - 12 = -$$

As ΔR_L is minus position 2 > position 3; the maximum left reaction is

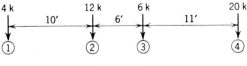

Fig. 4.14

$$R_L = 12 + \frac{6 \times 26 + 20 \times 15}{32} = 26.3 \text{ kips}$$

In considering the maximum value of the right reaction, load ④ is first placed at the right reaction and then the loads are moved to the right until load ③ is over the right reaction.

$$\Delta R_R = \frac{22 \times 11}{32} - 20 = -$$

As ΔR_r is minus the critical position is the first position, that is, with load ④ over the reaction.

$$R_R = 20 + \frac{6 \times 21 + 12 \times 15 + 4 \times 5}{32} = 30.2 \text{ kips}$$

EXAMPLE PROBLEM 4.3.   With the same loading as shown in Fig. 4.14 the maximum shear at a point 16 ft from the left reaction for a 50-ft simple beam will be determined. All the loads are on the span when load ① is at the point 16 ft from the left reaction.

$$\Delta V_{1\text{-}2} = \frac{Gb_1}{L} - P_1 = \frac{42 \times 10}{50} - 4 = +$$

$$\Delta V_{2\text{-}3} = \frac{Gb_2}{L} - P_2 = \frac{42 \times 6}{50} - 12 = -$$

As $\Delta V_{1\text{-}2}$ is (+) and $\Delta V_{2\text{-}3}$ is (−) the position for maximum shear is with load ② at the point of shear.

$$V = R_L - 4$$

$$R_L = \frac{4 \times 44 + 12 \times 34 + 6 \times 28 + 20 \times 17}{50} = 21.9 \text{ kips}$$

$$V = 21.9 - 4 = 17.9 \text{ kips}$$

There may be a greater value of shear at the point in question if the loads were reversed and moving from left to right. In a bridge structure the value of shear for this reversed loading condition would also have to be investigated and the maximum value of shear determined.

4.9   Moving Loads—Maximum Bending Moments at a Point on a Simple Beam

The influence diagram will give direction to the method for determining the position of load to produce the maximum moment at a given point on a simply supported beam. From the influence diagram for moment at a point it is readily seen that for a single concentrated load the position for maximum moment would be with the load at the point. For a uniform load the maximum moment would be with the uniform load over the entire beam. For a series of concentrated loads on a beam the criterion for maximum moment can be developed as follows.

In Fig. 4.15 a simple beam is shown. The influence diagram for moment at point A is drawn above the beam.

The maximum ordinate is ab/L. If a concentrated load P, resting at any point to the right of A, is moved a distance Δx to the left (but still remaining to the right of point A), the increase in moment at point A is

$$\Delta M_A = P\frac{ab}{Lb}\Delta x = P\,\Delta x\,\frac{a}{L}$$

If a concentrated load P, resting at any point to the left of point A, is moved a Δx distance to the left (but still remaining on the beam), the decrease in moment at point A is

$$\Delta M_A = P\frac{ab}{La}\Delta x = P\,\Delta x\,\frac{b}{L}$$

From these two equations it is seen that the change in moment at a given point on a simple beam produced by the movement of a concentrated load can be expressed by a simple equation. This increment equation involves the magnitude of the load, the distance moved, and the proportion of the span length from the point in question to the reaction.

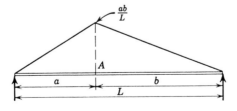

Fig. 4.15

Figure 4.16 shows a beam with a series of concentrated loads; G_1 is the resultant of the loads to the left of point A, and G_2 is the resultant of the loads to the right of point A. If all the loads are moved to the left a distance Δx, the change in moment at point A is

$$\Delta M_A = G_2 \, \Delta x \, \frac{a}{L} - G_1 \, \Delta x \, \frac{b}{L}$$

or

$$\frac{\Delta M_A}{\Delta x} = G_2 \frac{a}{L} - G_1 \frac{b}{L}$$

As Δx approaches zero as a limit

$$\frac{dM}{dx} = G_2 \frac{a}{L} - G_1 \frac{b}{L}$$

making $dM/dx = 0$, the condition for maximum

$$\frac{G_1}{a} = \frac{G_2}{b} \tag{4.5}$$

The criterion equation (Eq. 4.5) is general for all beams simply supported at their ends, whether the loading consists of concentrated loads or uniform loads. In fact, it can be applied to any part of any structure in which the influence diagram is a triangle. This is so for a great many structures.

The criterion for placement of loads to give the maximum moment at any point on a simply supported beam can be stated:

The maximum moment at a point on a simply supported beam will occur when the average loading on the portion of the beam to the left of the point will be equal to the average loading on the portion of the beam to the right of the point.

With a uniform load the average load on the left-hand portion can be made equal to the average load on the right-hand portion.

EXAMPLE PROBLEM 4.4.   Given a simply supported beam of 20-ft span, a uniform load of 12-ft length, and an intensity of 2.4 kips per foot, find

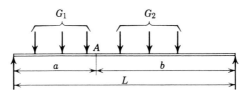

Fig. 4.16

the maximum moment at a point 8 ft from the left reaction. $G_1 = 2.4x$ and $G_2 = 2.4(12-x)$, where x is the length of the load to the left of the point in question.

$$\frac{G_1}{a} = \frac{G_2}{b}$$

$$\frac{2.4x}{8} = \frac{2.4(12-x)}{12}$$

$$12x = 8(12-x)$$

$$x = 4.8 \text{ ft}$$

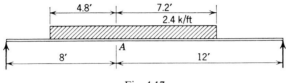

Fig. 4.17

$$R_L = \frac{2.4 \times 12 \times 10.8}{20} = 15.55 \text{ kips}$$

$$M_A = 15.55 \times 8 - \frac{2.4(4.8)^2}{2} = 124.4 - 27.6 = 96.8 \text{ ft-kips}$$

If a beam is subjected to a series of concentrated loads, it will not likely be possible to place the loads so that Eq. 4.5 is satisfied. The maximum moment at a point on a simple beam will occur when one of the concentrated loads is directly over the subject point. In order to ascertain which load should be placed at the point the procedure should be as follows:

Start by placing each load in order; first, just an infinitesimal distance to the right of the point and then an infinitesimal distance to the left of the point. Solve Eq. 4.5 in each case. The critical load will be the one where, with the load just to the right of the point $G_1/a < G_2/b$ and with the load just to the left of the point $G_1/a > G_2/b$.

EXAMPLE PROBLEM 4.5.   Given the same 20-ft beam as previously used, determine the maximum moment at point A for the following loading:

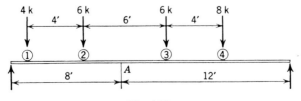

Fig. 4.18

With load ① just to the right of A

$$\frac{G_1}{a} = \frac{0}{8} < \frac{G_2}{b} = \frac{16}{12} \quad \text{(load ④ is off the span)}$$

With load ① just to the left of A

$$\frac{G_1}{a} = \frac{4}{8} < \frac{G_2}{b} = \frac{12}{12}$$

With load ② just to the right of A

$$\frac{G_1}{a} = \frac{4}{8} < \frac{G_2}{b} = \frac{20}{12}$$

With load ② just to the left of A

$$\frac{G_1}{a} = \frac{10}{8} > \frac{G_2}{b} = \frac{14}{12}$$

Load ② is the critical load.

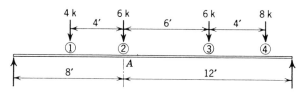

Fig. 4.19

The maximum moment is then found to be

$$R_L = \frac{8 \times 2 + 6 \times 6 + 6 \times 12 + 4 \times 16}{20} = 9.4 \text{ kips}$$

$$M_A = 9.4 \times 8 - 4 \times 4 = 59.2 \text{ ft-kips}$$

When the loading consists of several loads whose distance from the first load to the last load is greater than the length of the beam, there is the possibility of more than one load of the group satisfying the criterion. This situation can result from loads moving on and off the span. When this condition arises the critical load can possibly be ascertained by inspection or by comparing the computed values of moment for the several critical loads.

4.10   Moving Loads—Absolute Maximum Moment in a Simple Beam

In Section 4.9 the criterion for placement of loads to determine the maximum moment at a given point was developed. The prime requirement in the design and analysis of simple beams is usually determining the largest bending moment that can occur in a beam from the passage of a given loading. For a single concentrated load the maximum moment occurs under the load with the load at the midspan of the beam. With a uniform load of definite or indefinite length the maximum moment also occurs at mid-span with the load placed symmetrical with respect to the mid-span. A loading consisting of several concentrated loads at definite spacings requires a more involved criterion. The maximum moment does not occur at midspan. The influence diagram for moment at any point on a simple beam always consists of a triangle with the apex of the triangle at the point in question.

A group of concentrated moving loads is shown on the beam of Fig. 4.20. Since the maximum moment at any point on the beam will occur with one of the loads over that point, a general equation for moment at any point can be written. The value of the left reaction with the loads so placed that load P_3 is at a distance x from the left reaction is

$$R_L = \frac{G(L - x - s)}{L}$$

The value of G is the sum of all the concentrated loads on the span. The bending moment under load P_3 is

$$M = R_L x - P_2 b_2 - P_1(b_1 + b_2)$$
$$= \frac{G(L - x - s)}{L} x - P_1 b_1 - b_2(P_1 + P_2)$$

In order to determine the position on the beam at which the bending moment will be maximum under load P_3, the expression for moment is differentiated with respect to x and equated to zero.

$$\frac{dM}{dx} = \frac{G}{L}(L - 2x - s) = 0$$

$$x = \frac{L}{2} - \frac{s}{2}$$

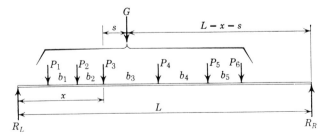

Fig. 4.20

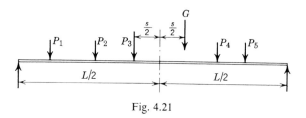

Fig. 4.21

From this expression for x it can be concluded that *the maximum moment under any particular load will occur when the center of the span is midway between that load and the resultant of all the loads on the span.*

Figure 4.21 shows the placement of the loads to obtain maximum moment under load P_3.

The question confronting the engineer is under which concentrated load will the absolute maximum moment occur. The bending moment under several loads could be determined by the foregoing method. However, this amount of work is not necessary. The absolute maximum moment occurs under one of the two concentrated loads adjacent to the resultant of all loads on the span. If the heavier of the two adjacent loads is nearer the resultant, then the absolute maximum moment occurs under this load. If the lighter of the two adjacent loads is nearer the resultant, then the moment under both loads, placed as just noted must be computed and the two values compared.

When a long multiple-axle vehicle crosses a short-span bridge, there may be several combinations of wheel loads on the span at particular times. The absolute maximum moment for a series of loads under this condition must be determined by comparing moments from the several possible combinations of wheel loads.

EXAMPLE PROBLEM 4.6.   Given the beam shown in Fig. 4.22 with the loads shown, determine the maximum absolute moment. With all four loads on the beam, the critical position is shown in Fig. 4.22.

$$R_L = \frac{28 \times 15.32}{32} = 13.41 \text{ kips}$$

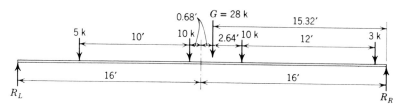

Fig. 4.22

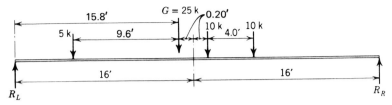

Fig. 4.23

$$M_{\max} = 13.42(16 - 0.68) - 5 \times 10$$
$$= 205.4 - 50 = 155.4 \text{ ft-kips}$$

There is a second possibility for maximum moment. With the first three loads only, the placement is as shown in Fig. 4.23.

For the critical position shown in this diagram the fourth load would be off the span.

$$R_R = \frac{25 \times 15.78}{32} = 12.33 \text{ kips}$$

$$M_{\max} = 12.33 \times 15.78 - 10 \times 4$$
$$= 194.6 - 40 = 154.6 \text{ ft-kips}$$

The moment with all four loads on the beam is the maximum.

EXAMPLE PROBLEM 4.7.   Determine the absolute maximum moment for the vehicle of Fig. 4.24 crossing a 36-ft simple beam bridge. From observation it appears that there are three possibilities for maximum moment: the first three loads, the first four loads, or the middle three loads. Each possibility will be checked.

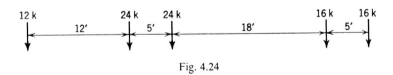

Fig. 4.24

First three loads
$$R_R = \frac{60 \times 17.8}{36} = 29.67 \text{ kips}$$
$$M_{\max} = 29.67 \times 17.8 - 24 \times 5$$
$$= 528 - 120 = 408 \text{ ft-kips}$$

First four loads. This case is not possible as the fourth load will be off the span.

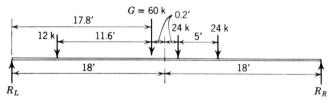

Fig. 4.25

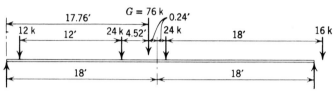

Fig. 4.26

Middle three loads

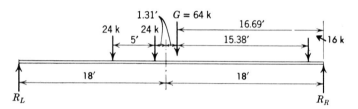

Fig. 4.27

$$R_L = \frac{64 \times 16.69}{36} = 29.67 \text{ kips}$$

$$M_{\max} = 29.67 \times 16.69 - 24 \times 5$$
$$= 495 - 120 = 375 \text{ ft-kips}$$

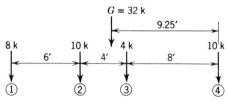

Fig. 4.28

The absolute maximum moment of 408 ft-kips is with the first three loads on the beam as shown in Fig. 4.25.

EXAMPLE PROBLEM 4.8. Given the loading shown in Fig. 4.28, find the absolute maximum moment in a simple beam span of 26 ft. The nearest load to the resultant is load ③. The maximum moment under this load is as follows:

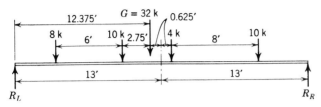

Fig. 4.29

$$R_R = \frac{32 \times 12.375}{26} = 15.23 \text{ kips}$$

$$M_{max} = 15.23 \times 12.375 - 10 \times 8$$

$$= 188.4 - 80 = 108.4 \text{ ft-kips}$$

As the other adjacent load [load ②] is greater in magnitude than the closest load, the moment with load ② in the critical position is checked.

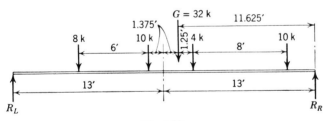

Fig. 4.30

$$R_L = \frac{32 \times 11.625}{26} = 14.32 \text{ kips}$$

$$M_{max} = 14.32 \times 11.625 - 8 \times 6$$

$$= 166.8 - 48 = 118.8 \text{ ft-kips}$$

This demonstrates that the critical load is not always the nearest load to the resultant.

Problems

1. Draw the shear and moment diagrams for the beam of Fig. 4.31.

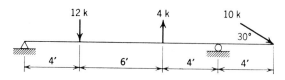

Fig. 4.31

Ans. $V_{max} = 6$ kips, $M_{max} = +24$ ft-kips.

2. Draw the shear and moment diagrams for the beam of Fig. 4.32.

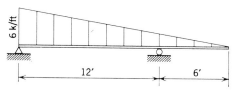

Fig. 4.32

3. Draw the shear and moment diagrams for the beam of Fig. 4.33.

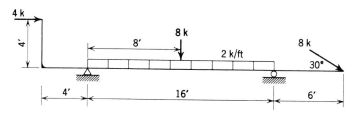

Fig. 4.33

Ans. $V_{max} = 22.5$ kips, $M_{max} = +92$ ft-kips.

4. Draw the shear and moment diagram for the frame ABC of Fig. 2.12.

5. Draw the shear and moment diagram for the beam shown below.

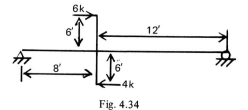

Fig. 4.34

6. (a) If the shear diagram between two points along a beam is a straight sloping line, what is the shape of load diagram and the bending moment diagram between these two points?

(b) If the bending moment diagram between two points along a beam is parabolic and concave downward, what is the shape of the shear diagram between these two points?

7. Draw the influence diagrams for a 20-ft simply supported beam for the following conditions: *(a)* shear just to the right of the left reaction, *(b)* shear at midspan, *(c)* moment at the quarter point, and *(d)* moment at the midspan.

8. Draw the influence diagrams for the beam of Fig. 4.35 for the following conditions: *(a)* reaction at *A*, *(b)* moment at *B*, *(c)* shear at *C*, *(d)* moment at *D*, and *(e)* shear at *E*.

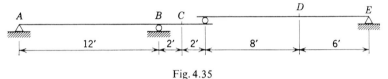

Fig. 4.35

9. For the beam of Fig. 4.36 draw the influence diagrams for moment at *B* and also for shear just to the left and right of *B* and from the influence diagrams determine the following:

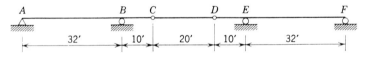

Fig. 4.36

(a) Values of shears and moments at *B* for a uniform dead load of 0.4 kips per foot.

(b) Live-load shears and moments at *B* for the H20–44 truck loading.

(c) Live-load shears and moments at *B* for the H20–44 equivalent lane loading.

Ans (a) $V_{BL} = -8.3$ kips, $V_{BR} = 8.0$ kips, $M_B = -60$ ft-kips;

(b) $V_{BL} = -36.5$ kips, $V_{BR} = +38.4$ kips, $M_B = -344$ ft-kips;

(c) $V_{BL} = -39.2$ kips, $V_{BR} = +38.8$ kips, $M_B = -276$ ft-kips.

10. For a simply supported beam of 36-ft span determine the crucial position and the following values for the loading of Fig. 4.37. *(a)* maximum shear at support, *(b)* maximum moment at the one-third point, and *(c)* maximum moment at midspan.

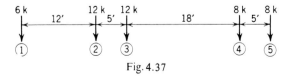

Fig. 4.37

11. Determine the same quantities as for Problem 10 but for a 60-ft beam span
    Ans. $V = 32.2$ kips, $M_{1/3} = 409$ ft-kips, $M_{\mathcal{C}} = 445$ ft-kips.

12. Given a simply supported beam of 80-ft. span, determine the absolute maximum moment and shear for the moving load of Fig. 4.38.

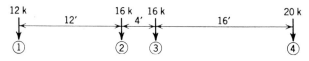

Fig. 4.38

Ans. $V = 52$ kips, $M = 992$ ft-kips.

13. A bridge consists of several steel beams of 44-ft. span. Assuming each beam supports one-half lane of traffic, compare the maximum moment and maximum shear for the H20-44 truck loading and the H20-44 equivalent lane loading.

14. A bridge is to be built. It will consist of two girders spanning 120 ft. It will carry two lanes of traffic but each girder will be designed for 1.15 lanes of traffic due to possible eccentricity of traffic with respect to bridge centerline. The design load is HS-20. Determine maximum shear and maximum bending moment for a girder. Calculate for both truck and equivalent lane loading and compare values.

15. At what span length of simply supported beam will the maximum shear be the same for the HS-20 truck loading and the equivalent lane loading? At what span length will the bending moments be equal?

16. Using the railroad loading of Fig. 3.10, determine the critical position and maximum values for *(a)* maximum moment at midspan of an 80-ft. simple beam and *(b)* maximum shear at the support of an 80-ft. simple beam.
    Ans. (a) 7264 ft-kips, *(b)* 447 kips.

17. Using the railroad loading of Fig. 3.10, determine the maximum shear and bending moment on a 160 ft. simple span.

.

TRUSSES

5.1 Introduction

Ever since Palladio started using frameworks in his buildings during the mid-sixteenth century, the truss has served man in many different types of structures. The truss has been employed in America from the historic old wooden bridges to the vehicles of air travel. The first trusses were utilized in wood for buildings. With the large-scale smelting of iron and the requirement of bridges of long span and large load capacity, occasioned by the introduction of the railroad, the truss gained great popularity among structural engineers. The railroads and highways of today are dotted with truss bridges fabricated from structural steel. The introduction of steel as a construction material gave the structural engineer an ideal medium for use in the fabrication of trusses. A material of high tensile strength as well as compressive strength, a material that could be easily and strongly fastened together, and a material that could be produced in sections of varying shape, different cross-sectional areas, and a multitude of lengths was the answer to truss construction. This combination of material and structural shape has resulted in great structures such as the Eiffel Tower, the Kill-van-Kull arch bridge, the Mackinac Straights Bridge, and many other monumental structures. In recent years, because of the increased labor costs in structural fabrication, the truss is used much less than the plate girder for medium span bridges. The plate girder has fewer parts and hence less labor for fabrication and erection.

The development of the box girder as a structural form for highway bridges has further reduced the use of trusses. The combination of box girder with supporting cable stays radiating from a tower has increased the economical span length for girder bridges to over one thousand feet. The high cost of fabricating trusses has reduced their use in highway bridges.

The single-span simply supported bridge truss is very rare in modern-day bridge engineering. However, it is treated here because the principles of truss analysis are easier to present and to understand when applied to a single-span, statically determinate truss. The step-by-step principles of determining bar forces are the same for a simple span structure as for a multiple span continuous bridge. Once the external reactions are determined, the procedure learned for a simple-span structure can be applied to the continuous structure.

A pin-connected truss is defined as a stable framework consisting of elements connected together at the ends so that no element is restrained from rotation about any axis perpendicular to the plane of the framework and passing through the ends of the elements. A truss should have all loads applied at the joints of the members. Loads applied to the truss at any point other than the joint introduces in addition to the axial loads, bending moments in the members. This additional bending moment results in a considerable increase in the required size of the member. It will be recalled from *Statics* that any member with a pin or hinge at the ends and with loads applied only at the pin or hinge is a two-force member; that is, the resultant force in the member must be along the axis of the member. A truss which has the loads applied at the joints and the members connected at their ends with frictionless pins is called an ideal truss. The configuration of the members of a truss conforms to sets of triangles. This is the only arrangement that is stable for members pinned at their ends. Trusses are usually composed of members that can resist both tensile and compressive forces. However, since many members of a truss are required to resist forces in only one direction during service, they can be of such a nature that they will only resist forces in one direction. An example is a cable, it will only resist tensile forces. It is impossible to apply a compressive force to a cable.

The early trusses were joined together at their ends by pins, resulting—except for minor friction—in an ideal truss. In modern practice it is not practical to join members by pins. The usual method of joining members is by riveting, bolting, or welding. Since a single rivet or bolt or a spot of weld is usually not sufficient to carry the load, groups of such fasteners are needed. Whenever more than one individual fastener is used at a joint, the joint becomes a moment resistant connection. However, if the elements of the fastener group are properly arranged and the stiffness (I/L) of the members is not large, the moments induced in the members because of the rigid connection is small and can be neglected. For large bridge trusses the analysis is performed by computer and programs are available for calculating these secondary bending moments as well as the primary axial forces. This chapter will only treat the calculation of axial forces.

Other arrangements are used in particular cases, but only the common ones are treated here. In bridge trusses the usual configuration is the Pratt truss or the Warren with verticals. The Warren is somewhat more popular with bridge designers as it gives a more pleasing appearance. In the Pratt truss, the diagonals are all in tension under dead load. However, the diagonals experience a reversal of stress because of the live load moving across the structure. The live-load tension force will be greater than the live-load compression force for simple-span Pratt trusses. The major force in the diagonals will thus be in tension, and compression will be possible only where the live-load force (plus any impact) is greater than the dead-load force. The end-post is of course in compression for all types. The Warren truss, noted by the alternating slope of the diagonals, has tension in one diagonal and compression in the adjacent diagonal under dead load. Under passage of the live load the force in the diagonals of the Warren truss will also reverse. However, the major force will be compression in one-half of the diagonals. Since the sense of the major force in all the diagonals of a Pratt truss is tension, there is some economy in this arrangement of the diagonal members.

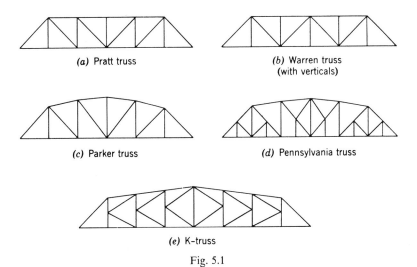

(a) Pratt truss

(b) Warren truss (with verticals)

(c) Parker truss

(d) Pennsylvania truss

(e) K-truss

Fig. 5.1

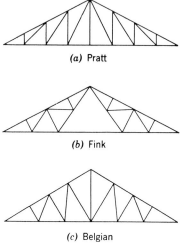

(a) Pratt

(b) Fink

(c) Belgian

Fig. 5.2

When the span of a bridge truss becomes quite large, it is more economical to increase the depth of the truss toward the center of the span, where the moment is greatest. Continuous and cantilever truss bridges almost always have a varying depth. The greatest depth will occur where the moment is greatest. When the depth of a truss is increased, it is necessary to increase the panel length so that the slope of the diagonals are at an economical angle. Long panel lengths (over about 25 ft) will result in a heavy floor system and thus large dead loads. In order to keep the weight of the floor system as small as possible, the panel length can be subdivided as in the Baltimore truss. The K-truss is another method of achieving short panel lengths with deep trusses.

The peaked roof truss (Fig. 5.2) is usually made in the configuration of a Pratt or Fink. Flat roof trusses are usually the Pratt or Warren. The Belgian truss has the struts perpendicular to the top chord. Because of this, there is a fixed slope of the roof for each particular number of panel points in the top chord. The Pratt and Fink can have any slope of the roof. The slope of the roof is determined by architectural or economic considerations.

Detail analysis of these roof trusses is treated in Chapter 6.

5.2   Stability and Determinancy of Trusses

In Section 4.1 an explanation was given regarding the procedure for determining whether a beam was statically determinate or indeterminate.

The same principle and procedure apply to the reactions of a truss. If there are more reaction components than there are applicable equations of equilibrium, the truss is statically indeterminate with regard to reactions, or as usually stated, the truss is statically indetermine *externally*. If there are fewer possible reaction components than there are applicable equations of equilibrium, the structure is unstable and undergoes excessive displacement under certain applications of loads. For instance, Fig. 5.3a shows a truss with only two possible reaction components, a vertical reaction only at each end. Under the action of vertical loads only the structure is in a state of unstable equilibrium. A horizontal load of the smallest amount causes the structure to move laterally. Figure 5.3b shows a truss that has three reactions—one more reaction than the equations $\Sigma F_y = 0$ and $\Sigma M = 0$—the only applicable equations of equilibrium for vertical loads. This structure is statically indeterminate with regard to vertical loads only. However, it is unstable with respect to horizontal loads as none of the supports is capable of resisting horizontal forces. This illustration introduces the basic requirement for external stability for any rigid body. All three equations of static equilibrium have to be possible if the structure is to be stable. In Figs. 5.3a and b the equation $\Sigma F_x = 0$ is not valid as there can be no reaction in the x direction. Figure 5.3c shows a structure that is unstable since the equation ΣM_o yields a finite value. Moments taken about point o have to equal zero since it is the intersection point of the reactions. A single external load applied anywhere to the structure will not yield a zero solution to the equation of moments taken about point o. The truss would then rotate about this point. Care should be taken in supporting a truss or any portion of a compound truss to see

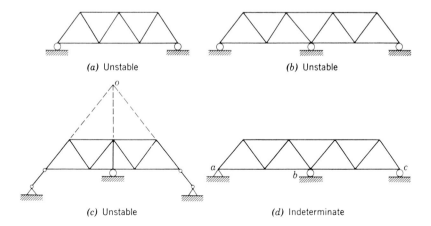

(a) Unstable   (b) Unstable

(c) Unstable   (d) Indeterminate

Fig. 5.3

that all three equations of static equilibrium can be applied to the structure. Figure 5.3*d* shows a truss that has four unknown quantities with regard to reactions; R_{ax} and R_{ay} (or the magnitude R_a and its direction), and the reactions R_b and R_c. These four unknowns exceed the three available equations of static equilibrium by one. The structure is statically indeterminate to the first degree with respect to external forces.

Another approach that can be used to determine the degree of indeterminacy is to remove reaction components one at a time and see if the structure is stable against all possible forces. The degree of indeterminacy is the number of reaction components that can be removed up to the point where the structure becomes stable and determinate. Removal of any reaction components from a structure that is statically determinate with respect to external forces results in an unstable structure.

Reaction forces that can be removed and still leave the structure stable and determinate are called *redundant* forces, that is, they are not needed to establish static equilibrium.

A truss can also be determinate, indeterminate, or unstable with respect to the system of bars. This is usually called the internal condition. A truss can be statically determinate internally and statically indeterminate externally, or the reverse of this can be true. As was previously stated, a truss consists of an assemblage of bars connected at their ends. In order for the bars to form a stable structure that can resist loads applied at the joints the bars must form triangular figures. Starting with a single figure composed of three bars as shown in Fig. 5.4*a*, it is seen that the figure contains three bars (*b*) and three joints (*n*). If a second triangle is formed to the first one as shown in Fig. 5.4*b*, there are now five bars *b* and four joints *n*. Adding still a third triangle, Fig. 5.4*c*, there are seven bars and five joints. The number of bars added for each joint after the first triangle was two. The number of bars as a function of the number of joints for any trussed structure consisting of an assemblage of bars forming triangles can be stated by the equation $b = 2n - 3$. Taking such a system of bars and writing the two equations of statics, $\Sigma F_x = 0$ and $\Sigma F_y = 0$, with respect to each joint there will be $2n$ independent equations. This means that the number of unknown forces that can be found by the equations of statics is $2n$.

In a structure such as shown in Fig. 5.4*c* there will be $2n - 3$ bar forces as unknown plus three necessary reaction components. This shows that the unknown forces will equal the available equation of statics. The general statement is that a trussed structure is statically determinate internally if the equation $b = 2n - 3$ is satisfied. Figure 5.5*a* shows a structure that is statically determinate as the number of bars *b* equals $2n - 3$. The structure in Fig. 5.5*b* has one more bar than $2n - 3$. This

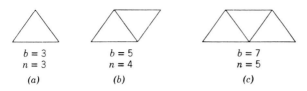

$$b = 3 \qquad b = 5 \qquad b = 7$$
$$n = 3 \qquad n = 4 \qquad n = 5$$
$$(a) \qquad (b) \qquad (c)$$

Fig. 5.4

structure is statically indeterminate to the first degree with respect to internal forces. Figure 5.5c shows a structure that is lacking one bar to equal the $2n - 3$ number. This structure is unstable. This is apparent as one panel in the assemblage is a rectangle instead of a triangle. In a framework of pin-connected members a rectangular figure is not stable. Caution should be applied in using the $b = 2n - 3$ equation as a rigid rule in determining if a structure is stable. Figure 5.5d shows a structure in which $b = 2n - 3$, but this structure is unstable as one panel has one member more than necessary, and the adjoining panel is lacking the diagonal member necessary to keep the structure from collapse. *The equation $b = 2n - 3$ is thus a necessary but not a sufficient condition for stability.*

Figure 5.6 shows a two-span truss. The number of equations of statics equals $2n = 36$. The unknown forces equal 32 bar forces plus 4 unknown reaction forces. Since the available equations equal the unknown forces, the structure is statically determinate. It should be noted here that the equation $b = 2n - 3$ does not apply to the structure as a whole. It can be applied to each rigid part of the compound truss of Fig. 5.6. A more general expression applicable to compound trusses as well as simple trusses is

$$b = 2n - r \tag{5.1}$$

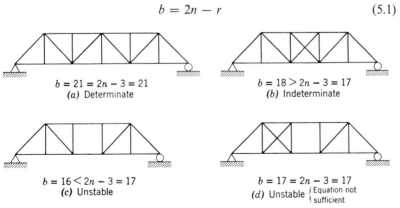

$b = 21 = 2n - 3 = 21$
(a) Determinate

$b = 18 > 2n - 3 = 17$
(b) Indeterminate

$b = 16 < 2n - 3 = 17$
(c) Unstable

$b = 17 = 2n - 3 = 17$
(d) Unstable $\begin{cases} \text{Equation not} \\ \text{sufficient} \end{cases}$

Fig. 5.5

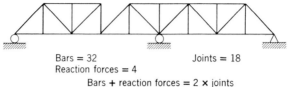

Bars = 32   Joints = 18
Reaction forces = 4
Bars + reaction forces = 2 × joints

Fig. 5.6

where r is equal to the number of components of reaction. This equation indicates that it is possible to decrease the number of bars in a truss if they are replaced by effective reaction components. *The value of r used in the equation should not exceed the minimum number of reactions required for external stability.*

5.3   Methods of Analysis for Statically Determinate Trusses

Method of Joints.   The first step in the determination of the bar forces in a truss is to determine all the external forces acting on the structure. After the applied forces are determined, the reactions are found by applying the equations of static equilibrium. This procedure is the same for trusses as for beams and is therefore not repeated here for simple-span structures. In a later section, the procedure for determining reactions of multiple trusses such as cantilever trusses is treated in more detail. At this point in a student's study, the computing of reactions for simple-span statically determine structures should be second nature to him.

In statics it was given that forces meeting at a point are called concurrent forces, and for any group of coplanar concurrent forces the magnitude of no more than two unknown forces can be determined by the equations of statics, the applicable equations being $\Sigma F_x = 0$ and $\Sigma F_y = 0$. Consider a typical joint in a truss as shown in Fig. 5.7a. The magnitude of two of the forces and the line of action of all four forces are known. The values of F_1 and F_2 can be quickly determined by employing the equations of statics

$$\Sigma F_y = 0 \qquad F_1 \sin \alpha - 30 = 0 \qquad F_1 = \frac{30}{\sin \alpha}$$

$$\Sigma F_x = 0 \qquad F_2 - 60 - F_1 \cos \alpha = 0 \qquad F_2 = 60 + F_1 \cos \alpha$$

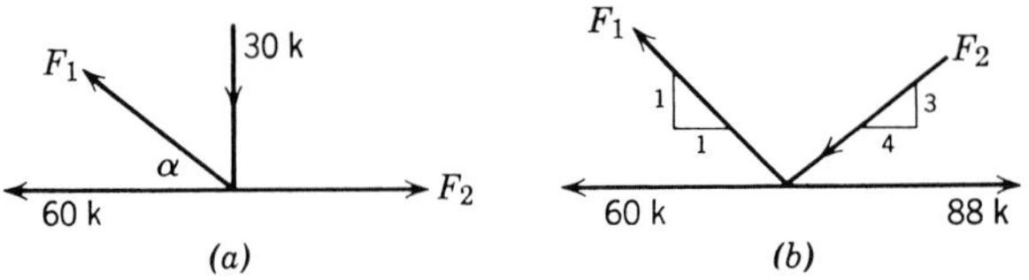

F_1 |30 k

60 k F_2

(a)

F_1 F_2
1
1   4
60 k   88 k

(b)

Fig. 5.7

The value of F_1 can be substituted directly into the expression for F_2 by solving first the equation for F_1. It is usual in truss analysis to use the slope of a member instead of the angle a member makes with respect to the reference member. This is illustrated by referring to Fig. 5.7b where the slope of F_1 is one horizontal to one vertical, and F_2 has a slope of four horizontal to three vertical. In the beginning the direction of F_1 and F_2 are usually unknown. When this is true, the directions are assumed and the correct sense found from the solution.

$$\sum F_y = 0 \qquad\qquad \frac{F_1}{1.414} - \frac{3F_2}{5} = 0$$

$$\sum F_x = 0 \qquad 88 - 60 - \frac{F_1}{1.414} - \frac{4F_2}{5} = 0$$

Substituting for F_1 the value of $0.85F_2$,

$$0.6F_2 + 0.8F_2 = 28$$
$$F_2 = 20 \text{ kips } F_1 = 0.85 \times 20 = 17 \text{ kips}$$

Since the values of F_1 and F_2 are positive, it means that the assumed directions are correct. If, for instance, they were assumed to be in the opposite directions (F_1 down to the right and F_2 up to the right), the first expression, relating the relationship of F_1 and F_2 would remain the same. However, the second equation would read

$$88 - 60 + \frac{F_1}{1.414} + \frac{4F_2}{5} = 0$$

and F_2 would equal -20 kips and $F_1 = -17$ kips

The minus signs indicate that the true direction of forces F_1 and F_2 are opposite to that assumed. Many engineers prefer in all cases to assume that the forces in all the unknown bars are tension. The resultant sign from the solution of the equations of equilibrium will conform to the usual sign convention; that is, plus for tension and minus for compression. The determination of the forces in all the bars of a truss can be a simple process of progressing from one joint to the next, performing a solution of the unknown forces for a system of concurrent forces. It is, of course, necessary to start at a joint that has no more than two bars with unknown forces. Progression then goes from joint to joint, always selecting a joint that has no more than two unknown bar forces. For a Pratt truss, as shown in Fig. 5.8, the method of joints solution can start at either end, for at these two points there are only two unknown bar forces. The lettering of the panel points is the usual method employed in bridge trusses. At L_0 the unknown forces are those in bars $L_0 U_1$ and $L_0 L_1$. It is usually easier to find the horizontal and vertical components in all the bars first

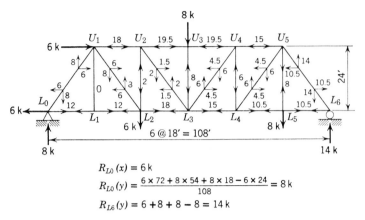

$$R_{L0}(x) = 6\,k$$

$$R_{L0}(y) = \frac{6 \times 72 + 8 \times 54 + 8 \times 18 - 6 \times 24}{108} = 8\,k$$

$$R_{L6}(y) = 6 + 8 + 8 - 8 = 14\,k$$

Fig. 5.8

and then find the actual bar forces after all the components are found. Since the slopes of all the bars are known as soon as the dimensions of the truss have been given, the magnitude of the second component can be quickly calculated as soon as one of the components is determined. In the given truss, Fig. 5.8, the slopes of the diagonal members are three horizontal to four vertical. This results in a 3:4:5 triangle. The loading given in the illustrated problem is not typical but is merely selected to illustrate the method. The first equation to consider at L_0 is $\Sigma F_y = 0$. In this equation there will only be one unknown, the y component of bar force $L_0 U_1$. If the process was started with $\Sigma F_x = 0$, the equation would yield two unknowns. If the y component is 8 kips the x component must be $\frac{8}{4} \times 3 = 6$ kips. This force must also be toward the joint. When we apply the equation $\Sigma F_x = 0$, the force in bar $L_0 L_1$ must equal 12 kips toward the right (away from the joint) to produce equilibrium of joint L_0. If the force in bar $L_0 U_1$ produces vertical and horizontal forces of 8 kips and 6 kips respectively acting toward joint L_0, it must also produce forces of like magnitude acting toward joint U_1. Since bar $L_0 L_1$ exerts a force of 12 kips away from joint L_0, it must also cause a force of 12 kips away from joint L_1. The forces are represented by arrows on the line diagram of the truss shown in Fig. 5.8. The forces as shown are those exerted by the bar on the joint. If the bar exerts a force toward the joint, the bar must be in compression. If it exerts a force away from the joint, the bar must be in tension. A free-body diagram of a bar and the joints at each end of the bar quickly proves this point.

With the forces at joint L_0 determined the operation moves to the next joint. There is the choice of moving to L_1 or U_1. The choice must be L_1, for at U_1 there are three unknown bar forces. At joint L_1 there is no external force, so if the equation $\Sigma F_y = 0$ is satisfied, there must be zero force in bar U_1L_1. With bar U_1L_1 determined the number of unknown bar forces at U_1 are now just two. Summing first the forces in the y direction gives a y component of 8 kips in bar U_1L_2 and thus a 6-kip horizontal component. With the horizontal component of both L_0U_1 and U_1L_2 equal to 6 kips to the right and an external force of 6 kips to the right, the force in U_1U_2 must be 18 kips to the left to balance the joint. The procedure to determine all the other bar forces is the same. The method is self-checking, for at the right support (L_6) the vertical component in U_5L_6 must be equal to and opposite to the reaction of 14 kips, and the horizontal component must be equal and opposite to the force in L_5L_6. Note that all the bars forming the top chord have compression and all the bars of the lower chord are in tension. This is so for simple-span trusses acted upon by gravity loads. It should also be observed that the diagonals—but not the end posts—are in tension, whereas the verticals U_2L_2 and U_4L_4 are in compression. This is typical for the Pratt truss. The hangers U_1L_1 and U_5L_5 are in tension if there is a load applied at the lower panel point of the hanger.

Graphical Solutions. The determination of bar forces in a truss can easily be made by a graphical solution. A force, as was shown in Chapter 2, is a vector quantity and can be represented by an arrow. If a system of concurrent coplanar forces are in equilibrium, the figure formed by the graphical representation of the force vectors will be a closed polygon. When we take the truss of Fig. 5.9 as an example, the graphical process can start with any point at which there are only two unknown magnitudes of force. Section 2.6 showed that only two unknown quantities can graphically (as well as numerically) be determined for any system of concurrent forces in equilibrium that are acting on a body. These two unknowns can be two magnitudes, one magnitude and one direction, or two directions. Since the directions will always be known for the bar forces of a truss, the unknowns will always be magnitudes. The unknown external forces (reactions) can also be found graphically as outlined in Section 2.6 or they can be determined algebraically. Graphical solutions do not have to be drawn to a very large scale to obtain the accuracy necessary for the usual design of an engineering structure.

The first necessary step in a graphical solution is to lay out the truss to a convenient scale, making sure to scale all dimensions accurately. Figure 5.9a shows a nonparallel chord bridge truss. Instead of following the procedure of lettering the joints, Bow's notation is followed by lettering and numbering the spaces. Each bar and force vector will be designated by the two letters or numbers it lies between.

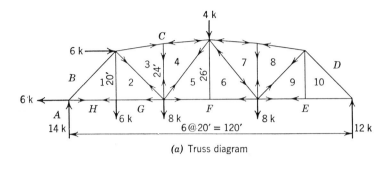

(a) Truss diagram

1H. 20 +
B1. 19.8 −
C3. 23.8 −
3,2. 4.7 +
2,1. 6 +
C4. 23.5 −
4,3. 2.3 +
4,5. 2.9 +
5F. 21.5 + = 6F
G2. 20 +
D7. 20.4 −
7,6. 2.5 −
D8. 20.4 −
7,8. 2.0 +
8,9. 11.3 +
9E. 12 +
D10. 17 −
10,9. 0
10E. 12 +

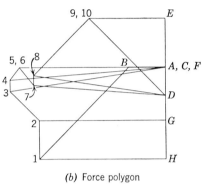

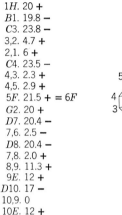

(b) Force polygon

Fig. 5.9

Graphical solutions of parallel chord bridge trusses are hardly worth-while as the algebraic solution is so quick and easy. However, the algebraic solution for nonparallel chord trusses becomes more involved and the graphical solution is usually a time saver. This is without a doubt true for sloping chord roof trusses. They are treated in a later chapter. In Fig. 5.9a the loads are not typical of loadings on bridge trusses but are selected to cover horizontal as well as vertical loads. In this manner the example solution will be more general. After the truss is laid out and the spaces between the bars are lettered as shown in Fig. 5.9a, the graphical solution can start. This solution is first started at a joint, taken as a free body, where only two unknown forces exist. This is usually at the end of the truss. In this example, the solution will start at the left end. At this joint the two unknown forces are *B1* and *1H*. Forces *HA* and *AB* are known. The procedure is to proceed around the joint, in this case clockwise, starting at *H* and intercepting first force *HA*. Force *HA* is a vertical force of 14 kips. This force is then drawn to scale in its correct direction in the force polygon. This is the line HA in Fig. 5.9b. This force polygon is usually referred to as a Maxwell diagram, after its originator, James Clerk Maxwell. In order to achieve the best results the force polygon is drawn to as large a scale as possible within the required accuracy. Experience will provide the rule. After HA has been drawn, the next force encountered in proceeding clockwise around the joint is then drawn in the force polygon. This is line AB representing a force of 6 kips to the left. It is immaterial which direction around the joint the process takes place as long as once a direction has been chosen this direction is not reversed. In circling the joint the starting point should begin at the place where the known forces will be encountered before the unknown. After AB has been drawn in the force polygon, the next force encountered is the unknown force $B1$. This force is known in direction but not in magnitude. From point B in the force polygon a line is drawn starting from B parallel to $B1$ in the truss diagram. The next force in proceeding around this first joint is $1H$. This is the force in the chord. Direction but not magnitude is known for this force. This fourth force completes all the forces acting on the joint, and if the joint is in equilibrium the force polygon comprising all the forces acting on the joint must close. In order for the force polygon to close, the line representing the force $1H$ which is drawn parallel to the lower chord must pass through the starting point H in the force polygon. This is accomplished by drawing a horizontal line through H until it intersects the line just previously drawn that was parallel to member $B1$ and passing through point B. The intersection of these two lines establishes the point 1 on the force polygon. The scaled distance from B to 1 and 1 to H represent the force in members $B1$ and $1H$ respectively. Scaling of vectors is usually held in abeyance until the entire force polygon is completed. The sense of the force, tension, or compression can be determined from the force polygon. For instance, force vector HA was drawn in the force polygon in the true direction of the force. The beginning of vector HA was labeled with the letter H (the first space encountered in proceeding clockwise around the joint), and the end was labeled A. Similarly, the end of AB was labeled B. In continuing clockwise around the joint, the procession was from B to 1. From B to 1 in the force polygon is in a direction down to the left. Force $B1$ must therefore be down to the left, which is toward the joint. A force toward the joint represents compression in the member. Force $1H$ is to the right, away from the joint; therefore there is tension in the member. The forces are now all determined for the first joint, and the solution can move on to any joint for which there are just two unknown bar forces. This will be the next panel point to the right, which has the forces GH and $H1$ known. The process of proceeding around the joint in a clockwise direction is repeated. The direction of the external force is drawn, and the magnitude of this force is scaled in the force polygon. The directions of the unknown forces are then drawn so that the lines representing all the forces around the joint form a closed polygon. The intersection of the lines representing *1-2* and *2G* will establish the point *2*. The process can now be repeated for the first top chord panel point on the left since bar forces *2-1* and *1B* have now been ascertained, leaving only the two bar forces *C3* and *3-2* to be found. This process repeats for all the joints of the truss until all the bar forces are determined. The method is self-checking since the last force to be drawn is the right reaction *DE*. Drawing this force in its true direction and magnitude should close the complete polygon if the solution has been accurately performed.

Care should be taken in transferring the direction of the bars in the truss diagrams to corresponding parallel lines in the force diagram. Sometimes a slight deviation can cause large errors in closure. After the force polygon has been successfully closed, the magnitude and sense of the bar forces can be tabulated (plus for tension, minus for compression as shown in Fig. 5.9*b*).

Method of Sections.   The third general method for finding bar forces in an internally determinate truss involves taking free-body sections of the truss. For instance, if a free-body section, consisting of joints L_0, U_1, and L_1 of the truss of Fig. 5.8, is formed by cutting the bars U_1U_2, U_1L_2, and L_1L_2 the result is as shown in Fig. 5.10*a*.

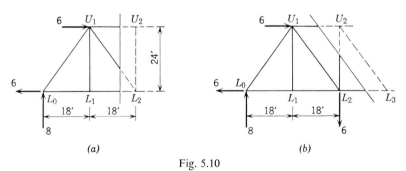

Fig. 5.10

The forces in the cut bars now change from internal forces to external forces and have to be considered in the equilibrium of the free body. As the external forces are not concurrent forces, the three equations of equilibrium, $\Sigma F_x = 0$, $\Sigma F_y = 0$, $\Sigma M = 0$, are valid instead of only the two equations in the method of joints. Because there are three available equations the free-body can be taken with three unknown forces acting. It is therefore important in cutting a section through the truss that no more than three bars are cut in which the forces in these bars are unknown. Cutting additional bars in which the forces have been previously determined is of course permissible. It is now possible to determine the forces in the cut bars by the static equilibrium equations. The force in bar U_1U_2 is found by taking moments about joint L_2. It must be remembered that moments can be taken about any point lying on the plane of the truss, but care should be taken to include only those forces that are acting on the free-body. The equation for moment at L_2 is

$$M_{L2} = 8 \times 36 + 6 \times 24 - U_1U_2 \times 24 = 0$$

$$U_1U_2 = \frac{432}{24} = 18 \text{ kips} \quad \text{(compression)}$$

In writing the equation, the moment caused by U_1U_2 was considered counterclockwise (force to the left). As the resultant value was positive, the assumed direction is correct.

The vertical component in U_1U_2 can be determined by summing forces in the y direction for the same free-body. The force in U_1L_2 is assumed to be tension.

$$\Sigma F_y = 8 - (U_1L_2)_y = 0 \qquad (U_1L_2)_y = 8 \text{ kips}$$

$$U_1L_2 = 8 \times \frac{5}{4} = 10 \text{ kips} \quad \text{(tension)}$$

The force in the remaining bar is found by employing the equation $\Sigma F_x = 0$.

$$\Sigma F_x = 6 - 6 - (U_1L_2)_x + 18 - L_1L_2 = 0$$

$$6 - 6 - \frac{10}{5} \times 3 + 18 - L_1L_2 = 0$$

$$L_1L_2 = 12 \text{ kips} \quad \text{(tension)}$$

The next section to cut would be one passing through U_1U_2, U_2L_2, and L_2L_3 as shown in Fig. 5.10b. The cut bars are three. However, one of the bar forces, U_1U_2, has previously been determined. The other two bar forces can be determined by the two equilibrium force equations

$$\Sigma F_x = 6 - 6 + 18 - L_2L_3 = 0$$

$$L_2L_3 = 18 \text{ kips} \quad \text{(tension)}$$

$$\Sigma F_y = 8 - 6 - U_2L_2 = 0$$

$$U_2L_2 = 2 \text{ kips} \quad \text{(compression)}$$

In the example just presented the direction of the unknown force at each

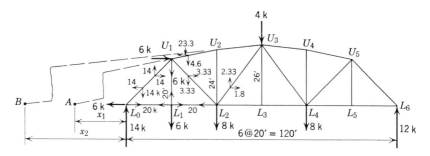

Fig. 5.11. Nonparallel chord truss.

joint was assumed. If the resulting answer from the algebraic equation was plus, the assumed direction was correct. A negative sign would indicate a force in the opposite direction. The sense of the bar force is then determined by the direction of the force with respect to the joint. Another procedure is to assume all unknown forces are acting away from the joint (tension). A plus sign for the answer then indicates tension and a minus sign compression.

The advantage of the method of sections is that it is not necessary to start at one end of the truss and find all the bar forces up to the desired bar force, if only one or two bar forces at an interior joint in the truss is desired. This method simplifies the determination of influence diagrams for bar forces in trusses as is treated in a later section.

With a bit of ingenuity the method of sections can be used to advantage in the determination of the forces in a nonparallel chord bridge truss. The truss of Fig. 5.9 is used to illustrate the procedure. The first step after determining reactions is to find the bar forces $L_0 U_1$ and $L_0 L_1$ by the method of joints. Figure 5.11 shows these forces.

The analysis of joint U_1 by the method of joints involves writing two simultaneous equations as both unknown bars have x and y components. However, by employing the method of sections, the simultaneous equations can be eliminated. A section is cut passing through bars $U_1 U_2$, $U_1 L_2$, and $L_1 L_2$. The force in bar $U_1 U_2$ is then projected to the left along its line of action until it intersects with the force in $L_1 L_2$ also projected to the left along its line of action. This point of intersection is point A (A has been moved to the right in the figure to keep it within the limits of the page). Since the slope of $U_1 U_2$ is five horizontal to one vertical point, A will be $5 \times 20 = 100$ ft to the left of L_1 or $x_1 = 80$ ft. If moments are now taken about A considering the free-body to the left of the cut section, the only unknown in the equation is the force in $U_1 L_2$. However, if the force in $U_1 L_2$ is taken as acting at L_2, the x component will pass through point A, and only the y component will enter into the computation of moments taken about A. This equation is

$$\Sigma M_A = 14 \times 80 - 6 \times 20 - 6 \times 100 - (U_1L_2)_y \times 120 = 0$$

$$(U_1L_2)_y = \frac{400}{120} = 3.33$$

$$U_1L_2 = 1.414 \times 3.33 = 4.7 \text{ kips}\quad\text{(tension)}$$

The force in bar U_1U_2 can now be found by the method of joints without solving simultaneous equations. The force in L_2U_3 can be handled the same way by projecting the force in bar U_2U_3 to the left along its line of action to the intersection of the projection of the lower chord. This point of intersection is labeled B in Fig. 5.11. Since the slope of U_2U_3 is one vertical to ten horizontal, point B is $10 \times 24 = 240$ ft to the left of L_2. The value of x_2 is then 200 ft.

$$M_B = 14 \times 200 - 6 \times 20 - 6 \times 220 - 8 \times 240 - 240(L_2U_3)_y = 0$$

$$(L_2U_3)_y = \frac{-560}{240} = -2.33$$

$$(L_2U_3)_x = \frac{-2.33}{26} \times 20 = -1.80$$

$$L_2U_3 = \sqrt{2.33^2 + 1.8^2} = 2.9 \text{ kips}\quad\text{(tension)}$$

The force in L_2U_3 was broken into its x and y components at L_2. This is necessary in order to have the x component pass through point B. In the solution given, the value of $(L_2U_3)_y$ is minus. This means that its true direction is opposite to the direction assumed in the moment equation. The direction assumed in the moment equation was down—causing a moment opposite to the 14-kip reaction. If $(L_2U_3)_y$ is up, the force in L_2U_3 is away from the joint. Forces acting away from a joint are, of course, tension.

The forces on the remaining right-hand half of the truss can be found in a similar manner. This exercise is left for the student.

Counter Systems. An arrangement of web members within a panel which was commonly used in the early truss bridges consists of crossed diagonals. Crossed diagonals are still used in the bracing systems of bridge trusses and towers. These crossed diagonals shown in Fig. 5.12 are given the name *counters*. Strictly speaking, such a system is statically indeterminate. In applying the $b = 2n - 3$ equation it is obvious that b is greater than $2n - 3$. Because of the physical nature of the diagonals, however, in some of these systems the structure "breaks down" into a

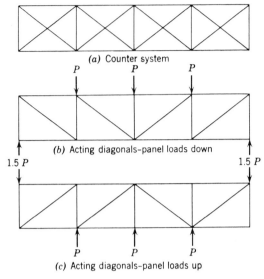

(a) Counter system

(b) Acting diagonals-panel loads down

(c) Acting diagonals-panel loads up

Fig. 5.12

statically determinate system. If the diagonals are fabricated of cross sections, such as round bars, flat bars, or cables, which have very small moments of inertia about at least one of the axes, the member will not be able to resist compressive forces. It is generally considered that a structural member with an L/r greater than about 200 is unable to carry a compressive load and will buckle if subjected to such a load. In a truss with a counter system the diagonal sloping in one direction would be subjected to a tensile force, whereas the one sloping in the other direction would have to resist a compressive force. If it cannot carry a compressive force, the entire shear in the panel is resisted by the diagonal that is being stressed in tension. The diagonals with the opposite slope are ineffective, and the analysis proceeds as if they were not present. This effect is shown in Fig. 5.12b. Figure 5.12c shows the opposite diagonals acting when the shear is from the opposite direction as it might be when the diagonal system is part of a wind-bracing truss. Counter systems are rare in main truss structures today.

5.4 Long-Span Trusses

K-Truss.   As was mentioned earlier, long-span trusses require a greater depth between chords than short-span trusses. With no increase in depth but with increase in span length, the required cross-sectional area of the chords becomes too large. As the depth increases, the panel point spacing increases if the diagonals remain at an economical slope. The K-truss is one means of keeping the panel point spacing within reason and also maintaining a proper slope of the diagonals.

Figure 5.13 shows a K-truss in which the upper chord panel points U_1 to U_9 lie on the arc of a parabola. The point of intersection of the diagonals (the M-points) are considered to be at the midheight of the verticals. This condition, of course, does not necessarily have to be so.

The bar forces in the K-truss can be found either by the method of sections or the method of joints or a combination of both. In employing the method of joints, the process could start at L_0 as soon as the reactions are found. Joint L_1 or U_1 is then considered next. The force in bar L_1M_1 is, of course, equal to and opposite to the vertical load at L_1. An analysis of joint U_1 determines the force in bar U_1M_1. The step next is to find the forces in the diagonals M_1U_2 and M_1L_2. Figure 5.13b shows a free-body diagram of joint M_1. If $F_1 > F_2$, F_3 and F_4 are in the direction shown. If $F_1 < F_2$, F_3 and F_4 are in the opposite direction. This can be easily proved by considering the equilibrium of joint M_1. The x components of F_3 and F_4 must be equal to and opposite to each other to maintain equilibrium at joint M_1. If the x components of these two bar forces are opposite, then the y components are in the same direction. The sum of the

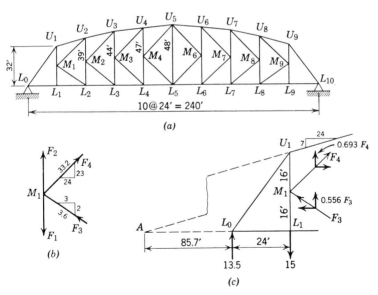

(a)

(b)

(c)

Fig. 5.13 K-truss.

y components must equal the difference in F_1 and F_2. Where F_3 and F_4 have the same slope, the y components of these two forces are equal to each other, and therefore each is equal to one-half the difference between forces F_1 and F_2. In Fig. 5.13b

$$\sum F_x = 0: \qquad 3 \times \frac{F_3}{3.6} = 24 \times \frac{F_4}{33.2} \quad \text{or} \quad F_3 = 0.87F_4$$

$$\sum F_y = 0: \qquad 2 \times \frac{F_3}{3.6} + 23 \times \frac{F_4}{33.2} = F_1 - F_2$$

$$0.556F_3 + 0.693F_4 = F_1 - F_2$$

$$0.556(0.87F_4) + 0.693F_4 = F_1 - F_2$$

$$F_4 = \frac{F_1 - F_2}{1.177} = 0.850(F_1 - F_2)$$

$$F_3 = 0.740(F_1 - F_2)$$

If the forces in the diagonal bars are desired without having to find any other bar forces, the method used in the sloping chord truss of Fig. 5.11 can be utilized. A section is cut vertically between panel points 1 and 2, severing bars U_1U_2, M_1U_2, M_1L_2, and L_1L_2. The left-hand portion of the truss is then drawn as a free-body diagram as shown in Fig. 5.13c. The chords are then projected to their point of intersection, point A. Then as was done for the previous truss, moments are taken about point A. To illustrate the procedure a problem is given with a single vertical load of 15 kips at panel point L_1. Point A is $\frac{3.2}{7} \times 24 = 109.7$ ft to the left of L_1. Taking moments about point A, the only unknown bar forces entering into the equation are the y components of F_3 and F_4. The moments of the x components of the two forces cancel each other, for they are equal in magnitude and have the same lever arm about A. The equation for moment about A is

$$15 \times 109.7 - 13.5 \times 85.7 = (0.556F_3 + 0.693F_4)109.7$$

$$15 - 10.55 = 0.556F_3 + 0.693F_4$$

Since $\qquad\qquad\qquad F_3 = 0.87F_4$

as previously determined

$$1.177F_4 = 4.45$$

$$F_4 = 3.78 \text{ kips} \quad \text{(tension)}$$

$$F_3 = 3.29 \text{ kips} \quad \text{(compression)}$$

The other diagonals of the truss can be handled in the same manner. The force in U_1U_2 can be determined by the method of sections as well as the method of joints. When we take moments about L_1 of the free-body diagram in Fig. 5.13c, the horizontal component of U_1U_2 is the only unknown force entering into the equation

$$\sum M_{L1} = 0 \qquad 13.5 \times 24 - \left(\frac{24U_1U_2}{25}\right)32 = 0$$

$$U_1U_2 = \frac{324}{30.7} = 10.6 \text{ kips} \quad \text{(compression)}$$

The forces F_3 and F_4 do not enter into the equation because the moments of the horizontal components cancel each other. The vertical components of course pass through point L_1.

Subdivided Panel Truss.   Another type of long-span bridge truss is the subdivided truss of Fig. 5.14. Here again the determination of bar forces at joints L_0, L_1, and U_1 is performed in this order in the same manner as for the previous trusses. However, in proceeding to joints U_2 or L_2 it is seen that there will be three unknown bar forces at each joint. The inexperienced might get the idea that the structure is statically indeterminate, but by applying the $b = 2n - 3$ equation of Section 5.1 we see that the equation is satisfied for static determinacy. There are a number of ways to proceed with the solution. One method would be to cut a vertical section 1–1 as shown in the figure. By taking moments about L_4 the force in U_2U_4 can be found. Then proceeding to joint U_2 the remaining two unknowns at this joint, bar U_2M_1 and U_2L_2, can be determined. The force in bar U_1U_2 would be determined previously by an analysis of joint U_1. Moving down to joint L_2, bar forces in L_2M_1 and L_2L_3 can be found. The remaining forces at M_1 can be ascertained if bar force U_3M_1 is known. This bar will be subjected to a force only if there is a panel point load at U_3. By the approach used here it appears that the bar forces could not be found by the method of joints alone. This, however, is not so. The method of joints could be used exclusively, but it would result in a set of simultaneous equations that are usually time consuming

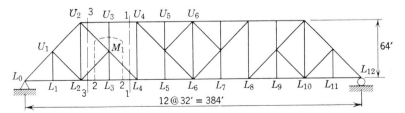

Fig. 5.14   Subdivided panel truss.

in their solution. The method just described illustrates a combining of both the standard methods for a quicker solution.

It may be desirable to find the forces in the bars joining at M_1 without having to determine the forces in other bars of the truss. This can be done by cutting a section around joint M_1 as shown by the dashed line 2–2 of Fig. 5.14. In cutting section 2–2 there are six bars cut. By considering only the part of the truss inside the section lines 2–2 as a free-body the force in member M_1L_2 can be found. First, however, the force in bar M_1U_3 must be determined. This is easily done by an analysis of joint U_3. Bar force M_1U_3 is equal in magnitude to the value of the panel point loading at U_3. Next, moments are taken about panel point L_4. It is observed that the line of action of the forces in all of the cut bars except M_1L_2 and M_1U_3 passes through panel point L_4. Since force M_1U_3 has already been ascertained, the equation for moments about L_4 contains only one unknown, the force in M_1L_2. It is necessary, however, that there be an external load at L_3 or U_3 in order for there to be a force in M_1L_2. If there is no load at L_3 or U_3, the equation for moments about L_4 contains only one term (force in M_1L_2), which, of course, must equal zero. In taking moments about L_4, it is best to resolve M_1L_2 into its x and y components at a point along its line of action where the x component passes through L_4. This point will, of course, be at panel point L_2. If panel point L_3 is midway between L_2 and L_4, as it usually is, then the y component of the force in M_1L_2 is equal to one-half the sum of the external panel point loads at U_3 and L_3. This is true when the external loads at U_3 and L_3 are both in the same direction, which is true for all bridge trusses. The y component of the force in M_1L_4 can be found by cutting the section 1–1 and considering the portion of the truss to the left of the section line as a free-body. In order to satisfy the equation $\Sigma F_y = 0$ for this free-body the y component of the force in bar M_1L_4 must be equal to the algebraic sum of the y components of all the external panel point loads on the free-body. The force in M_1U_2 can be found in like manner after the force in M_1L_2 has been found. Cut a vertical section 3–3 through the truss. Bar force M_1U_2 is now found in the same way as M_1L_4. It is necessary that the y component of the force in M_1L_2 be considered as an external force in the y direction.

5.5   General Features of Truss Bridges

Bridge trusses can be divided into two general classifications based upon the location of the roadway. If the roadway frames into the top chord panel points, the bridge is called a deck-type truss, whereas if the roadway

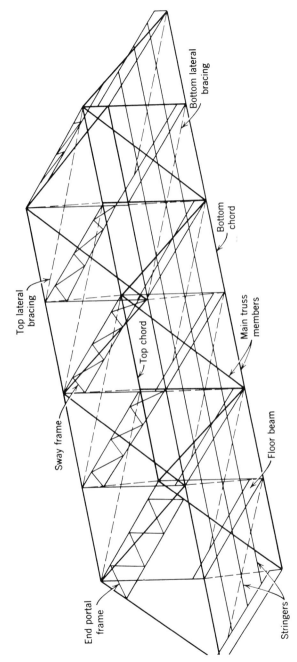

Bottom lateral
bracing

Bottom
chord

Top lateral
bracing

Top chord

Main truss
members

Sway frame

Floor beam

End portal
frame

Stringers

Fig. 5.15 Through truss highway bridge.

is supported at the lower chord panel points, the bridge is classified as a through-type bridge. The trusses and bracing are very much in evidence when crossing a through bridge, whereas the trusses are not in view during passage over a deck bridge.

A truss bridge is in reality a space frame consisting of several structural systems all working together to give strength and stability to the bridge. However, for purposes of analysis, each truss can be considered as a separate unit and analyzed as a two-dimensional planar structure. It is important that the way in which loads are imparted to all parts of the structure be fully understood. The live load has to be traced from the wheels of the vehicle to the supporting earth. The component parts of a highway truss-bridge are shown in Fig. 5.15. The first item in the support of the vehicle is the roadway deck. In modern highway bridges the deck is usually reinforced concrete or an open steel grid floor. The roadway deck acts as a beam and has for its supports the stringers. The stringers are longitudinal beams spanning between larger transverse beams called floor beams. The stringers are designed as simple beams if they are attached to the web of the floor beams by standard web angles. However, where the spacing of floor beams is not large, the stringers are sometimes designed to pass over the top of the floor beams and are made continuous for two or three span lengths. When this is so, the stringers should be analyzed as continuous beams. The ends of the floor beams are attached to the main bridge trusses at the panel points. There is a floor beam for each panel point. This meets one of the conditions of an ideal truss in that all loads have to be applied to the joints. The trusses then carry the load to the substructure, which in turn is supported by the earth.

In addition to the floor system and the main trusses there are usually four distinct bracing systems in a truss bridge. There are two horizontal trusses, one in the plane of the top chord and one in the plane of the bottom chord, that transmit any lateral forces to the foundation. The usual lateral force is the wind load. As the top lateral truss cannot continue on down to the bridge bearings in a through truss bridge, the forces in the top lateral truss have to be carried through a frame at each end of the structure. This frame is called the portal frame. The analysis of portal frames is treated in Chapter 7. In addition to the portal frame in through trusses there is usually a frame system at each panel point. These intermediate frames are usually not designed for any specific loading, but the members are selected on the basis of meeting L/r requirements only. These frames add general lateral stability to the structure. Deck-type bridges do not have a cross frame at each panel point, but can use a simpler system of crossed diagonals whose L/r requirements have to conform to that for tension members only.

5.6   Dead-Load Forces

The nature of the dead loads for highway bridges has been discussed in Section 3.3 and for railroad bridges in Section 3.4. The floor system is designed before the truss itself, so that only the weight of the truss need be estimated to make a determination of the dead-load forces in the members of a truss. The weight of the floor system is applied to the panel points of the truss by the floor beams. The dead load of the floor system will thus

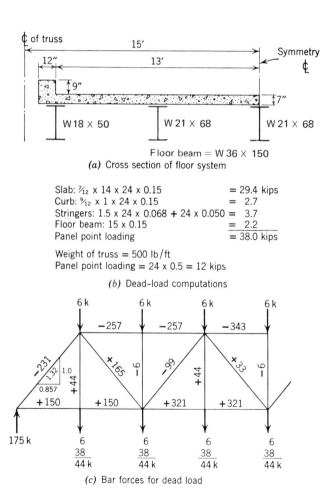

(a) Cross section of floor system

Slab: $\frac{7}{12}$ x 14 x 24 x 0.15   = 29.4 kips
Curb: $\frac{9}{12}$ x 1 x 24 x 0.15   =   2.7
Stringers: 1.5 x 24 x 0.068 + 24 x 0.050 =   3.7
Floor beam: 15 x 0.15   =   2.2
Panel point loading   = 38.0 kips

Weight of truss = 500 lb/ft
Panel point loading = 24 x 0.5 = 12 kips

(b) Dead-load computations

(c) Bar forces for dead load

Fig. 5.16 Dead-load analysis.

be applied to the top chord panel points for deck trusses and bottom chord panel points for through trusses. The dead weight of the main truss and bracing system should be correctly applied to all the top and bottom chord panel points of the truss. However, some designers apply the truss weight entirely to the same panel points that the floor weight is applied.

Which system is used makes no difference in the dead-load forces of most of the members of a truss. For instance, in a simlpe-span Pratt or Warren truss, the verticals will be the only members affected by the two different methods of distribution of the dead load. The effect on the verticals will usually be slight. After the panel point dead loads have been determined, the forces in all of the members are computed by either one or a combination of the three methods explained in Section 5.3. A complete analysis of a highway bridge truss for dead load is shown in Fig. 5.16.

5.7 Live-Load Forces

Influence Line Method. The live load on a bridge truss is always a moving load. A moving load presents a more complex problem to an engineer than does a static load. Before the bar forces can be determined it is necessary to find the position of the live load that will result in the greatest force in the member under consideration. Otherwise, the vehicle could be placed in such a way that the force calculated would not be the maximum. In addition, the position for maximum effect, which is called the *critical position*, may be different for nearly each member of the truss. It is therefore apparent that methods must be established by which this *critical position* can be found. First, however, the development of influence lines for bridge trusses is presented.

Influence diagrams for beams were covered in Section 4.5. The same definition for an influence diagram applies to trusses. However, for trusses, the influence diagram is drawn for the force in a particular member of a truss. Therefore, it appears necessary to draw an influence diagram for each member of the truss. As most bridge trusses are symmetrical about the midspan of the truss, the number of influence diagrams will be approximately one-half the total number of members in a truss. Even this number can usually be reduced because several members will have similar influence diagrams. The influence diagram is plotted with position of load on the structure as abscissa, and the force in the member due to a unit load on the structure as ordinate. Warning is again given to the student to learn the principles of determining influence diagrams and not try to remember assorted shapes and sizes that match the various members of numerous types of trusses.

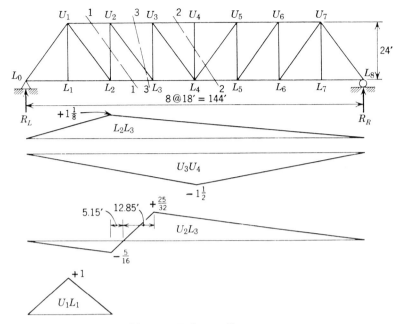

Fig. 5.17 Influence diagrams.

Several influence diagrams are plotted for the simple through truss in Fig. 5.17. The first diagram, that for lower chord member L_2L_3, is a simple triangle with apex at panel point L_2. By employing the method of sections, the force in member L_2L_3 can be found by cutting section 1–1 through members U_1U_2, U_2L_2, and L_2L_3 and taking moments about panel point U_2. With the unit load anywhere between L_2 and L_8, the moment about U_2 (due to external loads) is simply the value of the reaction at L_0 times the lever arm of 36 ft. Since the portion of the truss to the left of the section line is the free body under consideration, the moment of the reaction must be balanced by the moment due to the force in member L_2L_3. The force in L_2L_3 is then equal to the moment of the reaction about U_2 divided by the perpendicular distance between the chords. The value of the vertical reaction at L_0 is a straight line function of the distance from L_8 to the unit load. Therefore the moment about U_2 and consequently the force in L_2L_3 is a straight line function of the position of the load from L_8 to L_2. The influence diagram for this member is a straight line from zero at L_8 to $\frac{6}{8} \times \frac{36}{24} = 1\frac{1}{8}$ at L_2. When the unit load is moved to the left of L_2, the moment about U_2 reduces, for the moment of the unit load has to be considered. With the unit load at L_1 the moment about U_2 is $\frac{7}{8} \times 36 - 1 \times 18 = +13.5$. The force in L_2L_3 and thus the ordinate to the influence diagram at L_1 equals $13.5/24 = +0.5625$. This is just one-half of the value of the ordinate at L_2, proving that the influence line is a straight line between L_2 and L_0. The plus values of the ordinate indicate tension in the member.

The influence diagram for U_3U_4 is constructed similarly to L_2L_3. The section is cut along line 2–2, and moments are taken about panel point L_4. With the unit load between L_8 and L_4, the moment about L_4, due to the forces on the left portion, is $72R_L - 24\,(U_3U_4) = 0$. The value of U_3U_4 is a direct function of R_L, and thus the influence line ordinates are a straight line function from L_8 to L_4. When the unit load is between L_4 and L_0, the right-hand free body can be utilized and the moment about panel point L_4 can be written as $72R_R - 24(U_3U_4) = 0$. The influence line will then be a straight line from the maximum at L_4 to zero at L_0.

The third influence diagram drawn in Fig. 5.17 shows a configuration different from that of the previous two. This diagram indicates the force in member U_2L_3. This member has a $3:4:5$ slope. In order to determine the force in member U_2L_3, section 3–3 is cut. The y component of the force in U_2L_3 must then be equal to the algebraic sum of all the vertical forces acting on either the left-hand or right-hand free body. With the unit load between L_8 and L_3, the only external vertical forces acting on the left-hand portion is the reaction R_L. The force in U_2L_3 is then equal to $\frac{5}{4}R_L$. As R_L is a straight line function of position of load between L_8 and L_3, the line varies from zero at L_8 to $\frac{5}{4} \times \frac{5}{8} = +\frac{25}{32}$ at L_3. The plus value indicates tension. When the unit load is at L_2, the sum of the vertical forces on the left-hand free body is $\frac{3}{4} - 1 = -\frac{1}{4}$, making the force in U_2L_3 equal to $-\frac{1}{4} \times \frac{5}{4} = -\frac{5}{16}$ (compression). Placing the unit load at L_1, it is readily seen that the ordinate is just one-half of the value obtained with the unit load at L_2. The diagram is then a straight line to a value of zero at L_0. At this stage of the development, it is well to consider the diagram between L_2 and L_3. As all bridge trusses have stringers spanning between the floor beams, all loads must be applied to the truss by the floor beams; that is, all loads are applied to the truss at the panel points. If the unit load is placed midway between L_2 and L_3, for purposes of analysis it is necessary to replace this unit load by loads of 0.5 applied simultaneously at L_2 and L_3. The force in U_2L_3 is therefore then (considering the left-hand free body) equal to $(\frac{11}{16} - \frac{1}{2})\frac{5}{4} = +\frac{15}{64}$. A check of a point on a straight line midway between a value of $+\frac{25}{32}$ and $-\frac{5}{16}$ proves to be $+\frac{15}{64}$. The influence diagram is therefore a straight line between L_2 and L_3.

The bottom diagram of Fig. 5.17 shows the influence line for the hanger U_1L_1. The only way for this member to be subjected to a force is to have an applied load at panel point L_1. In order to have an applied load at panel point L_1 there must be a load on the floor beam at L_1. As the unit load moves to the left from L_8 there is no load applied to floor beam L_1 until the unit load passes to the left of L_2. As the unit load moves to the left of L_2 the load on the floor beam at L_1 increases until it is equal to unity when the load is at L_1. The diagram then decreases to zero again at L_0.

Section 4.4 developed the principles that the effect of a concentrated load on a structure was equal to the magnitude of the load times the ordinate of the influence diagram at the point of loading. The sum effect of several loads on a structure is then the algebraic sum of the individual loads multiplied by their respective ordinates of the influence diagram. It was also proved that the effect of a uniform load on a structure could be found by multiplying the intensity of the uniform load by the area of the influence diagram between the limits of the uniform load. These same rules, of course, apply to trusses.

If the truss of Fig. 5.17 were to be designed for the H20-44 loading (Figs. 3.7 and 3.8), the question would be which loading should be used, the truck loading or the equivalent lane loading. As the AASHTO Specification states that "the loading which produces the maximum stress shall be used," it probably will be necessary to try both loadings. The AASHTO Specification contains tables that indicate which loading controls for shear and moment on simple beam spans. These tables can be utilized to a great extent to indicate the maximum loading for the bar forces of trusses. However, in case of doubt the forces due to both loadings should be computed. First, the force in bar L_2L_3 is determined for one lane of H20-44 loading. The truck loading is used first. Since the truck consists of two axles, with 8 kips on the front axle and 32 kips on the rear axle, the question is where to place the truck. From the influence diagram for L_2L_3 it is apparent that for the maximum force the 32-kip axle should be placed at the apex of the triangle. This point is at L_2. As the distance between axles is a fixed distance of 14 ft, it is also noted that the ordinate of the influence diagram 14 ft to the right of L_2 is greater than the ordinate 14 ft to the left of L_2. With the vehicle on the bridge so that the 32-kip axle is at L_2 and the 8-kip axle 14 ft to the right of L_2, the resulting force in L_2L_3 is

$$F_{L_2L_3} = 32 \times 1.125 + 8\left(1.125 \times \frac{94}{108}\right) = 36 + 7.8 = 43.8 \text{ kips} \quad \text{(tension)}$$

The term in parentheses is the ordinate of the influence diagram at the 8-kip load. The force due to the lane loading will be calculated as a comparison. The lane loading consists of a uniform load of 0.64 kips per foot and a concentrated load of 18 kips for moment and 26 kips for shear. Since the chords resist the bending moment in the structure the 18-kip concentrated load is used. The force is found by multiplying the 0.64-kip uniform load by the area of the influence diagram and the 18-kip concentrated load by the maximum ordinate and then adding the two values. The computation is as follows:

$$F_{L_2 L_3} = 0.64 \times 0.5 \times 1.125 \times 144 + 18 \times 1.125 = 51.9 + 20.2$$
$$= 72.1 \text{ kips} \quad \text{(tension)}$$

It is seen that the lane loading is considerably more severe than the truck loading for this member. This will be true where the base of the influence diagram is long (greater than approximately 56 ft for moment resistant members with the H20 loading) and consists of a single triangle. The force in $U_3 U_4$ due to one lane of traffic per truss can be found in a similar manner.

TRUCK LOADING

$$F_{U_3 U_4} = 32 \times 1.5 + 8\left(1.5 \times \frac{58}{72}\right)$$

$$= 48 + 9.7 = 57.5 \text{ kips} \quad \text{(compression)}$$

LANE LOADING

$$F_{U_3 U_4} = 0.64 \times 0.5 \times 1.5 \times 144 + 18 \times 1.5 = 69.1 + 27$$

$$= 96.1 \text{ kips} \quad \text{(compression)}$$

From the experience with member $L_2 L_3$ it would not have been necessary to make the calculation for force in $U_3 U_4$ due to the truck loading.

Member $U_2 L_3$ is a web member and as such resists the shear forces. Therefore, in applying the lane loading to this member, the 26-kip concentrated load is used instead of the 18-kip load. The influence diagram for this member shows what is termed a stress reversal in the member. That is, the stress in this member due to a load passing across the bridge would go from compression to tension or tension to compression depending on the direction of travel. In this case it is necessary to determine both the maximum tensile and compressive forces in the member. The same procedure is used to determine the forces due to both the truck and lane loading.

TRUCK LOADING
Tension:

$$F_{U_2 L_3} = 32 \times \frac{25}{32} + 8\left(\frac{25}{32} \times \frac{76}{90}\right) = 25 + 5.3 = 30.3 \text{ kips}$$

Compression:

$$F_{U_2 L_3} = 32 \times \frac{5}{16} + 8\left(\frac{5}{16} \times \frac{22}{36}\right) = 10 + 1.5 = 11.5 \text{ kips}$$

LANE LOADING
Tension:

$$F_{U_2 L_3} = 0.64 \times \frac{1}{2} \times \frac{25}{32} \times 102.85 + 26 \times \frac{25}{32} = 25.7 + 20.3 = 46 \text{ kips}$$

Compression:

$$F_{U_2L_3} = 0.64 \times \frac{1}{2} \times \frac{5}{16} \times 41.15 + 26 \times \frac{5}{16} = 4.1 + 8.1 = 12.2 \text{ kips}$$

It is seen that the lane loading is also the maximum condition for this member; however, the compression values for the two loading conditions are quite close.

The last member for which an influence diagram is drawn in Fig. 5.17 is the hanger U_1L_1. This member is subjected to tensile forces only. Since it neither resists shear nor moment but just carries the floor beam reaction, it is not designated by the AASHTO Specification which concentrated load is to be used when applying the lane loading. The writer has always used the 26-kip value, for this is the load that would be used in determining the maximum reaction of the stringer and thus the maximum floor beam loading.

TRUCK LOADING

$$F_{U_1L_1} = 32 \times 1 + 8 \times \tfrac{4}{18} = 32 + 1.8 = 33.8 \text{ kips} \quad \text{(tension)}$$

LANE LOADING

$$F_{U_1L_1} = 0.64 \times 0.5 \times 36 + 26 \times 1 = 11.5 + 26 = 37.5 \text{ kips} \quad \text{(tension)}$$

If the 18-kip concentrated load had been used, the truck loading would have been maximum.

These examples demonstrate the advantage in the use of influence diagrams for determining forces in trusses due to live loads. The influence diagram can also be used in determining dead-load forces provided all the dead load is considered to be acting along the loaded chord (the loaded chord is the one to which the floor beams are attached). Here the dead-load forces are determined by multiplying the total area of the influence diagram by the magnitude of the dead load per unit length. For member U_2L_3 of Fig. 5.17 the dead-load intensity would be multiplied by the net area of the influence diagram. The dead-load forces for the four members of this truss for a unit dead load of 1.5 kips per foot are computed as follows:

$$F_{L_2L_3} = 1.5 \times 0.5 \times 1.125 \times 144 = 121.5 \text{ kips}$$

$$F_{U_3U_4} = 1.5 \times 0.5 \times 1.5 \times 144 = 162 \text{ kips}$$

$$F_{U_2L_3} = 1.5 \times 0.5 (0.781 \times 102.85 - 0.312 \times 41.15) = 50.6 \text{ kips}$$

$$F_{U_1L_1} = 1.5 \times 0.5 \times 1 \times 36 = 27 \text{ kips}$$

Maximum Force Criterion. There is usually no question of where to place the AASHTO H-Loading vehicles to obtain the maximum force in the members of a bridge truss. With a two-axle vehicle the heavier axle is placed at the apex of the influence diagram triangle and intuitive reasoning indicates the location of the other axle. Even with the three-axle H-S truck there is usually nothing more necessary than simple reasoning for the proper placing of this vehicle. However, with railroad loadings and specific multiaxle trucks simple reasoning will not usually give a ready answer to the proper placement for maximum bar forces. Simple mathematical methods will, however, give the solution. Sections 4.5 and 4.6 developed methods for determining the proper location to cause maximum effect in simple beams. The methods developed for beams can be readily applied to simple-span trusses.

First to consider is the chord members of a simple-span truss. The chord members resist the bending moments to which the truss is subjected. The resulting influence diagram is a triangle whose base is the full span of the truss. The apex of the triangle is at the point on the truss where moments are taken (by the method of sections) to obtain the force in the member. This influence diagram is then of the same nature as the influence diagram for moment in a beam. In fact, the force in a horizontal chord member, or the horizontal component in a sloping chord member, can be found by dividing the equivalent simple beam moment, taken at the appropriate point, by the perpendicular distance between chords. Since the criterion for maximum moment at a point in a simple beam, Section 4.9, was developed on the basis of the shape of the influence diagram, the criterion there developed is applicable for determining the position for the maximum force in the chords of a simple-span truss. The criterion is given by Eq. 4.5, which is $G_1/a = G_2/b$, where G_1 is the total load on segment a of the influence diagram and G_2 is the total load on segment b of the diagram (Fig. 4.16). The critical position is obtained when G_1/a equals G_2/b. This equality can be obtained for uniform loads, but it is not usually possible for a set of concentrated loads. Then the concentrated loads are moved in turn from the left to the right of the location of the apex of the influence diagram triangle. When the equation reverses from $G_1/a < G_2/b$ to $G_1/a > G_2/b$ by moving a particular load across the point of the apex, the critical position is with that load at the apex of the triangle. This method is demonstrated by determining the maximum force in member L_2L_3 of the truss of Fig. 5.17 for the four-axle vehicle shown in Fig. 5.18.

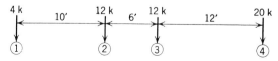

Fig. 5.18   Vehicle load.

The a and b portions of the influence diagram for L_2L_3 (Fig. 5.17) are 36 and 108 respectively. It is obvious that axle number ① is not the critical one, so the process will start with axle number ②. Moving this axle from just to the right of panel point L_2 to just to the left gives

$$\frac{G_1}{a} = \frac{4}{36} < \frac{44}{108} = \frac{G_2}{b} \qquad \text{Axle ② to right}$$

$$\frac{G_1}{a} = \frac{16}{36} > \frac{32}{108} = \frac{G_2}{b} \qquad \text{Axle ② to left}$$

As the equation changed from $<$ to $>$, axle ② is the critical axle. The force in L_2L_3 with the vehicle in this position can be found by again employing the influence diagram.

$$F_{L_2L_3} = 1.125\left(12 + 4 \times \frac{26}{36} + 12 \times \frac{102}{108} + 20 \times \frac{90}{108}\right)$$

$$= 1.125 \times 42.9 = 48.3 \text{ kips} \quad \text{(tension)}$$

The preceding computations were for the vehicle proceeding from right to left across the structure. It may be possible that a greater force will result when the vehicle is reversed and is proceeding from left to right. Sometimes for particular members it will be obvious which direction of vehicle movement will be the most critical, but in most cases it will not be. The only solution then is to determine the forces under both conditions. With the load moving from left to right it is again apparent that axle ① is not critical. In placing axle ② from just to the left of L_2 to just to the right and keeping a and b the same as before.

$$\frac{G_1}{a} = \frac{44}{36} > \frac{4}{108} = \frac{G_2}{b} \qquad \text{Axle ② to left}$$

$$\frac{G_1}{a} = \frac{32}{36} > \frac{16}{108} = \frac{G_2}{b} \qquad \text{Axle ② to right}$$

Axle ② is not critical in this case.

$$\frac{G_1}{a} = \frac{32}{36} > \frac{16}{108} = \frac{G_2}{b} \qquad \text{Axle ③ to left}$$

$$\frac{G_1}{a} = \frac{20}{36} > \frac{28}{108} = \frac{G_2}{b} \qquad \text{Axle ③ to right}$$

Axle ③ is also not the one,

$$\frac{G_1}{a} = \frac{20}{36} > \frac{28}{108} = \frac{G_2}{b} \qquad \text{Axle ④ to left}$$

$$\frac{G_1}{a} = \frac{0}{36} < \frac{48}{108} = \frac{G_2}{b} \qquad \text{Axle ④ to right}$$

Of course, the last exercise of moving axle ④ was not necessary because if the first three axles were not critical, axle ④ had to be. But it is pleasing to see that the method works. The force in member L_2L_3 is

$$F_{L_2L_3} = 1.125\left(20 + 12 \times \frac{96}{108} + 12 \times \frac{90}{108} + 4 \times \frac{80}{108}\right)$$

$$= 1.125 \times 43.6 = 49.1 \text{ kips} \quad \text{(tension)}$$

In this case the force is nearly the same for both directions of movement.

Member U_3U_4 will be subject to the same criterion. It is not necessary, however, to consider the passage of the vehicle in both directions since the influence diagram is symmetrical. Since a is equal to b, it is apparent that the division of the loads is at axle ③. Therefore this axle is placed at L_4. The force is

$$F_{U_3U_4} = -1.5\left(12 + 12 \times \frac{66}{72} + 4 \times \frac{56}{72} + 20 \times \frac{60}{72}\right)$$

$$= -1.5 \times 42.8 = -64.2 \text{ kips} \quad \text{(compression)}$$

The web member U_2L_3 has an influence diagram consisting of two triangles. If the maximum tension force is required for this member, it is desirable to have as many loads as possible within the length of the positive portion of the diagram. Since each portion of the diagram is a triangle, the same criterion can be applied as long as all the loads remain within the length of the same triangle. This will usually be the case as one of the segment lengths will be quite small in proportion to the other segment. Therefore, the number of loads within the short segment will usually be only one. Consider the vehicle of Fig. 5.18 moving from right to left and consider axle ② as a possible candidate for critical honors.

$$\frac{G_1}{a} = \frac{4}{12.85} < \frac{44}{90} = \frac{G_2}{b} \qquad \text{Axle ② to right}$$

$$\frac{G_1}{a} = \frac{16}{12.84} > \frac{32}{90} = \frac{G_2}{b} \qquad \text{Axle ② to left}$$

This is the critical axle for right to left movement. However, it is seen that if we reverse the loading, axle ④ will play the same role, and as ④ is much heavier than ② this latter position will produce the maximum force. The force is

$$F_{U_2L_3} = 0.781\left(20 + 12 \times \frac{78}{90} + 12 \times \frac{72}{90} + 4 \times \frac{62}{90}\right)$$

$$= 0.781 \times 42.8 = 33.5 \text{ kips} \quad \text{(tension)}$$

By similar circumstances axle ④ is placed at panel point L_2 to give the ᴍnum possible compressive force. The vehicle is positioned so that other axles are to the left of L_2. The negative force is

$$F_{U_2L_3} = -0.313\left(20 + 12 \times \frac{24}{36} + 12 \times \frac{18}{36} + 4 \times \frac{8}{36}\right)$$

$$= -0.313 \times 34.9 = -10.9 \text{ kips} \quad \text{(compression)}$$

The last example for this truss is member U_1L_1. The same criterion holds in this case, and since the influence diagram is symmetrical the position of loading will be the same as was found for member U_3U_4; that is, axle ③ is placed at L_1. Since the base of the triangle is relatively short, it should be checked to see that all the axles are within the limits of the influence diagram. Any axle loads falling out of the limits of the diagram should not be included in the criterion check or force computations.

$$F_{U_1L_1} = 12 + 12 \times \frac{12}{18} + 4 \times \frac{2}{18} + 20 \times \frac{6}{18} = 27.1 \text{ kips} \quad \text{(tension)}$$

The railroad loading involves more work in locating the critical position as well as the final bar forces. The E72 railroad loading of Fig. 3.10 is applied also to the truss of Fig. 5.17. If the structure is a single track bridge, each truss carries one-half of the loading. If it is a double track structure, each truss is designed for a full train loading if regulations permit two trains on the bridge at one time. If railroad regulations permit only one train at a time, the truss is designed for a train loading somewhere between one-half to a full train, depending on the location of the tracks with respect to the two trusses. In the example to follow, the truss is considered as carrying one full train load.

The first member considered for this example is L_2L_3 of the truss of Fig. 5.17. Since the influence diagram for force in this member is a simple triangle, the criterion equation $G_1/a = G_2/b$ can be applied. The value of a is 36 ft and b is 108 ft. However, since this is a proportionality equation a can be taken as equal to 1 and b equal to 3. It is quite obvious that the first or second axles of the E72 loading (Fig. 3.10) is not the critical axle. The procedure will start with the number ③ axle. With this axle just to the right of L_2 there are 12 ft of uniform load on the bridge. Therefore

$$G_1 = 108 \text{ kips} \qquad G_2 = 1022.4 - 108 + 12 \times 7.2 = 1000.8 \text{ kips}$$

$$\frac{G_1}{1} = \frac{108}{1} < \frac{1000.8}{3} = \frac{G_2}{3}$$

With axle ③ just to the left of panel point L_2,

$$G_1 = 108 + 72 = 180 \text{ kips} \qquad G_2 = 1000.8 - 72 = 928.8 \text{ kips}$$

$$G_1 = 180 < \frac{928.8}{3} = \frac{G_2}{3}$$

Since G_1 remains $< G_2/3$ during the movement of axle ③, this is not the critical axle.

With axle ④ to right of L_2,

$$G_1 = 180 < \frac{964.8}{3} = \frac{G_2}{3}$$

With axle ④ to left of L_2,

$$G_1 = 252 < \frac{892.8}{3} = \frac{G_2}{3}$$

This is not the critical axle.

With axle ⑤ to right of L_2,

$$G_1 = 252 < \frac{928.8}{3} = \frac{G_2}{3}$$

With axle ⑤ to left of L_2,

$$G_1 = 324 > \frac{856.8}{3} = \frac{G_2}{3}$$

Since the left-hand side of the equation (G_1) changes from $<$ to $>$ in the movement of axle ⑤, this axle is the critical load. With this axle placed at panel point L_2 the maximum force in member L_2L_3 will be obtained. The force in this member can be found by the usual procedure of summing up the product of the axle loads times their respective influence line ordinates.

Moving Loads on K-Trusses.   Influence diagrams for selected members of the K-truss of Fig. 5.19 are now developed. To determine the influence diagram for any member, the unit load is moved across the structure and the force versus load position relations are established. The first member considered is the upper chord U_1U_2. Section 5.4 showed that the force in this member could be established by cutting a section down through the truss between panel points 1 and 2 (section 1–1 of Fig. 5.19), and taking moments about L_1 of the forces acting on the free body to the left of the section line. The horizontal component of the force in U_1U_2 is equal to the moment of the original external forces about L_1 divided by the distance of 32 ft. The influence diagram for this member will then be a triangle whose base is the length of the truss and whose apex is at L_1. The maximum ordinate is then computed as follows for a unit load at L_1

$$F_{U_1U_2} = \frac{9}{10} \times \frac{24}{32} \times \frac{25}{24} = 0.703 \quad \text{(compression)}$$

The influence diagram is plotted in Fig. 5.19.

The next member to be considered is M_1L_1. The force in this member is found by the summation of F_y at joint L_1. By this method it is obvious that $F_{M_1L_1}$ has to be equal to the floor beam reaction at L_1. The influence diagram is then the same as for any hanger.

It is not necessary to know the value of M_1L_1 in order to determine the influence diagram for M_1U_1. A section is cut as shown by the dashed line 2-2 of Fig. 5.19, and moments are taken about the intersection point of the projection of chord U_1U_2 and the lower chord. This is point A. When

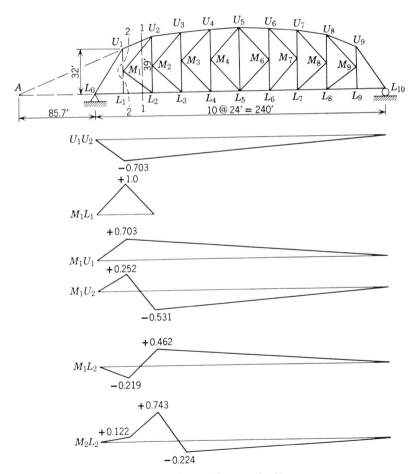

Fig. 5.19 Influence diagrams for K-truss.

the unit load is between L_2 and L_{10}, the moment about point A of the forces acting on the free body only involves the reaction at L_0 and the bar forces in M_1U_1. The influence diagram is a straight line with values varying from zero at L_{10} to $0.8 \times 85.7/109.7 = +0.624$ at L_2. When the unit load is between L_0 and L_2 there is a panel point load at L_1. Since the force in L_1M_1 is always equal and opposite to this panel point load, the moment about point A of the panel point load at L_1 and the force in L_1M_1 cancel each other. The ordinate to the influence diagram at L_1 is then $0.9 \times 85.7/109.7 = +0.703$.

To determine the forces in bars M_1U_2 and M_1L_2 the relationship between the two forces is determined by an analysis of the joint M_1 as a free body. This was done in Section 5.4, and for this truss it was found that $F_{M_1L_2} = 0.87F_{M_1U_2}$. The value of either bar force for a unit load at any point can now be found by taking the free body to the left of section lines 1-1 and taking moments again about point A. The vertical components of these two bar forces are the two unknowns in the equation. Then by substituting for one unknown in terms of the other the solution is obtained. With a unit load at L_2 the equation for moments about point A is

$$0.8 \times 85.7 - (0.556F_{M_1L_2} + 0.693F_{M_1U_2})109.7 = 0$$

Substituting $F_{M_1L_2} = 0.87F_{M_1U_2}$

$$1.177F_{M_1U_2} = 0.624$$
$$F_{M_1U_2} = 0.531 \quad \text{(compression)}$$
$$F_{M_1L_2} = 0.462 \quad \text{(tension)}$$

With a unit load at L_1 the moment equation is

$$0.9 \times 85.7 - 1 \times 109.7 + 1.177F_{M_1U_2} \times 109.7 = 0$$
$$1.177F_{M_1U_2} = 0.297$$
$$F_{M_1U_2} = 0.252 \quad \text{(tension)}$$
$$F_{M_1L_2} = 0.219 \quad \text{(compression)}$$

The results indicate a reversal of stress in these two members when the load moves from L_1 to L_2. The influence diagrams for M_1U_2 and M_1L_2 are given in Fig. 5.19.

One other bar is analyzed here to give a coverage of all types that will be encountered in this truss. For illustration bar M_2L_2 is investigated. The analysis for this bar is different from M_1L_1 since at panel point L_1 there is no diagonal present such as there is at panel point L_2. The analysis of all the bars at similar locations would follow the same procedure. Probably, the best way to determine the influence diagram for bar M_2L_2

is to determine the influence diagram for M_1L_2 first. The value of M_2L_2 is then equal to the vertical component of the force in M_1L_2 plus or minus the panel point loading at L_2. With a unit load at L_3 the ordinate to the influence diagram for M_1L_2 is $+0.462 \times \frac{7}{8} = +0.404$. The force in M_2L_2 is $0.404 \times 2/3.6 = -0.224$ (see Fig. 5.13b). With the unit load at L_2 the force in M_2L_2 is

$$1.00 - 0.462 \times \frac{2}{3.6} = +0.743$$

and with the unit load moved to L_1 the force in M_2L_2 is $0.219 \times 2/3.6 = +0.122$. The plotting of these values shows that the tension portion of the influence diagram is not a triangle. This is the first such case that has been encountered in the examples up to this point. This should stress the point that the shape of an influence diagram should not be assumed, but should always be determined by analytical reasoning.

Moving Loads on Subdivided Panel Truss.   The next truss for which influence diagrams will be drawn is the subdivided panel truss of Fig. 5.20. The influence diagrams for the members between panel points 2 and 4 will be drawn as typical for this truss. In order to find the force in U_2U_4 section 1–1 is cut and the free-body to the left of this section is analyzed. The force is constant throughout the length of U_2U_4 and is found by dividing the moment of the original external forces about panel point L_4 by the depth between trusses. The force in L_4L_6 will always be equal and opposite to that in U_2U_4. This is proved by applying the equation $\Sigma F_x = 0$ to the free body. With a unit load at L_4 the maximum force in U_2U_4 is obtained. This value for a unit load is

$$F_{U_2U_4} = -\frac{2}{3} \times \frac{80}{40} = -1.33$$

The force in U_4L_4 can also be found by cutting section 1–1 and taking the equation $\Sigma F_y = 0$ for either free body. Considering the left-hand free body, the force in the bar will be toward the joint L_4 when the unit load is between L_5 and L_{12}. However, the force will be away from joint L_4 when the load is between L_0 and L_4.

The force in M_1L_4 can easily be determined once the influence diagram for U_4L_4 has been drawn. The equation $\Sigma F_y = 0$ at joint L_4 is used to find the force in M_1L_4. With a unit load at L_3,

$$F_{M_1L_4} = -0.25 \times 1.414 = -0.354$$

With a unit load at L_4, the force in M_1L_4 is

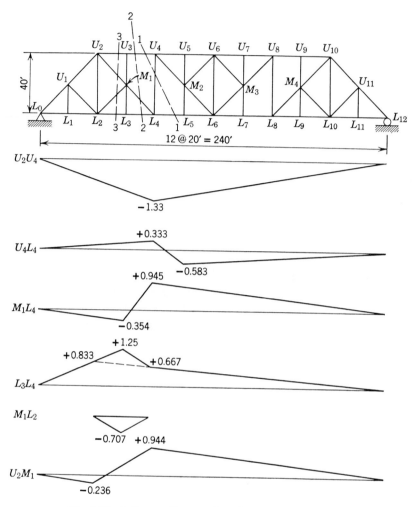

Fig. 5.20. Influence diagrams for subdivided panel truss.

$$F_{M_1L_4} = +\tfrac{2}{3} \times 1.414 = +0.945$$

The force in L_3L_4 can be determined by taking $\Sigma F_x = 0$ at joint L_4 or by cutting section 2-2 and taking $\Sigma M = 0$ about U_2 of the left-hand free body. With a unit load at L_2, the force in L_3L_4 is

$$F_{L_3L_4} = \frac{10}{12} \times \frac{40}{40} = +0.833$$

With a unit load at L_3,

$$F_{L_3 L_4} = \left(\frac{9}{12} \times 40 + 1 \times 20 \right) \frac{1}{40} = +1.25$$

With a unit load at L_4,

$$F_{L_3 L_4} = \frac{2}{3} \times \frac{40}{40} = +0.667$$

The influence diagram for this member $(L_3 L_4)$ is a different shape from any presented previously This is characteristic of chord members within the subdivided panels. The dashed line between points L_2 and L_4 of the influence diagram is the shape of the diagram if the panel were not subdivided. This indicates that although there is a saving in the floor system in the subdivided truss there will be an increase in live-load force in some of the chord members. However, there would no doubt be an over-all saving in a truss of this size by making the panel points at 20-ft spacing instead of 40 ft.

In Section 5.4 the method for finding the force in $M_1 L_2$ was outlined. It was shown that the magnitude of this force is equal to one-half of the panel point loading at L_3 multiplied by the slope of the member. Under live-load conditions there is no load at U_3. The influence diagram is then simply an isosceles triangle with base from L_2 to L_4 and height equal to 0.707. The force in this member is, of course, compression.

The final bar to be considered for this truss is $U_2 M_1$. When we cut a section 3–3 as shown in Fig. 5.20 and take $\Sigma F_y = 0$ for either free body, there are two unknown forces in the equation. However, having previously determined the influence diagram for $M_1 L_2$ as outlined in the previous paragraph, the force in $U_2 M_1$ can be readily obtained. With a unit load at L_2 the force in $U_2 M_1$ is equal to the reaction at L_{12} multiplied by the slope of the member, $M_1 L_2$ being zero for this position of load. This value is 0.236 compression. With the load at L_3 the reaction at L_0 is 0.75 upward and the vertical component of $M_1 L_2$ is 0.50 downward. The difference is the vertical component in $U_2 M_1$. The force in $U_2 M_1$ must then be 0.354 tension. With the unit load at L_4, $M_1 L_2$ is again zero and $U_2 M_1$ is the reaction at L_0 (0.667) multiplied by 1.414. This force of 0.944 is tension also. These three values indicate that the influence diagram is a straight line between L_2 and L_4.

The procedures already developed for the previous trusses can be used in determining the position of live load for maximum effect.

Problems

1. State whether the following structures (Fig. 5.21) are stable or unstable, and if stable whether they are statically determinate or indeterminate both externally and internally. Consider all bars are pinned at their ends.

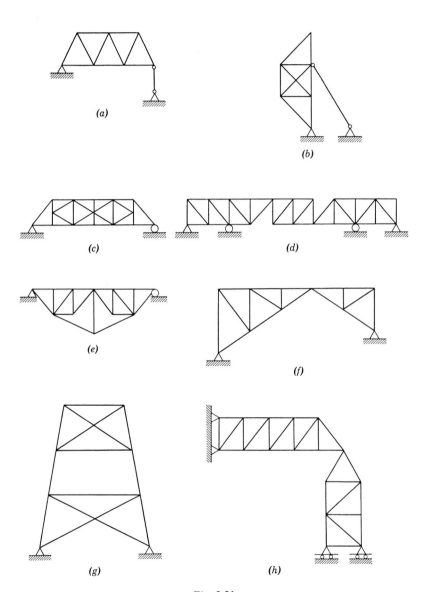

Fig. 5.21

2. Determine all the bar forces in the truss of Fig. 5.22.

3. Change the truss of Fig.5.22 to a Warren truss with verticals and determine the new force in those bars which change.

4. Find the forces in the members of the truss of Fig. 5.23.

5. Determine the bar forces in the truss and supporting bars of Fig. 5.24.
Ans. $F_{ab} = +26.4$ kips, $F_{bc} = +4.8$ kips, $F_{bd} = +26.8$ kips,

6. Determine the bar forces in the tower of Fig. 5.25.

7. Determine graphically the bar forces in the truss of Fig. 5.26.

8. For the truss of Fig. 5.26 plot the influence diagram for the end post (L_0U_1) the first diagonal (U_1L_2) and the chord U_2U_3.

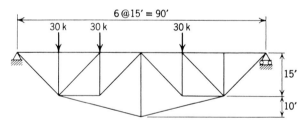

Fig. 5.22

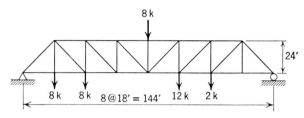

Fig. 5.23

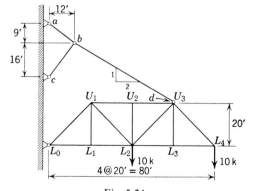

Fig. 5.24

9. Draw an influence diagram for members a, b, and c of the cantilever truss of 5.28.

10. For the truss of Fig. 5.27 plot the influence diagram for members $U_1 M_1$, $M_1 L_1$, $M_1 U_2$, and $U_2 U_3$. M_1 and M_2 are at midpoints of the verticals.

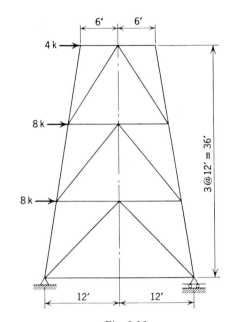

Fig. 5.25

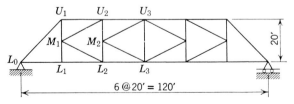

Fig. 5.26

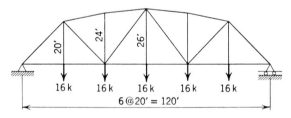

Fig. 5.27

11. Given the truss of Fig. 5.22, find the bar forces in the following members for a uniform load of 1 kip per foot over the whole span and a concentrated load of 24 kips at panel point L_4. Use influence diagrams. L_0U_1, U_1L_1, U_2L_2, U_3U_4, L_3L_4.

Ans. $L_0U_1 = -93.8$ kips, $U_1L_1 = +18$ kips, $U_2L_2 = -39$ kips, $U_3U_4 = -144$ kips, $L_3L_4 = +128$ kips.

12. For the bars named in Problem 10 determine the following:
(*a*) Dead-load forces for a uniform dead load of 1.5 kips per foot per truss.
(*b*) Live-load plus impact forces for one H20–44 truck.
(*c*) Live-load plus impact forces for one lane of H20–44 equivalent lane loading. *Ans.* (*b*) $U_1M_1 = +38.9$ kips, $M_1L_1 = +44.7$ kips, $M_1U_2 = -35.2$ kips or $+8.3$ kips, $U_2U_3 = -61.9$ kips.

13. Find the forces in members of the truss of Fig. 5.29. Consider the crossed diagonals as counters.

14. For the truss of Fig. 5.17 determine the influence diagrams for members U_1U_2, L_3L_4, and U_3L_3. Determine the maximum forces in these three members for the HS-20 truck. Consider the truck can be proceeding across the bridge from either direction.

15. For the truss shown in Fig. 5.30 determine the forces in bars *a.* *b* and *c.*

16. Determine the bar forces for the same members as for Problem 14 for the vehicle of Fig. 5.18.

17. Determine the maximum bar forces in the following members with the E72 railroad loading of Fig. 3.10 crossing the truss of Fig. 5.19. Assume each truss carries one-half train load. (*a*) U_1U_2, (*b*) M_1L_1, (*c*) M_1U_2, (*d*) M_2L_2.

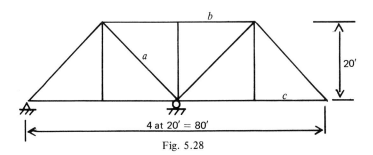

Fig. 5.28

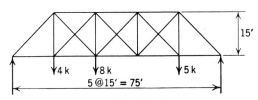

Fig. 5.29

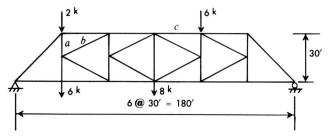

Fig. 5.30.

18. Determine the forces in the following bars of the truss in Fig. 5.20 for 5^k downward loads at panel points L_1 through L_{11}.

(a) U_1U_2
(b) U_1L_2
(c) L_2M_1
(d) U_4M_2
(e) U_5U_6

19. For the truss of Fig. 5.20 plot the influence diagram for members U_1L_2, U_4M_2, and M_2L_6.

20. For the following members of the truss of Fig. 5.20 determine the maximum bar forces for the AASHTO HS-20 loading. Investigate both the truck and lane loading.

(a) U_4L_4, (b) L_3L_4, (c) M_2L_6, (d) M_2L_5.

Ans. $U_4L_4 = -51.5$ kips or $+26.2$ kips, $L_3L_4 = +105.6$ kips, $M_2L_6 = +60.5$ kips or -36.8 kips, $M_2L_5 = +57.2$ kips.

6

ROOF TRUSSES AND THREE-HINGED ARCHES

6.1 Roof Trusses—Introduction

Roof trusses are designed in a multitude of shapes and sizes from the simple wooden rafter truss of residential buildings to the massive steel truss of the airplane hangar. The basic function of the roof truss is to support the roofing material which keeps the elements from entering the inside of the building. Therefore the loads that the roof truss has to support, besides its own weight, are the roofing materials, snow, ice, and wind. In addition to these loads, the truss may have to support loads suspended from it such as hoists or utility lines. All possible loadings should be given careful consideration. A detail treatment of building loads was given in Chapter 3.

In selecting the particular type of roof truss, many factors must be taken into consideration. The slope of the roof will be governed by architectural considerations and possibly also by the snow and wind loads. The configuration of the truss will be dependent on many factors such as roof slope, material of the truss, use of the building, and other factors.

Roof trusses can be divided into peaked roof trusses and flat roof trusses. Peaked roof trusses have been used extensively for industrial buildings in which a relatively short span is required. For longer spans the flat roof truss is more common. Recent awareness of esthetic considerations, even in industrial buildings, has reduced the use of the peaked roof truss,

Competition from new roof shapes in concrete has also reduced the application of the peaked roof truss.

Four basic configurations of the peaked roof truss are used in modern-day engineering. These four trusses together with the two standard flat roof trusses are shown in Fig. 6.1. In addition to these trusses there are others that have been used in past structures. Many are variations of these basic trusses. The fan truss is used for relatively short spans. The Belgian truss has the struts perpendicular to the top chord, producing a fixed slope of the roof for each particular number of panel points in the top chord. The Fink truss is usually limited to three panel points between the peak and the eave, whereas the Pratt truss can have any desired number of panel points.

All the roof trusses shown in Fig. 6.1 are statically determinate. The application of the equation $b = 2n - 3$ will prove this. Roof trusses have been supported in a number of different ways. Supporting the ends by concrete or steel columns is the most common method except for low-height structures with very short spans.

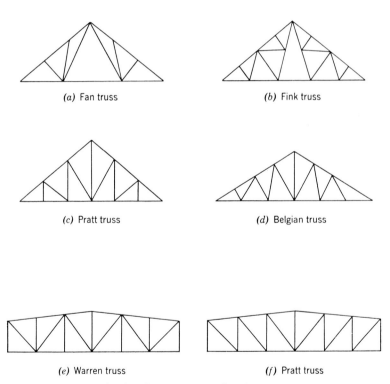

(a) Fan truss   (b) Fink truss

(c) Pratt truss   (d) Belgian truss

(e) Warren truss   (f) Pratt truss

Fig. 6.1 Common types of roof trusses.

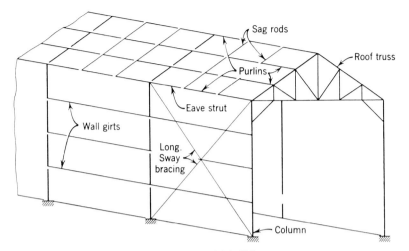

Fig. 6.2 Industrial building.

In all roof trusses, except for residential or small commercial buildings, the roof loads are applied to the top chord panel points. This is accomplished by attaching small beams to the trusses at the top chord panel points. These beams span between trusses, and the roofing material is attached to these beams. They are called purlins and are illustrated in the picture of the structural parts of a building in Fig. 6.2. Because the gravity loads applied to the purlins are at an angle to the principal axes of the cross section of the purlin, bending takes place about both axes. In order to reduce the bending moment about the Y–Y axis, and thus reduce the required size of beam, the span of the purlin with respect to the Y–Y axis of the beam is reduced. This is accomplished by attaching sag rods to the purlins at points between the trusses. These sag rods as shown run parallel to the slope of the roof. Whether more than one row of sag rods for each bay is used depends on the distance between trusses. If the distance between purlins is too great for the roofing material, rafters running perpendicular to the purlins are used. The roof load is then considered to be carried from the roofing to the rafters, to the purlins, to the trusses, and then from the trusses to the columns or walls.

6.2 Roof Trusses—Analysis

A Fink-type roof truss is shown in Fig. 6.3. The pitch is $\frac{1}{5}$ (height over total span). The angle formed by the two chords is then 21° 50′. The

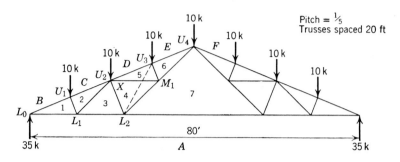

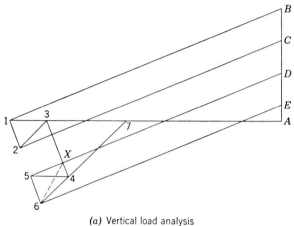

(a) Vertical load analysis

Fig. 6.3(a)

spacing of the trusses is 20 ft center-to-center. The following loads are used in the analysis of the truss.

Vertical Loads

1. DEAD LOAD

$$
\begin{aligned}
\text{Roofing} &= 8 \text{ psf of roof surface} \\
\text{Purlins} &= 4 \text{ psf of roof surface} \\
\text{Truss + bracing} &= \underline{8} \text{ psf of roof surface} \\
\text{Total} &= 20 \text{ psf of roof surface}
\end{aligned}
$$

Panel point load $= 20 \times 20 \times 10.77 = 4.3$ kips

2. SNOW LOAD

25 psf of roof surface $= 5.4$ kips per panel point

The total dead load is usually applied as panel point loads along the top

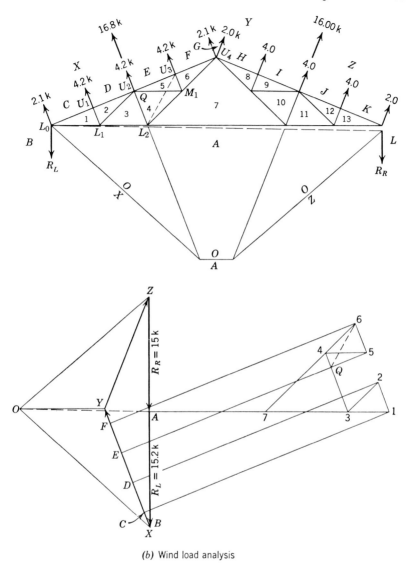

(b) Wind load analysis

Fig. 6.3   Forces in members of a Fink roof truss.

chord for roof trusses. For short-span roof trusses resting on walls or concrete columns, the ends of the trusses usually rest on steel plates and are anchored down by bolts passing through slotted holes. The horizontal component of the wind force is considered to be divided equally between supports. For longer span trusses one end of the truss is fixed against lateral movement and the other end supported on a rocker or on rollers.

With trusses attached to steel columns it is usual to divide the horizontal force equally to the two columns. A further treatment is given in Section 7.5.

The vertical load analysis in Fig. 6.3*a* is made for a panel point loading of 10 kips. The bar forces due to dead load and snow load are 43 % and 54 % respectively of the forces found in the vertical load analysis. The dead- and snow-load forces are tabulated separately so that they can be used in various combinations with the wind loading. The analysis of roof trusses is quicker by the graphical method than by any of the algebraic methods. Reactions can be found graphically for the wind loads, but there is no point in this procedure for dead or snow loads.

Many designers of roof trusses recommend a procedure of adding a concentrated live load to one or two bottom chord panel points. The total bar forces should include this load when attachment of utility lines, monorail hoists, or other items is specified in the design. Even when bottom panel point loadings are not specified, the user of the building could add these extra loads later. The analysis in Fig. 6.3, however, does not provide for such a loading. It would have to be provided for by a separate analysis.

The vertical load analysis of Fig. 6.3*a* follows the usual graphical solution as described in Section 5.3. Since both the loading and the truss are symmetrical about the centerline of the truss, the bar forces in only one-half of the truss need be determined. Completing the solution for the whole truss provides a check on the accuracy of the solution. The drawing of the last reaction force in the force polygon should close the diagram if the solution has been correctly and accurately performed.

The graphical analysis of a Fink truss presents one unique problem. As the analysis proceeds from joint L_0 to U_1 to L_1 there are only two unknown bar forces at each joint. However, in proceeding from these three joints to the next joint, there are three unknown bar forces at either joint U_2 or L_2. The simplest method the writer has found for handling this impasse is to calculate the force in the long lower chord, bar *A*-7, by the method of sections. This is always a simple matter. For the vertical loading shown, the force in *A*-7 is

$$F_{A\text{-}7} = \frac{35 \times 40 - 10(30 + 20 + 10)}{16} = 50 \text{ kips} \quad \text{(tension)}$$

With this force known, there are only two unknown bar forces at L_2, and the graphical solution can proceed at this joint. With bar 3-4 determined at joint L_2, only two unknown bar forces remain at U_2. The analysis can then proceed in order through the remainder of the structure.

There is a graphical method whereby all the bar forces in a Fink truss can be determined without any numerical analysis. This method consists of replacing members 4-5 and 5-6 by a single member from L_2 to U_3. This would reduce the bars intersecting at panel point U_2 from five to four and the graphical analysis could proceed from L_1 to U_2 to U_3 without encountering any more than two unknown bar forces at each joint. It can easily be proved that the substituting of the member connecting panel points L_2 and U_3 will not change the force in member E-6. After the force in member E-6 is determined by the graphical analysis of joint U_3, the substitute member can be removed and the truss returned to its true condition. It will then be necessary to reanalyze joint U_3 to determine bar forces D-5 and 5-6 and then go back to joint U_2 and find the bar forces at this joint for the true condition of the truss. The analysis can then proceed to joints L_2 and M_1.

Wind Load

$$\theta = 21.8 \text{ degrees}$$
$$\text{Openings} = 20\%.$$

External force on windward side $= 1.20(21.8) - 36 = -9.84$ psf (suction)

External force on leeward side $= -9$ psf (suction)
Internal pressure $= 4.5 + 0.25(20) = 9.5$ psf (pressure)
  or $-[4.5 + 0.15(20)] = -7.5$ psf (suction)

The panel point loads for the wind loading are

External force, windward side $= 9.84 \times 20 \times 10.77 = 2120$ lb (suction)
External force, leeward side $= 9 \times 20 \times 10.77 = 1940$ lb (suction)
Internal pressure, both sides $= 9.5 \times 20 \times 10.77 = 2050$ lb
Internal suction, both sides $= 7.5 \times 20 \times 10.77 = 1615$ lb

We now analyze the truss for the maximum condition, which is the external forces combined with the internal pressure. This combination gives panel point loads as follows:

Windward side $= 2120 + 2050 = 4170$ lb
Leeward side $= 1940 + 2050 = 3990$ lb

Both these loads are away from the roof.

The analysis of the truss under the action of the wind loads is shown in Fig. 6.3b. The analysis is based on the left end fixed and the right end on bearings. This support condition makes the truss unsymmetrical with respect to the wind loads, and the truss is usually analyzed twice: wind from the left and wind from the right. However, in the illustrated example, the wind loads on the windward side and the leeward side are practically equal and are both in an outward direction. This results in a horizontal

Table 6.1 Forces in Roof Truss

Bar	Force due to 10 kips Vertical Panel Load (kips)	Dead Load (kips)	Snow Load (kips)	Forces in Members Wind Load (kips)	$DL + SL$ (kips)	$(DL + WL)3/4$
L_0L_1	+88	+38	+47.5	−32	+85.5	Less Than Dead Load Only
L_1L_2	+75	+32	+40.5	−26.5	+72.5	
L_2L_3	+50	+21.5	+27	−16	+48.5	
L_0U_1	−94	−40.5	−51	+35	−91.5	
U_1U_2	−91	−39	−49	+35	−88	
U_2U_3	−86	−37	−46.5	+35	−83.5	
U_3U_4	−82	−35.5	−44.2	+35	−79.7	
U_1L_1	−9.5	−4.1	−5.1	+4.2	−9.2	
L_1U_2	+13	+5.6	+7	−6.8	+12.6	
U_2L_2	−18.5	−8.0	−10	+10	−18	
U_2M_1	+12	+5.2	+6.5	−6.8	+11.7	
U_3M_1	−9.5	−4.1	−5.1	+4.2	−9.2	
L_2M_1	+25	+10.8	+13.5	−13	+24.3	
M_1U_4	+37	+16.0	+20	−20	+36	

+ = tension
− = compression

reaction of an insignificant amount. As seen from the graphical determination of the reactions, the left and right reactions are practically equal. The bar forces for the left half of the truss are practically the same as those on the right half. For the sake of simplicity in Fig. 6.3*b*, only the analysis of one-half of the truss is shown. As a check on the accuracy of the analysis the usual procedure would be to carry the graphical solution through the entire truss. The force polygon should close if the work has been accurate.

The alternate methods of either using the substitute member L_2U_3 or the numerical determination of the force in *A*-7 can be used here again.

Load Combinations. After the forces in all the members for the individual load conditions have been found, the bar forces are combined into design forces, as shown in Table 6.1. As seen from this table the forces due to wind in all the members is of opposite sense to that caused by the dead or snow load. The maximum force in all the members is obtained only under dead and snow loading.

The wind loads do not affect the required size of any of the members. This is usually true for long-span, low slope trusses because the dead load is high and the wind load is away from the roof. If it is physically possible to have either suction or pressure inside a building (depending on direction of wind or which side of the building has openings) then there would be two loading systems on the roof structure that could be possible. The roof would have to be designed using both load systems and then each member would be selected on the basis of which loading required the greater size member. It is possible that due to one set of wind loads a member of a truss would be in compression while under the second set of wind loads it would be subjected to a tension force.

Also there is the added danger of not adequately fastening the roof to the supporting structure if uplift forces were neglected. Most failures of roof structures result in a blowing off of the roof and not a blowing in of the structure. With roofs having an angle to the horizontal greater than 30 degrees, the wind loading on the windward side will be toward the roof and, therefore, the wind load can add to the forces due to dead and snow loads. With short-span trusses and low roof slopes the load due to wind may cause a greater force in the member than the dead load force, but of opposite sense. Then the member would have to be designed both for tension and compression.

The American Institute of Steel Construction Specification for the Design of Steel Buildings states that for members subject to wind forces only, or those members that are designed for a combination of wind and other forces, may be proportioned for an allowable unit stress $33\frac{1}{3}\%$ greater than that used for dead and live load only. Another way of saying the same thing is that the member may be designed for three-fourths of the sum of the wind load combined with other loads but using the normal allowable stress.

6.3 Arches—General

A structure that is suitable for long spans is the arch. An arch is a structural unit that is supported by horizontal reactions as well as vertical reactions. The horizontal reactions must be capable of exerting thrusts (forces towards each other). Without the capability of horizontal thrusts the arch degenerates to a curved beam. Whereas the curved beam would have large bending moments, the arch is subjected to small bending moments. The internal force condition on an arch rib is primarily an axial load. The horizontal thrust causes moments that counteract the

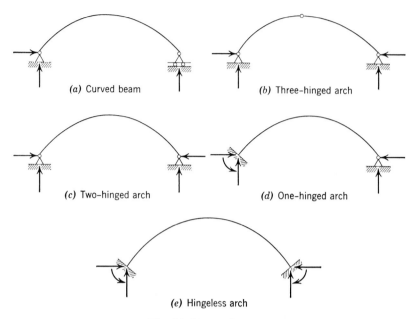

(a) Curved beam   (b) Three-hinged arch

(c) Two-hinged arch   (d) One-hinged arch

(e) Hingeless arch

Fig. 6.4 Types of arches.

moments because of the vertical reactions. There will be no bending moment in the arch if the arch is in the form of a solid rib whose axis forms a parabola and it is subjected to a uniform vertical load. Because of the characteristic of having small bending moments, the arch was ideal for masonary construction as practiced by the ancient builders.

Since the arch requires horizontal reactions, the foundations must be capable of furnishing this thrust. Good rock foundations prove ideal for resisting the arch thrust. The two supports have sometimes been tied together with cables or rods when rock foundation material is lacking.

The arch can be in the form of a solid rib as previously mentioned or of a trussed structure. The solid arch rib has been constructed in timber, stone, concrete, steel, and aluminum. The solid arch rib bridge can be a beautiful structure as attested by the Santa Trinita Bridge of Florence, Italy (1569), and the Sando Bridge over the Angerman River, Sweden (1942).

The characteristics of the support further divides the arch into different classes. These are shown in Fig. 6.4 along with the curved beam. The hingeless, one hinged and two-hinged arches are statically indeterminate structures. These structures have more reaction components than available equations of equilibrium. The single-hinge arch is rarely used.

Since the three-hinged arch is the only statically determinate type, it is the only one discussed further in this chapter.

The three-hinged arch has been used for both bridges and buildings. When used for bridges it has usually been in the form of a trussed arch. The three-hinged arch can undergo settlements of the supports without inducing dangerous bending moments in the structure. This is the primary advantage of the three-hinged arch over the two-hinged structure. The two-hinged arch, when on solid foundation, is more economical and has less deflection than the three-hinged structure.

6.4 Analysis of Three-Hinged Solid Rib Arch

Analytical Method. The three-hinged arch can be analyzed by either analytical or graphical methods. The structure has four components of reaction. Available for the determination of the reactions are the three equations of equilibrium and one equation of condition. The latter equation is that the moment at the central hinge is zero. These four equations are

$$\sum F_x = 0$$
$$\sum F_y = 0$$
$$\sum M = 0$$
$$\sum M_{\text{hinge}} = 0$$

The application of these equations to the determination of the reactions for the arch of Fig. 6.5 is as follows:

$$\sum F_x = 0 \qquad H_L - H_R + 8 = 0$$
$$H_L = H_R - 8$$

$$\sum F_y = 0 \qquad V_L + V_R - 12 - 16 - 4 = 0$$
$$V_L = 32 - V_R$$

$$\sum M_a = 0 \qquad 80V_R + 5H_R - 40 - 360 - 960 - 280 = 0$$
$$H_R = 328 - 16V_R$$

$$\sum M_b \qquad 40V_R - 15H_R - 320 - 120 = 0$$
$$H_R = 2.67V_R - 29.3$$

Equating the last two equations yields

$$2.67V_R - 29.3 = 328 - 16V_R$$
$$V_R = 19.1 \text{ kips}$$

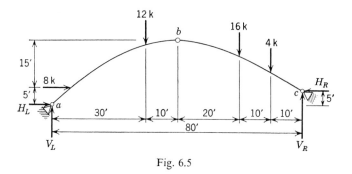

Fig. 6.5

Solving for the other three reactions gives

$$H_R = 21.8 \text{ kips}$$
$$V_L = 12.9 \text{ kips}$$
$$H_L = 13.8 \text{ kips}$$

With the reactions known, the bending moment and shears at all points along the arch axis can be determined in the same manner as was employed for beams. Determining shears for the arch rib, of course, involves more computation than for a straight beam. The required shear for design purposes is the force perpendicular to the axis of the arch. If the shape of the arch can be defined by a continuous mathematical expression, the slope of the perpendicular to the axis can be expressed by a single equation. As most arches are parabolic or circular, this is usually true. To find the shear at a point on the arch rib, the resultant of the forces to the left or the right of the section must be resolved into components perpendicular to the axis of the arch.

The magnitude of the axial thrust at specific points along the arch axis is also a design requirement. The axial thrust is the normal component of the resultant force acting on the arch cross section. If the slope of the arch axis is known at a particular point, the axial thrust at this point can be easily determined. The axial thrust will always be at right angles to the shear.

The moment at any point along the arch can be found simply by the usual analytical methods. The moments of all forces either to the right or left of the section are computed. It should be remembered that the moments are taken about an axis perpendicular to the plane of the arch and passing through the point on the axis of the arch. The axis of the arch corresponds to a line passing through the centroid of the cross section at all points along the arch.

The bending moments, thrusts, and shears will be calculated at several points on the arch of Fig. 6.5. The arch axis forms a second-degree parabolic curve. The equation of the parabolic curve, where the origin of coordinates x and y are at the springing (point of support), is

$$y = 4h\left(\frac{x}{L} - \frac{x^2}{L^2}\right)$$

The constants h and L are the total rise and the span of the arch respectively. The arch in Fig. 6.5 is unsymmetrical so the arch from a to b would have a different equation from that from c to b.

Inserting the constants from a to b

$$y = 4(20)\left(\frac{x}{80} - \frac{x^2}{6400}\right) = x - \frac{x^2}{80}$$

Note: L would be double the horizontal distance from a to b. The slope and thus the angle of the tangent to the arch axis at any point can be determined by taking

$$\frac{dy}{dx} = \tan\alpha = 1 - \frac{x}{40}$$

at the springing $x = 0$ and $\tan\alpha = 1.0$ and $\alpha = 45$ degrees. The thrust (T) and shear (S) at the springing is then determined by finding the components of V and H in a direction parallel and perpendicular to the tangent. The components are then summed in these two directions.

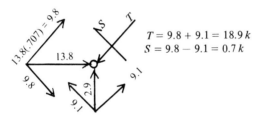

$$T = 9.8 + 9.1 = 18.9\,k$$
$$S = 9.8 - 9.1 = 0.7\,k$$

It is now desired to calculate the thrust, shear and bending moment at a point on the arch just above the point of application of the 8 kip horizontal load. The first step is to determine the x coordinate of this point on the arch.

$$y = x - \frac{x^2}{80} = 5$$

$$(x - 40)^2 = -400 + 1600$$

$$x = \pm\sqrt{1200} + 40 = 5{,}36$$

$$\tan\alpha = 1 - \frac{5.36}{40} = 0.866$$

$$\alpha = 40.9°$$

Now both *H*, *V*, and the 8 kip load will have to be resolved into components parallel and perpendicular to the arch axis at the 8 kip load point.

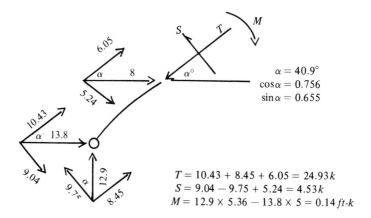

$$T = 10.43 + 8.45 + 6.05 = 24.93\,k$$
$$S = 9.04 - 9.75 + 5.24 = 4.53\,k$$
$$M = 12.9 \times 5.36 - 13.8 \times 5 = 0.14\,ft\text{-}k$$

It is seen that the bending moments in arches are quite low in value.
Moments, shears, and thrusts at a point along the arch axis can be determined by the same procedure as that used for finding these quantities
at the point of application of the 8 kip load.

6.5   Analysis of Three-Hinged Truss Arch

Reactions.   Three-hinged arches can be utilized for trusses as well as for
solid ribs. Such structures can be utilized for both bridges or buildings.
    The determination of the reactions can be made by employing analytical
or graphical methods. Since there are four reactions, the same four
equations as was used in determining the reactions for the arch rib can be
used. The three-hinged arch of Fig 6.7 is hinged at U_3 as well as the two
supports. The three equations of equilibrium and the one equation of
condition are:

(1) $F_x = 0$ $H_L + H_R = 0$
(2) $F_y = 0$ $V_L + V_R - 32 = 0$
(3) $M_{L_6} = 0$ $120V_L + 10H_L - 10(80 + 100) - 12(20) = 0$
(4) $M_{U_3} = 0$ $60V_L - 30H_L - 10(20 + 40) = 0$

Multiplying Eq. 4 by 2 and dividing the last two equations by 10,
Eqs. 3 and 4 become

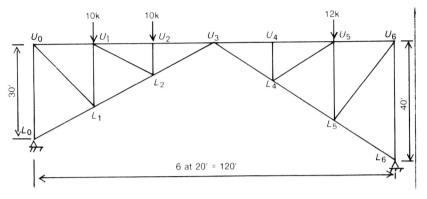

Fig. 6.6

$$12V_L + H_L = 204$$
$$12V_L - 6H_L = 120$$

Subtracting
$$7H_L = 84$$
$$H_L = 12 \text{ kips} = H_R$$

$$V_L = \frac{204 - 12}{12} = 16 \text{ kips}$$

$$V_R = 32 - 16 = 16 \text{ kips}$$

$$R_R = R_L = \sqrt{(16)^2 + (12)^2} = 20 \text{ kips}$$

It is noted that H_L will always equal H_R provided there are no applied horizontal loads. In this problem R_L is equal to R_R, but it should be apparent that this is true only for this particular loading.

Bar Forces. The determination of the bar forces in a trussed arch can best be performed by a graphical solution. The procedure, once the reactions have been determined, is the same as for any truss (see Section 5.3). The graphical solution for the bar forces of the truss of Fig. 6.6 is shown in Fig. 6.7.

The solution starts at panel point L_0 as there are only two unknown bar forces at this joint. After forming the force polygon at this joint, the analysis moves to joint U_0. At each joint the force polygon is drawn. The known force vectors are first drawn, and then the two unknown vectors have to close the polygon. Since the directions of the unknown vectors are parallel to the respective bars of the truss, only the magnitude of the vectors is unknown. All the vectors representing all the bar forces are determined by drawing a closed force polygon for each joint. Combining

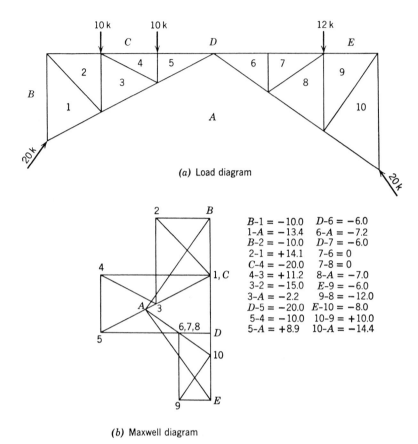

(a) Load diagram

B-1 = -10.0 D-6 = -6.0
1-A = -13.4 6-A = -7.2
B-2 = -10.0 D-7 = -6.0
2-1 = $+14.1$ 7-6 = 0
C-4 = -20.0 7-8 = 0
4-3 = $+11.2$ 8-A = -7.0
3-2 = -15.0 E-9 = -6.0
3-A = -2.2 9-8 = -12.0
D-5 = -20.0 E-10 = -8.0
5-4 = -10.0 10-9 = $+10.0$
5-A = $+8.9$ 10-A = -14.4

(b) Maxwell diagram

Fig. 6.7   Forces in three-hinged arch.

all the polygons into one Maxwell diagram saves drawing each vector twice. Bow's notation has been used in the graphical solution.

Influence Diagrams.   The influence diagrams for a three-hinged arch truss shows some shapes quite different from a simple-span truss. The influence diagrams for several typical bars of the truss of Fig. 6.6 are shown in Fig. 6.8.   The influence diagrams will be drawn first for the two components of the left reaction. Once these are drawn it would be a very simple matter to draw the influence lines for the other reactions. For this problem it is probably best to derive expressions for the reactions in terms of the distance the unit load is from the left reaction. This will be done as a variation of the approach that has been used on the trusses of Chapter 5, in which the load was placed at the various panel points and an independent solution was made for each load position.

With a unit load at a distance x from U_0 the three equations of static equilibrium and the one equation of condition are written as follows:

(1) $\Sum F_x = 0$ $H_L = H_R$

(2) $\Sum F_y = 0$ $V_L + V_R = 1$

(3) $\Sum M_{L_0} = 0$ $120V_R - 10H_R - 1(x) = 0$

(4) $\Sum M_{U_3} = 0$ $60V_R - 40H_R = 0$

$$V_R = \tfrac{2}{3}H_R$$

$$80H_R - 10H_R = x$$

$$H_R = \frac{x}{70} = H_L$$

$$V_R = \frac{2}{3}\left(\frac{x}{70}\right) = \frac{x}{105}$$

$$V_L = 1 - \frac{x}{105}$$

These solutions are valid only for a unit load between U_0 and U_3. This is because Eq. (4) as written is only valid for the load positions as stated. Writing the same equations for $x = {>}60$, there will be no change in the first three equations. The fourth equation can now be written considering the left-hand portion of the truss.

(4) $\Sum M_{U_3} = 0$ $60V_L - 30H_L = 0$

$$V_L = \tfrac{1}{2}H_L$$

$$V_R = 1 - V_L = 1 - \tfrac{1}{2}H_L$$

$$120(1 - \tfrac{1}{2}H_L) - 10H_R = x$$

$$H_L = \frac{12}{7} - \frac{x}{70} = H_R$$

$$V_L = \frac{6}{7} - \frac{x}{140}$$

$x = 0$	20	40	60	80	100	120
$V_L = 1$	0.810	0.619	0.429	0.286	0.143	0
$H_L = 0$	0.286	0.572	0.857	0.572	0.286	0

These values are plotted in Fig. 6.8. Positive values indicate that the reactions are in the direction shown on the diagram. It should be especially noted that the influence diagram for V_L is not a single straight line but consists of two straight lines. If both reactions were on the same level, it

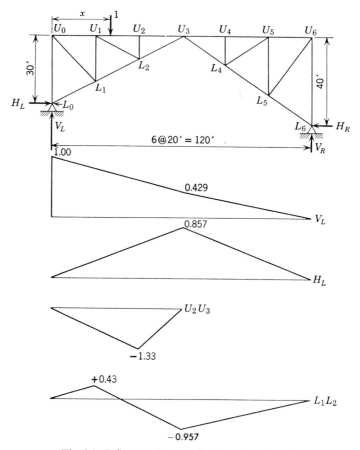

Fig. 6.8   Influence diagrams for three-hinged arch.

would be a single straight line. The apex of the triangle for the horizontal reaction will always be at the point beneath the center hinge.

The influence diagram for any of the members can be readily developed once the reactions are known. As an example, the influence diagram for chord member U_2U_3 can be found by writing the equation for moments about L_2 considering only forces to the left of L_2.

$$F_{U_2U_3} = \frac{40V_L - 20H_L - (40 - x)}{10} \quad \text{(for values of } x = 0 \text{ to } x = 40)$$

When $x = >40$, the expression becomes

$$F_{U_2U_3} = \frac{40V_L - 20H_L}{10}$$

A little reasoning indicates that the last expression is unnecessary, since with the unit load anywhere between U_3 and U_6 the force in the upper chord U_0U_3 must be zero. With no load on the left-hand portion the resultant reaction at L_0 must pass through U_3 in order to maintain equilibrium. Substituting the previously found values of V_L and H_L for loads on the right-hand portion will give $F_{U_2U_3} = 0$.

The influence diagram for member L_1L_2 indicates a stress reversal for this member. The force in this member with loads from U_3 to U_6 is simply $\frac{1}{2}H_L\sqrt{5}$.

Problems

1. Determine the bar forces by graphical methods for the Fink roof truss of Fig. 6.9.

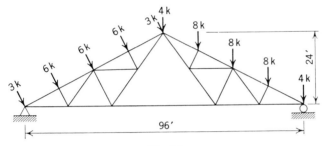

Fig. 6.9

2. Given the Pratt roof truss of Fig. 6.10, determine the bar forces in all the members for the following conditions:
  (*a*) ASCE 1940 Committee wind loads with $n = 10\%$.
  (*b*) Dead load of 24 lb per square foot of roof surface.
  (*c*) Snow load of 30 lb per square foot of horizontal projection.
  (*d*) Combine these loadings and determine the maximum design forces in accordance with AISC procedure.

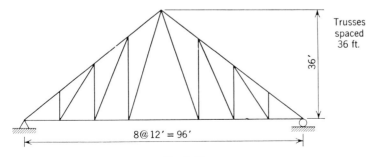

Fig. 6.10

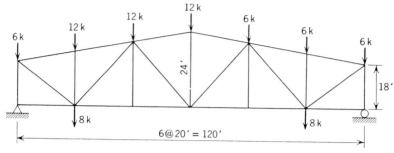

Fig. 6.11

3. Determine the bar forces for the Warren roof truss of Fig. 6.11.

4. Consider the three-hinged arch of Fig. 6.8 used as a highway bridge. Determine the force in member U_2U_3, U_0L_0, and L_1L_2 for one H-20 truck on the bridge.

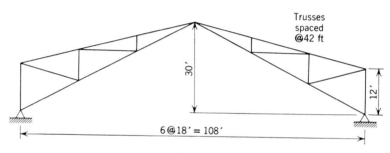

Fig. 6.12

5. Determine the forces in the three-hinged arch roof truss of Fig. 6.12 for the following loading:
(a) Roof load of 16 lb per square foot of roof surface.
(b) A snow load of 25 lb per square foot of horizontal projection.

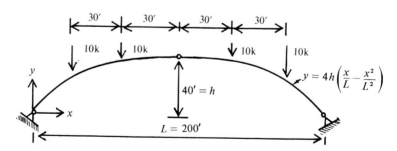

Fig. 6.13

6. The roof truss of Fig.6.12 has purlins at each roof panel point. Determine the bending moment about each axis of the purlin section: (*a*) without sag rods, and (*b*) with sag rods at the third points. The purlins are not continuous over the trusses.

7. Determine the thrust and shear at the supports for the three-hinged parabolic arch of Fig.6.13. Also determine the thrust shear and bending moment at each load point.

8. Determine the reactions, and the thrust, shear and bending moment at the load point for the three-hinged circular arch of Fig. 6.14.

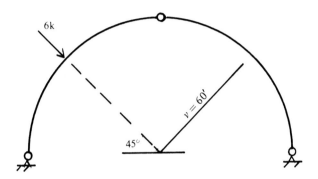

6k

$v = 60'$

45°

Fig. 6. 14

9. Determine the reaction and the thrust and shear at the supports for the unsymmetrical parabolic three-hinged arch of Fig. 6.15. Also determine the thrust, shear, and bending moment at a point 20 ft. from the left support. The load is considered as uniformly distributed with respect to span.

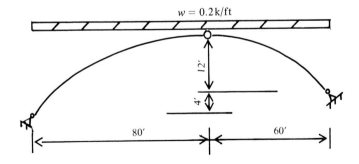

$w = 0.2 \text{k/ft}$

12'

4'

80'

60'

Fig. 6.15

7

FRAMES AND BENTS

7.1 Introduction

Frames or bents (the two terms are used synonymously) can be divided into many categories. The first distinction is between a braced bent, Fig. 7.1a, and a nonbraced bent, Fig. 7.1b. The braced bent is simply a vertical truss, and if the diagonals are tension-resistant members only, the structure is statically determinate and all the forces can be found by the principles already covered in this text. The cross-bracing is in action only when the bent is subjected to horizontal loads such as wind or inertia forces from earthquake motion, or for unsymmetrical vertical loads. When the bent is subjected to vertical loads along the horizontal member bc, a contradiction to the requirements of a truss appears; that is, the load is not applied at a joint. If the member bc in Fig. 7.1a were pinned at the ends, it would serve as a beam in transferring vertical loads to the columns. This member could also act as part of the truss in carrying horizontal loads. If, however, the member bc in Fig. 7.1b were pinned at the ends, the structure would be unstable. It is then apparent that an unbraced bent has to have joints that are capable of resisting moments. Such moment-resistant joints have to be properly designed. An assumption that the joint will transmit moments from the end of one member to the end of the attached member is not in itself sufficient. A pitfall that the inexperienced engineer can fall into is to make an assumption with regard to a condition of the structure when the actual physical condition is not compatible with the

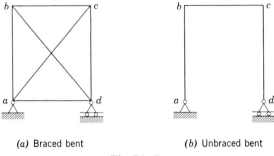

(a) Braced bent   *(b)* Unbraced bent

Fig. 7.1 Bents.

assumption. Complete followup of design assumptions and final construction details are necessary in engineering design. The members of a bent are not two-force members such as a truss, but are multiforce members. Multiforce members have shear forces and bending moments as well as axial forces.

Unbraced bents can be classified into categories such as number of stories, number of bays, or method of support at the base of the columns. The three single-storey, single-bay bents of Fig. 7.2 are different with respect to the manner of the base support of the columns. Bent *(a)* is statically determinate as it has only three reaction components, a vertical and a horizontal reaction at *a* and a vertical reaction at *d*. Of course, the joints *b* and *c* must be moment resistant. Bent *(b)* has four components of reaction, the vertical and horizontal reactions at each support. Bent *(c)* has both column bases fixed. In addition to the vertical and horizontal reactions at both bases, there is also a moment at the base giving six unknown reaction components. This bent is then said to be statically indeterminate to the third degree. There is the possibility of a bent being statically indeterminate to the second degree, but this is unusual. Such a bent would have one column base pinned and the other fixed.

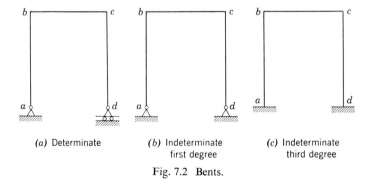

(a) Determinate   *(b)* Indeterminate   *(c)* Indeterminate
                            first degree   third degree

Fig. 7.2 Bents.

7.2 Statically Determinate Bents

The reactions for a statically determinate bent are found by first taking moments about the pin at the base of the column which is not on rollers. A bent with unequal length legs is shown in Fig. 7.3. Taking moments about the pin at A, the reaction at D can be determined.

$$\sum M_A = 0 \qquad 12 \times 12 + 4 \times 9 - 12R_D = 0$$

$$R_D = \frac{180}{12} = 15 \text{ kips}$$

$$\sum F_y = 0 \qquad R_{Ay} - 9 + 15 = 0$$

$$R_{Ay} = -6 \text{ kips} \quad \text{(down)}$$

The minus sign indicates a force in the opposite direction to the arrow.

$$\sum F_x = 0 \qquad R_{Ax} - 12 = 0$$

$$R_{Ax} = 12 \text{ kips}$$

The moment and shear diagrams can be determined once the reactions are found. The value of shear at any point on a bent is determined in the same manner as for beams; that is, the shear at any point is the algebraic summation of the forces normal to the axis of the member taken either to the left or right of the point. The shear at any point along member AB is is equal to $R_{Ax} = 12$ kips. This is the only force normal to the axis of the

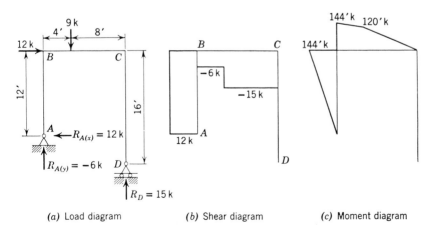

(a) Load diagram   (b) Shear diagram   (c) Moment diagram

Fig. 7.3 Statically determinate bent.

member considering forces below any respective point. Since this force is to the left, the shear diagram is drawn on the left of member AB. The shear along member BC is handled the same way as for any beam. The shear from B to the 9 kips load is equal to $R_{Ay} = 6$ kips. Since this reaction is down, the shear is negative according to the beam sign convention. From the 9-kip load to C the shear is equal to a negative 15 kips. Since there is no horizontal reaction at D, there can be no value of shear anywhere along CD. This is also proven by taking the summation of the horizontal forces from A around the bent to any point below C. This summation is zero. The moment diagram is determined by starting at A and proceeding around the bent to D, taking moments at pertinent points. The moments in a bent are determined by using the same method as for beams. The moment at any point from A to B is equal to the value of R_{Ax} times its lever arm. Since this is a first-order function, the moment diagram is a straight line from A to B, with a moment of $12 \times 12 = 144$ ft-kips at B. Moving from B to C, the moments are due not only to R_{Ax} but also R_{Ay}. At the 9-kip load $M = 12 \times 12 - 6 \times 4 = 120$ ft kips. It is not possible to apply the beam sign convention for moments to the vertical members of a bent, so instead of using signs for moments, the moment diagram is always plotted on the compression side of the member. Member AB has compression on the outside and tension on the inside. At point C the moment is zero. This value is arrived at whether the forces from A to C are considered or the forces from D to C. If the bent is thought of as a nonstraight beam, there should be no difficulty in determining shears and moments at any point on the structure.

7.3   Approximate Analysis of Single-Story Statically Indeterminate Bents

The bents of Fig. 7.2b and c are statically indeterminate. Additional equations beyond the three equations of static equilibrium are necessary in order to determine the shears and moments in the members. One additional equation is needed for the bent of Fig. 7.2b and three additional equations for Fig. 7.2c. These equations will have to be based on slope or deflection relationships. Any equations of slope or deflection involve the modulus of elasticity of each member and also the moment of inertia. It is important to recognize the fact that the shears and moments in a statically determinate structure are independent of the cross-sectional dimensions of the members or the material of which they are made. For instance, the shears and moments in the bent of Fig. 7.3 for the given

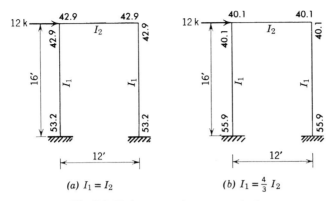

(a) $I_1 = I_2$ (b) $I_1 = \frac{4}{3} I_2$

Fig. 7.4   End moments by exact methods.

loading are the same regardless of whether the bent is of steel, aluminum, or wood, or regardless of the size of the members providing the stresses nowhere are beyond the elastic limit. This statement cannot be made for a statically indeterminate bent. For instance, the bents of Fig. 7.4a and b have the same overall dimensions and the same loading. For bent (a) the moment of inertia of the horizontal member is the same as the vertical member, whereas for bent (b) the moment of inertia of the horizontal member is three-fourths of that of the vertical member. The numbers at the ends of the members are the end moments as determined by a standard method of indeterminate structural analysis. It is seen that the variation of end moments is slight even with a change in ratio of moments of inertia between the horizontal and vertical members.

The moments in a single-story, single-bay bent can be approximated quite accurately for lateral loads by simple assumptions. This is particularly true for bents with columns of equal length and equal moment of inertia. For a bent with the bottom of the columns pinned only one assumption is necessary since there is only one more unknown reaction than equations of statical equilibrium. This assumption, where the moments of inertia of both columns are equal, is that the horizontal shear is equally divided between the columns. Figure 7.5 is the analysis of a bent with the same dimensions and loads as Fig. 7.4. Since the shears are equally divided between the two columns, the horizontal reaction in each is 6 kips. The vertical reactions are found by taking moments about the pins at the bottom of the columns. It is apparent that the approximate method is sufficiently exact for this type of bent when subjected to lateral loads.

When the footings of the columns rest on a very firm foundation and are rigidly attached to the columns, the columns may be considered as fixed

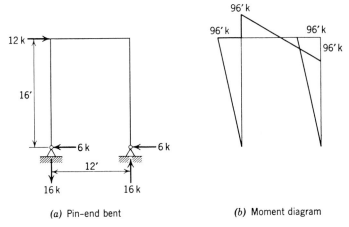

(a) Pin-end bent   (b) Moment diagram

Fig. 7.5 End moments by approximate methods.

at the base. With footings on soil, however, the true end condition may be closer to a pin-end condition than to a fixed-end. It only takes a slight rotation of the footing to approach a pin-end condition. This rotation can take place by a settlement of the soil on one side of the footing.

With a fixed-end condition at the base of the columns, the bent has three more unknown reactions than equations of statical equilibrium. An approximate method will then call for three simplifying assumptions. The first is the same as was used in the bent with a pinned base; that is, the horizontal shear is equally divided between the two columns. The second and third assumptions are that a point of contraflexure (point of zero moment) is at the midheight of each column. This is a double assumption, for it relates to both columns. Some engineers have recommended assumed points of contraflexure slightly different from the midpoint. If the footing is only partially fixed against rotation, the point of contraflexure moves down. Also, the greater the ratio of relative stiffness (I/L) between the beam and the column, the lower is the point of contraflexure. The contraflexure point for an infinitely stiff beam and fixed footings is at the midheight of the column.

The bent in Fig. 7.6 is fixed at the base and has the same loading and dimensions as the bent of Fig. 7.5. If a point of contraflexure is assumed at midheight of the columns, the condition would be the same as if there were pins at these points. The columns are then cut at these points, and the portion of the frame above the points of contraflexure is taken as a free body. By taking moments about one of the contraflexure points of a column, the axial load in the column can be determined. The reactions at the base of the columns are determined by taking the lower half of each

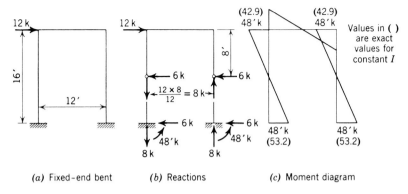

(a) Fixed-end bent   (b) Reactions   (c) Moment diagram

Fig. 7.6 Bent-fixed at base—approximate method.

column as a free body and applying the shears and axial loads at the point of contraflexure to this portion of the column. It is well to remember that the reactions at the bottom of the upper half of the column will be the loads that are acting on the top of the lower half of the column but in the opposite direction. Part (c) of Fig. 7.6 shows the moment diagram with the point of contraflexure assumed at midheight of the column. The values in parentheses are the exact values when the column and the beam have equal moments of inertia. From the exact values it is seen that the true point of contraflexure for this bent is about 0.9 of a foot above the midpoint. A slight rotation of the footing would bring the point of contraflexure down so that eventually the midpoint may be very near the true value for the location of the point of contraflexure. The moments, shears, and reactions between the pin-end column bent of Fig. 7.5 and the fixed-end case of Fig. 7.6 should be compared. It is observed that the maximum moments in the fixed-end bent are just one-half (by the approximate method) of those of the pin-end bent. The axial load in the columns are in the same ratio, whereas the shears are the same for both bents.

The examples treated so far in this chapter have been for bents with columns of equal length and equal moments of inertia. Where this solution is not so, adjustments in the methods previously given have to be made for indeterminate bents. From moment area principles it can be proven that if a constant section beam is fixed at both ends and one end is given a displacement Δ perpendicular to the axis of the member, the moment induced at the ends of the beam is equal to $6EI\Delta/L^2$. If an unequal leg bent of Fig. 7.7 is subjected to a lateral load in line with the horizontal member, the upper ends of the columns are displaced laterally an equal amount Δ. The moment induced in the ends of the columns, if the beam BC were infinitely stiff so that no rotation of the joints could take place, is proportional to the ratio of I/L^2 of the columns. If we call this quotient by the letter S, the ratio of S_1/S_2 is

$$S_1 = \frac{I_1}{L_1{}^2} = \frac{1}{(15)^2} \qquad S_2 = \frac{8}{(30)^2}$$

$$\frac{S_1}{S_2} = \frac{1/(15)^2}{8/(30)^2} = \frac{1}{2} = \frac{M_{BA}}{M_{CD}}$$

where M_{BA} is the moment at end B of member BA, and M_{CD} is the moment at end C of the member CD. This indicates that the moment induced in column BA is one-half of the moment in CD. An approximate method of solution would have to result in ratios of end moments in the columns of one-half. If the point of contraflexure is assumed at the midpoint of the column, the shears in the columns would have to be equal for $2M_{BA} = M_{CD}$. This relationship between shears in the columns can be stated in equation form as follows:

$$\frac{H_1}{H_2} = \frac{I_1(L_2)^3}{I_2(L_1)^3} \tag{7.1}$$

Applying this equation to the bent of Fig. 7.7, the ratio of shears is

$$\frac{H_1}{H_2} = \frac{1(30)^3}{8(15)^3} = 1$$

The shears and moments in the bent with the column shears equally divided in accordance with Eq. 7.1 are shown in Fig. 7.7b and c. A comparison of the approximate moments with the exact moments can be made by noting the exact moments given in parentheses. A value of $I_3 = \frac{8}{3}$ was used for the beam in the exact analysis. The comparison is quite close for this bent. Larger errors are possible for such bents, and

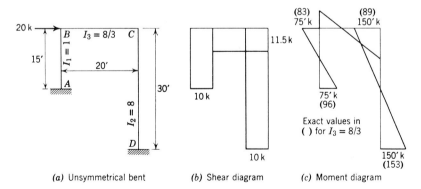

(a) Unsymmetrical bent   (b) Shear diagram   (c) Moment diagram

Fig. 7.7 Approximate solution for unsymmetrical bent.

the approximate method should be used only for very preliminary calculations for the unsymmetrical type of bent. For single-story bents, the exact solution by methods of indeterminate analysis takes only a little longer than the approximate solution.

Approximate methods are usable for preliminary design. For final design more exact methods are usually required. These more exact methods are presented in later chapters on the subject of analysis of statically indeterminate structures. For design of multi-story and multi-bay frames, a method of analysis employing a computer would most likely be used.

7.4 Bridge Portals

As was explained in Chapter 5, the end portal of a bridge truss transfers the wind loading on the upper portion of the truss to the substructure. This portal frame consists of the two end posts of the main trusses as columns and an overhead truss work. One such frame is shown in Fig. 7.8. Many different arrangements of members can and have been used for the horizontal truss joining the two end posts. The best arrangement is usually the one that is simplest and most pleasing in appearance.

The analysis of a portal frame is treated the same way as the single-bay symmetrical bent of this chapter. The lateral load is equally divided between the columns, and the point of contraflexure is assumed at the midpoint between the bottom of the end post and the intersection of the lower chord of the cross truss with the axis of the end post. The axial load in the columns (end posts) is determined by cutting the frame at the points of contraflexure and treating the upper portion as a free body. This free body is shown in Fig. 7.8b. Moments can then be taken about either point of contraflexure to find the axial load in the column. The next step is to find the force in the lower chord member m_1m_2. This is done by passing a vertical section line through the members U_1n_1, U_1m_2, and m_1m_2. The resultant free body is the portion of the left column from the point of contraflexure to the upper panel point U_1. By taking moments about U_1 of this free body, the only unknown in the equation is the force in m_1m_2. This value of bar force is

$$Fm_1m_2 = \frac{8 \times 19}{8} = 19 \text{ kips} \quad \text{(tension)}$$

With this force known the shear and moment diagram in the column can be drawn. The result is shown in Fig. 7.8c and d. The moment in the

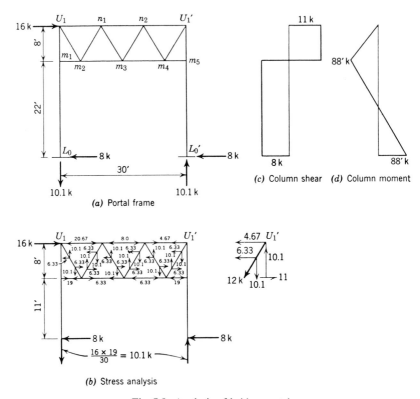

(a) Portal frame

(b) Stress analysis

(c) Column shear   (d) Column moment

Fig. 7.8 Analysis of bridge portal.

column must be considered along with the direct force in the determination of the required cross section. The direct force has to include the primary force in the end post due to dead load plus live load plus impact as determined from the analysis of the main bridge truss combined with the 10.1-kip axial force due to the portal action. As bridge specifications allow a higher working stress when wind loading is combined with dead load plus live load plus impact, the cross section also has to be determined for the forces without wind using the basic working stress. Whichever cross section is the greater as determined for the two loading conditions is, of course, the one to be used. The other members of the cross truss can be determined by either the method of joints or the method of sections. The analysis by the method of joints is shown in Fig. 7.8b. Care has to be taken at joint U_1 or U'_1 to include the column shear in the summation of F_x at these two joints.

The forces in the chords of the truss are of the opposite hand with the wind from the other direction. However, the chords are of one constant section throughout their lengths, and as such only the maximum tension and compression force are of interest to the designer. The diagonals all have the same slope, and therefore they have equal forces. With wind from one direction the force is of opposite sense from that when the wind is from the other direction. The magnitude of force is the same.

7.5 Industrial Building Bents

Many buildings have a steel roof truss supported by steel columns. If the chords of the roof truss along with the column axis intersect at a point, there will be little lateral rigidity unless a knee brace is used to connect the second lower chord panel point of the truss to the column. The knee brace is arranged to form an angle of approximately 45 degrees with the column. Figure 7.9 shows this arrangement. With the knee brace in place the truss and columns form a rigid bent. The analysis can proceed in a similar manner as the bridge portal of the preceding section.

The same assumptions, such as equal shear in the two columns and the location of the point of contraflexure in the columns, are made. The columns will no doubt be supported by concrete spread footings. Usually the footings will rest on soil. Slight soil settlements on one side of the footing, when there is a moment at the bottom of the column, results in a point of contraflexure closer to the footing than to the knee brace connection. A point of contraflexure about one-third the distance from the top of the footing to the point of intersection of the column and knee brace is the usual assumption. Although the Fink-type roof truss probably lends itself the most readily to this type of construction, the example problem of Fig. 7.9 utilizes a Pratt-type truss. A Fink roof truss has already been analyzed in Section 6.2. There will be no moments in the columns under dead load; so the bent is only analyzed for wind loads in the example presented here. The dead load, or any other vertical loads, will, of course, cause axial loads in the columns. This axial load together with the axial loads and bending moments from the wind loading have to be combined before a steel section can be selected as a column.

The analysis of the building bent is as follows. Refer to Fig. 7.9.

The pitch of the roof $= \frac{1}{3}$
Height of peak $= \frac{60}{3} = 20$ ft
Angle of roof with horizontal $= \tan^{-1}(\frac{20}{30}) = 33°-43'$
Per cent of openings $= 10\%$ (given)
Design wind load $=$ in accordance with 1940 ASCE Committee

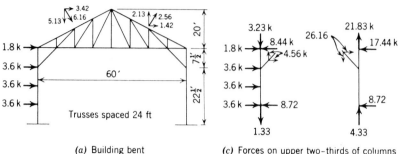

(a) Building bent

(c) Forces on upper two-thirds of columns

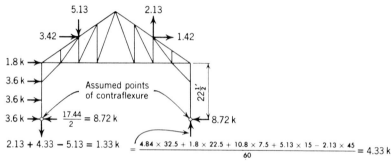

(b) Determination of reactions

Pressure on windward side of roof $= 0.3(33.72) - 9$

$$= 10.1 - 9 = 1.1 \text{ psf}.$$

Pressure on leeward side of roof $= -9$ psf (suction)

Internal force on both sides of roof:

Pressure $= 4.5 + 0.25(10) = 7$ psf

Suction $= -[4.5 + 0.15(10)] = -6$ psf

Panel point wind loads

Outside pressure windward side $= 1.1 \times 9 \times 24 = 238$ lb $\searrow$

Outside suction leeward side $\quad = 9 \times 9 \times 24 = 1942$ lb $\nearrow$

Inside pressure—both sides $\quad = 7 \times 9 \times 24 = 1512$ lb $\nwarrow \nearrow$

Inside suction—both sides $\quad = 6 \times 9 \times 24 = 1296$ lb $\searrow \swarrow$

The structure is designed for the most critical loading. Combining the outside loads with the loads produced by the internal effects gives the following panel point loads.

A. Outside loads and internal pressure

Panel point loads on windward side $= 0.24 - 1.51 = -1.27$ kips $\nwarrow$

Panel point loads on leeward side $= -1.94 - 1.51 = -3.45$ kips $\nearrow$

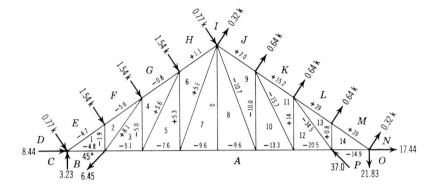

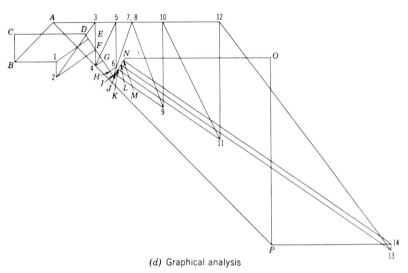

(d) Graphical analysis

Fig. 7.9 Wind analysis of building bent.

B. Outside loads and internal suction

Panel point loads on windward side = 0.24 + 1.30 = +1.54 kips↘

Panel point loads on leeward side = −1.94 + 1.30 = −0.64 kips↗

These two combinations give different forces in the bars, and a complete design would involve a truss analysis for both loadings. However, as the procedure is the same for both, only the analysis for loading B is demonstrated in Fig. 7.9.

The wind load on the wall is taken as 20 psf in accordance with the 1940 ASCE Report. Wind forces are transferred from the wall surface material to the columns through girts running perpendicular to the axis of the columns. The bent of Fig. 7.9 has girts spaced at 7.5 ft up the column. The concentrated load applied to each column at the position of the girts is

Column wind loading $= 20 \times 7.5 \times 24 = +3600$ lb/girt

This load is all applied to the windward side.

The first step in the analysis of the frame is to determine all the loads and their positions on the structure, as in Fig. 7.9a. The total wind load on the roof is shown instead of the individual panel point loads. The next step is to determine the reactions at the points of contraflexure. The point of contraflexure was chosen as one-third the distance between the top of the footing and the point of attachment of the knee brace to the column. This is the usual assumption when the bottom of the column is rigidly attached to the footing and when the footing is resting on reasonably firm earth. If the footing rested on more rigid material such as rock, then the point of contraflexure would be assumed to be somewhere between the third point and the midpoint. After the reactions are determined, the columns are taken as free-body diagrams and all the forces acting on the columns are found. This step is shown in Fig. 7.9c. The horizontal component of the force in the knee braces is determined by taking moments about the tops of the columns, considering only that portion of the column from the point of contraflexure to the top of the column. The last and final step is to determine the forces in all the members of the truss. The forces acting at the top of the columns as well as the forces in the knee braces are applied as reactions to the truss. Figure 7.9d shows the graphical analysis of the truss. It is recommended that the analysis start at the end of the truss which has the greatest reactions. Here it is the right end. By following this procedure more accuracy will ordinarily be obtained. With the wind blowing from right to left the forces in the members would be of opposite hand to that shown. For instance member 2–3 has a force of 8.1 kips tension when the wind is moving from left to right, and it will have a force of 34.5 kips compression (same as member 12–13) when the wind is moving from right to left. The two forces would have to be combined with the dead load, snow load, live load, etc., to determine the design force.

7.6 Multiple Story, Multiple Bay Frames

For the final analysis of a building frame of several stories and bays a solution by a method of statically indeterminate structures is usually used. By employing an electronic digital computer much time can be saved. However, before a final statically indeterminate solution can be made, the elastic properties of the beams and columns have to be known. This means that the sizes of these members have to be known or assumed. Approximate methods of frame analysis have been developed for lateral loads whereby the members can be sized for use in a statically indeterminate solution. Many times these methods are used as the terminal solution. The most used of the lateral load approximate methods for building frames are the cantilever, portal, Bowman method C,[1] factor method,[2] and two-cycle moment distribution method. The first three methods can be applied without recourse to previously determined sizes of the members. Certain assumptions are made in these three methods with respect to distribution of shears and points of contraflexure. After these assumptions are made, the solution can be carried out by the standard equations of statics. The last two methods require a previous knowledge of the elastic relationship of all the members meeting at a point. The factor method requires nothing beyond the equations of statics once the elastic relationships of the members have been selected. However, this method is considerably more involved than the first three methods. The factor method or the two-cycle moment distribution method, which is a method of analysis of indeterminate structures, is not discussed further here.

It is difficult to say which of the three methods considered here is the most accurate. Generally speaking it is thought that for tall, narrow buildings, the cantilever method is preferred, whereas for shorter, wider buildings the portal method is used. The Bowman method C is usually more exact than the portal method, especially for unsymmetrical frames. Reference (1) compares the results of the three methods with an exact solution by the slope deflection method for a five-story, three-bay frame. In this example the Bowman method C was closer than the other two methods to the values as determined by the slope deflection method. In practice, it is possible to proportion the sizes of the members so that the moments and shears given by a particular approximate method fit very closely to those given by a so-called exact solution. However, the final criterion in the design of a structure is that it carry the loads with a required safety factor and that the structural elements be the most economical size possible.

The building frame of Fig. 7.10 is analyzed to demonstrate the three methods, and the results are compared with an "exact" solution as determined by the moment distribution method. In the "exact" solution the moment of inertia of the beams and columns was the same throughout the structure.

Cantilever Method. In this method the following assumptions are made:

1. The points of contraflexure are located at the midpoint of the columns.

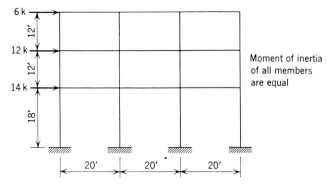

Fig. 7.10 Building frame.

2. The points of contraflexure are located at the midpoints of the girders.

3. The unit direct stresses in the columns vary as the distances of the columns from the center of gravity of the column cross sections. This would indicate a prior knowledge or assumption of the cross-sectional areas of the columns. This is not usually necessary since with the common case of symmetrical frames the center of gravity of the columns is at midpoint, and it is usually assumed that the columns have equal cross-sectional areas. The analysis by this method is shown in Fig. 7.11.

The first step in this method is to locate the points of contraflexure and take a free-body diagram of the joints of the top floor and the column segments above the points of contraflexure of the top story columns. Assuming the columns have equal area and utilizing the third basic assumption of the method, we find that the axial load in the interior and exterior columns is V and $3V$ respectively. Moments are then taken about the point of contraflexure in the right exterior column, considering the top half of the top story as a free body. The numerical value of V is thus determined. The second step is to take free-body sections of each joint, cutting the columns and girders at the points of contraflexure as shown in Fig. 7.11. With the use of the equations of statics the shears and axial loads in the columns and girders are determined. The third step is to find the moments in the columns and girders at the joints by multiplying the shear in the member by the distance from the point of contraflexure to the joint. In Fig. 7.11 the sign convention is plus for a clockwise moment on the joint and negative for a counterclockwise moment. This is the usual practice in frame analysis. If the shear in the lower column section is determined by taking moments about the point of contraflexure of the girders, the analysis will be self-checking by proceeding from the left exterior joint to the right exterior joint.

Portal Method.   In this method the following assumptions are made:

1. The points of contraflexure are located at the midpoints of the columns.

2. The points of contraflexure are located at the midpoints of girders.

3. The shear in each exterior column is the same and equals one-half the shear in an interior column.

This method is the easiest of the three methods to apply. Figure 7.12 is an analysis by the portal method of the frame of Fig. 7.10. The first step is to locate the points of contraflexure and draw free-body diagrams of each joint and attached column and girder sections in the same manner as for the previous method. The shears are then divided among the columns in accordance with the third assumption. For the given frame, the shear in each exterior column for the top floor is 6 kips/6 = 1 kip, and the interior columns carry shears of 2 kips each. The second step is to take moments about the midpoints of the girders to determine the axial forces in the columns. The process should start from the top line of girders and proceed in orderly fashion down to the column foundations. It is seen that the moments in the interior columns are twice the moment in the exterior columns, and the moments at the ends of all girders of a given floor are equal.

Bowman Method.   The assumptions of this method are somewhat more involved than the other two methods. They are as follows:

1. In frames of one or more stories, the points of contraflexure in the bottom story columns are at 0.60 of the column height from the base of the column. In frames of two or more stories, the points of contraflexure in the top story columns are at 0.65 of the column height from the top of the column. In frames of three or more stories, the points of contraflexure in the columns of the story next to the top story are at 0.60 of the column height from the upper end. In frames of four or more stories, the points of contraflexure in the columns of the second story from the top story are at 0.55 of the column height from the upper end. In frames of five or more stories the points of contraflexure of all story columns not covered by any of these assumptions are at midheight.

2. The points of contraflexure in the exterior girders are at 0.55 of the center-to-center distance of columns from the outside column. In all other girders the points of contraflexure will be at the midpoints. In a single or double bay frame, the points of contraflexure, of course, have to be at the midpoint from a standpoint of symmetry and equilibrium.

3. An amount of shear equal to $\dfrac{\text{Number of bays} - \frac{1}{2}}{\text{Number of columns}}$ times the total

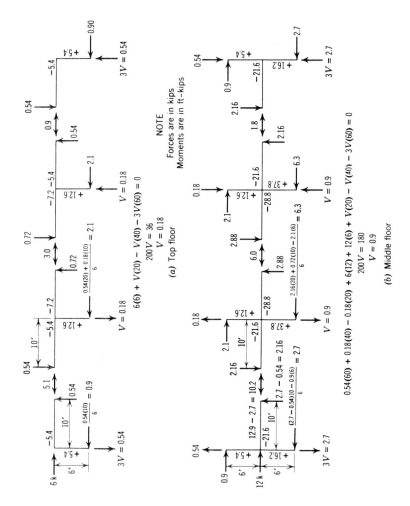

NOTE

Forces are in kips
Moments are in ft-kips

$6(6) + V(20) - V(40) - 3V(60) = 0$

$200V = 36$

$V = 0.18$

(a) Top floor

$0.54(60) + 0.18(40) - 0.18(20) + 6(12) + 12(6) + V(20) - V(40) - 3V(60) = 0$

$200V = 180$

$V = 0.9$

(b) Middle floor

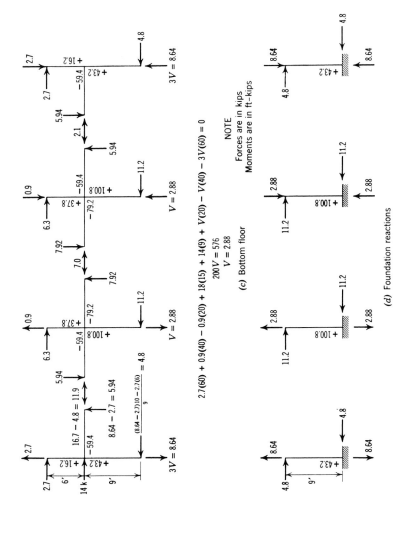

$$2.7(60) + 0.9(40) - 0.9(20) + 18(15) + 14(9) + V(20) - V(40) - 3V(60) = 0$$

$$200V = 576$$
$$V = 2.88$$

(c) Bottom floor

NOTE
Forces are in kips
Moments are in ft-kips

(d) Foundation reactions

Fig. 7.11 Frame analysis by cantilever method.

195

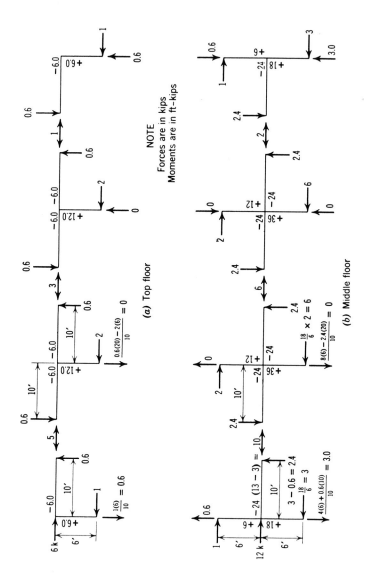

NOTE
Forces are in kips
Moments are in ft–kips

(a) Top floor

(b) Middle floor

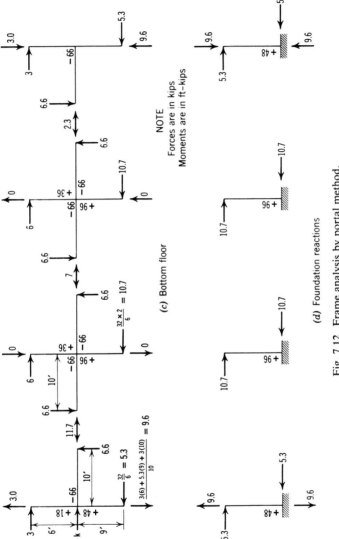

NOTE
Forces are in kips
Moments are in ft–kips

(c) Bottom floor

(d) Foundation reactions

Fig. 7.12 Frame analysis by portal method.

197

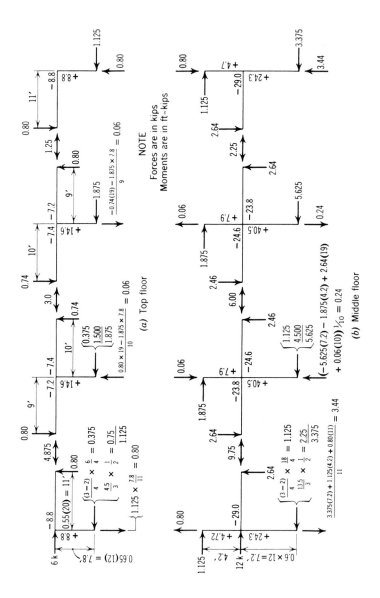

NOTE
Forces are in kips
Moments are in ft-kips

(a) Top floor

(b) Middle floor

198

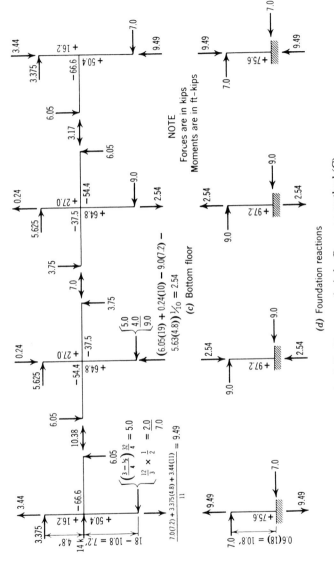

NOTE
Forces are in kips
Moments are in ft-kips

(c) Bottom floor

(d) Foundation reactions

Fig. 7.13 Frame analysis by Bowman method (C).

199

shear in the bottom story is divided among the columns of the bottom story in proportion to the moments of inertia of the column cross sections. The remaining shear in the bottom story is divided among the bays of this story inversely as the widths of the bays. This shear in each bay is divided equally among the columns of each bay.* For other stories, an amount of shear equal to $\dfrac{\text{Number of bays} - 2}{\text{Number of columns}}$ times the total shear in the story is divided among the columns of that story in proportion to the moments of inertia of the column cross sections. The remaining shear in the story is divided among the bays inversely as the widths of the bays. This shear in each bay is divided equally among the columns of that bay.*

In the example frame used in the illustration of the two previous methods, the moments of inertia of the girders and columns are considered equal. This condition might be assumed in the first trial design of a frame and then adjusted later for other trials.

The first step in this method as shown in Fig. 7.13 is to proportion the column shears in accordance with assumption 3. For the top story an amount of shear equal to 6[(3 − 2)/4] = 1.5 kips is equally divided among the columns. This gives a shear of 0.375 kip per column. The remaining shear in the story (6 − 1.5) = 4.5 kips is divided equally among the bays, giving a shear of 4.5/3 = 1.5 kips per bay. This shear per bay is equally divided among the columns of the bay. The inside columns will therefore receive twice as much of this shear as the outside columns. After the shear in each top story column is determined, moments are taken about the points of contraflexure of each girder to find the axial loads in the columns. With the axial loads in the columns known, a $\Sigma F_y = 0$ equation will yield the girder shears. The end moments in the columns and girders are then found easily by multiplying the shears by the distance from the respective points of contraflexure to the joints of the frame. After the analysis of the top story, the same procedure is applied to the next floor below and so on down to the foundations of the columns.

Figure 7.14 gives a comparison of the end moments in the members of the frame for the three approximate methods with an "exact" solution. The "exact" solution was made by the moment distribution method. For a majority of the members the Bowman method gives end moments closer to those of the "exact" method than do the other two approximate methods. The approximate solutions all gave base moments in the outer columns less than the moments by the "exact" method, and therefore higher moments at the base of the inner columns. There is one other aspect pertaining to the

---

* When the moments of inertia of the girders above a story are not equal, this remaining shear is divided among the bays directly as the I/L of the girders above the bay.

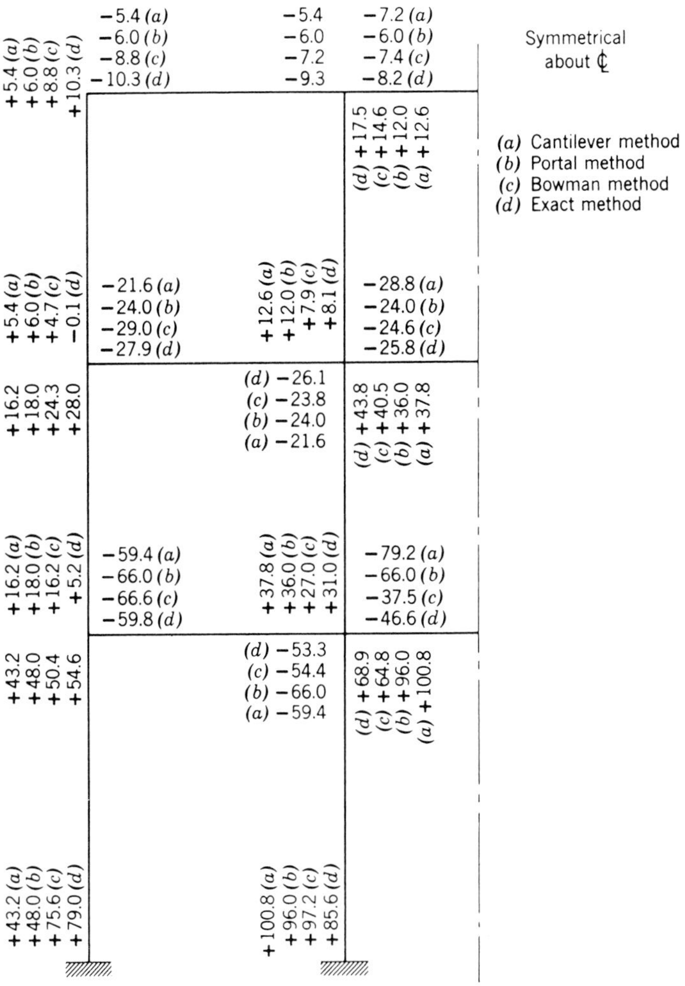

Top member:

−5.4 (a)   −5.4   −7.2 (a)
−6.0 (b)   −6.0   −6.0 (b)
−8.8 (c)   −7.2   −7.4 (c)
−10.3 (d)   −9.3   −8.2 (d)

Symmetrical about ₵

Upper-left column: +5.4 (a), +6.0 (b), +8.8 (c), +10.3 (d)

Upper-right column: +17.5 (d), +14.6 (c), +12.0 (b), +12.6 (a)

(a) Cantilever method
(b) Portal method
(c) Bowman method
(d) Exact method

Second level:

Left column: +5.4 (a), +6.0 (b), +4.7 (c), −0.1 (d)

−21.6 (a), −24.0 (b), −29.0 (c), −27.9 (d)

Middle: +12.6 (a), +12.0 (b), +7.9 (c), +8.1 (d)

Right: −28.8 (a), −24.0 (b), −24.6 (c), −25.8 (d)

Left column: +16.2, +18.0, +24.3, +28.0

Middle: (d) −26.1, (c) −23.8, (b) −24.0, (a) −21.6

Right: +43.8 (d), +40.5 (c), +36.0 (b), +37.8 (a)

Third level:

Left column: +16.2 (a), +18.0 (b), +16.2 (c), +5.2 (d)

−59.4 (a), −66.0 (b), −66.6 (c), −59.8 (d)

Middle: +37.8 (a), +36.0 (b), +27.0 (c), +31.0 (d)

Right: −79.2 (a), −66.0 (b), −37.5 (c), −46.6 (d)

Left column: +43.2, +48.0, +50.4, +54.6

Middle: (d) −53.3, (c) −54.4, (b) −66.0, (a) −59.4

Right: +68.9 (d), +64.8 (c), +96.0 (b), +100.8 (a)

Bottom:

Left column: +43.2 (a), +48.0 (b), +75.6 (c), +79.0 (d)

Middle: +100.8 (a), +96.0 (b), +97.2 (c), +85.6 (d)

Fig. 7.14 End moments by various methods.

comparison. The "exact" solution was based on the moments of inertia of the columns and girders being equal. A comparison of the resultant end moments suggests that this equality of moments of inertia would probably not exist if lateral forces dictated the design of the structural members. A change in relationship of the stiffness of the members would change the end moments and perhaps bring them even closer to an approximate solution. It should be remembered that there is also no *exact* solution. Slight rotation of footings when end fixity is assumed, approximation of loads, and other factors may render an approximate solution as near the values that will exist in the actual structure as a so-called exact solution. This should be kept in mind when selecting a method of analysis.

Most of the approximate methods for vertical loads are based upon statically indeterminate methods and will not be covered here. For a detailed treatment of this subject, the reader is referred to Reference (3).

References

1. Sutherland, H. and H. L. Bowman, *Structural Theory*, John Wiley, 4th Edition, pp. 295–300, 1954.
2. Wilbur, J. B. and C. H. Norris, *Elementary Structural Analysis*, McGraw-Hill, 2nd Edition, pp. 305–315, 1960.
3. Benjamin, J. R., *Statically Indeterminate Structures*, McGraw-Hill, 1959.

Problems

(When plotting the moment diagram for bents, plot the diagram on the compression side of the member.)

1. Determine the forces in the members of the braced bent of Fig. 7.15. Assume both supports can have reactions both up or down.

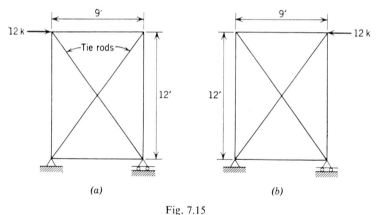

(a)   (b)

Fig. 7.15

2. Draw the shear and moment diagrams for the bents of Fig. 7.16. All bents have constant E and I.

3. Draw the shear and moment diagrams for the bents of Fig. 7.17 using approximate methods. Members have constant E and I.

4. Draw the shear and bending moment diagrams for the bent of Fig. 7.18. Use approximate methods.

Ans. $M_A = -M_B = 44.1$ ft-kips, $M_C = -M_D = 37.2$ ft-kips.

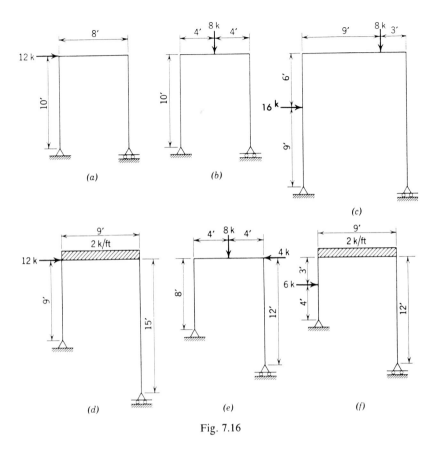

Fig. 7.16

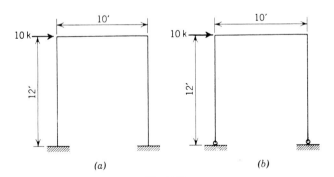

Fig. 7.17

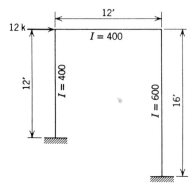

Fig. 7.18

5. Draw shear and bending moment diagrams for the legs, and determine the forces in the bars of the bridge portal frame of Fig. 7.19.

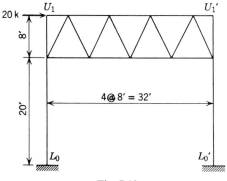

Fig. 7.19

6. Make an analysis of the building bent of Fig. 7.20 using the 1940 ASCE Committee wind loading. Wind on the columns is applied through the girts. Use panel point loadings to nearest 100 lb; $n = 20\%$, spacing of trusses $= 30$ ft.

7. Analyze the building bent of Fig. 7.21 for the following conditions:

(a) 1940 ASCE Committee wind loading. $n = 15\%$.

(b) Total dead load of 18 lb per square foot of horizontal projection of roof surface.

(c) Snow load of 20 lb per square foot of horizontal projection of roof surface.

8. The building frame of Fig. 7.22 is subject to earthquake forces. The dead load of each floor per frame is as given. Determine the moments in the columns and girders under earthquake loads in accordance with the recommendations of the SEAOC. Assume the moments of inertia of all the columns are equal. Use $Z = \frac{3}{4}$, $I = 1.5$, $K = 0.67$, $T_s = 1.5$ sec.

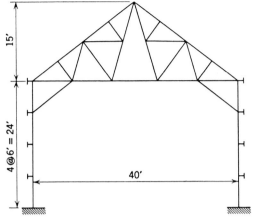

Fig. 7.20

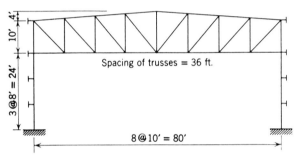

Fig. 7.21

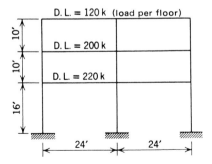

Fig. 7.22

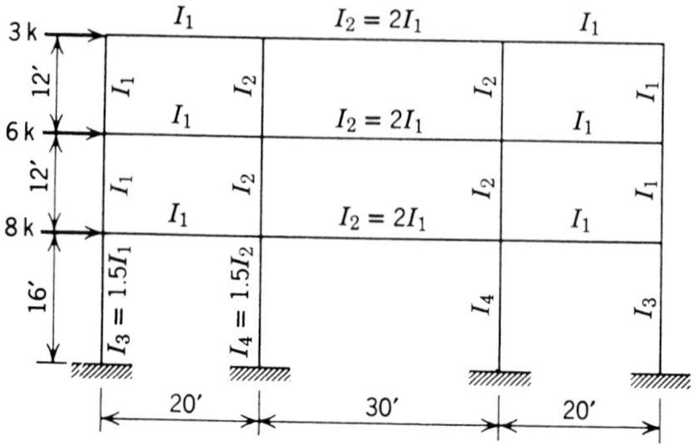

Fig. 7.23

9. The building frame of Fig. 7.23 has wind loads at each floor as shown. Determine column and beam moments using the following approximate methods: (a) portal method, and (b) cantilever method.

10. Determine the column and beam moments and shears for the frame of Fig. 7.23 using the Bowman Method.

STATICALLY DETERMINATE
SPACE STRUCTURES

8.1 Introduction

Not all structures can be broken down into two-dimensional problems. Chapter 2 reviewed the basic principles of three-dimensional statics. Towers, domes, and like structures usually have to be treated in their three-dimensional shape. These structures are statically determinate if all the bar forces can be determined by the six equations of statical equilibrium.

$$\sum F_x = 0 \qquad \sum M_x = 0$$
$$\sum F_y = 0 \qquad \sum M_y = 0 \qquad (8.1)$$
$$\sum F_z = 0 \qquad \sum M_z = 0$$

The moment equations, of course, apply to moments about three orthogonal axes.

For three-dimensional concurrent forces, the moment equations are not applicable and only three unknowns can be determined. This is so when we have a space truss in which all members intersect at joints. Three unknown bar forces are then the maximum number at any one joint that can be determined by applying Eqs. 8.1 only. In a space framework, as in a planar truss, the following conditions define such a structure:

1. The members are straight between joints.
2. The center of gravity lines of all members meeting at a joint intersect at a common point.

3. All loads are applied at the joints.
4. The ends of all members are free to rotate in any direction.

If these conditions are met, the forces in all the members will be axial, that is, all members will be two force members only. Condition (4) may not be met exactly in most space frames because of practical fabrication problems. However, if the member does not have too small an L/I ratio there will be very little error by assuming that the ends of the members are free to rotate.

As with a planar truss, the triangle formed by three members is the simplest stable configuration. If to the three joints just formed, Fig. 8.1a, three additional bars are connected and these bars joined at a common point (A), Fig. 8.1b, a space framework is formed consisting of four joints and six bars. With the addition of each joint after the first three, three bars are added. The equation of static determinacy can then be derived by noting the following conditions:

when $\qquad\qquad n = 4 \qquad b = 6$

when $\qquad\qquad n = 5 \qquad b = 9$

$$b = \text{bars}$$

$$n = \text{joints}$$

The relationship between b and n is a straight-line function and can be expressed by the equation

$$b = kn - r$$

Substituting these two relationships of b and n into this straight-line equation,

$$6 = 4k - r$$
$$9 = 5k - r$$

Solving these two equations for k and r, the resulting expression is

$$b = 3n - 6 \qquad\qquad (8.2)$$

This equation can be applied to any space framework to determine if the

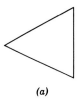

(a)

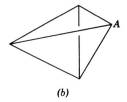

(b)

Fig. 8.1

structure is internally unstable, statically determinate, or statically indeterminate. If

(1) $b < 3n - 6$,   structure is unstable
(2) $b = 3n - 6$,   structure is statically determinate
(3) $b > 3n - 6$,   structure is statically indeterminate

Two conditions must be noted in applying Eq. 8.2. The first is that the structure could be stable even if $b < 3n - 6$, provided additional reaction components, beyond the usual six required for external equilibrium, are at the proper points. The second condition is that all the bars meeting at a joint *do not* lie in a common plane. Consider the six-bar frame of Fig. 8.2 with bar m inserted as shown by the dashed line. This additional bar has added one more joint (E) and increased the number of bars by just two instead of the three that were added for each new joint in Fig. 8.1. All the bars meeting at joint (E) lie in a common plane. The addition of two bars for each joint takes place in all planar structures. Equation 8.2 can then be amended to take care of

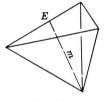

Fig. 8.2

space frameworks in which some joints have intersecting members lying in a common plane. This equation is

$$b = 2t + 3n - 6 \qquad (8.3)$$

where b = number of bars in the structure
   t = number of joints where all the intersecting members lie in a common plane
   n = number of joints where all the intersecting members do not lie in a common plane

A space framework can be considered as a rigid body. The number of reactions acting on a rigid body that can be determined by the six static equilibrium equations (Eq. 8.1), are of course, six. This means that a space frame that has more than six reaction components will normally be externally statically indeterminate. However, as previously stated, bars can sometimes be removed if replaced by adding the same number of reaction components without changing the state of static determinancy. This can be stated by a revision of Eq. 8.3.

$$b + r = 2t + 3n \qquad (8.4)$$

where r = number of reaction components for stability; all other terms are the same as for Eq. 8.3.

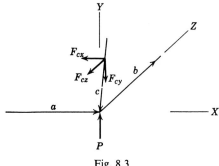

Fig. 8.3

Certain rules applying to three-dimensional frameworks can be used in simplifying the analysis. If a single joint of a space framework, as shown in Fig. 8.3, is analyzed by the method of Section 2.7, the component lengths of all the members meeting at a joint are first determined. The axes are so chosen that bar a lies along the X axis and bar b along the Z axis. The external load at the joint is perpendicular to the XZ plane. The lengths of the bars are indicated as:

	X	Y	Z	d	F_x	F_y	F_z	F
Bar a	a	0	0	a	1	0	0	F_a
Bar b	0	0	g	g	0	0	1	F_b
Bar c	$-c_x$	$-c_y$	$-c_z$	c	$-c_x/c$	$-c_y/c$	$-c_z/c$	F_c
					0	P	0	P

$$\Sigma F_y = 0 \qquad (c_y/c)F_c = P$$

$F_c(c_y/c)$ = component of force in bar c in the Y direction

The analysis proves that the component in the Y direction of the force in bar c is equal to the external load P. This can be stated in a rule as follows:

RULE 1.   If all the bars intersecting at a joint lie in a common plane, with the exception of bar c, the component normal to the plane of the force in bar c is equal to the component normal to the plane of any external force applied at that joint.

From this rule it is obvious that if there is no component of the external force applied at the joint that is normal to the plane, there can be no force in the intersecting bar that does not lie in the plane.

If in the joint of Fig. 8.3 there were no external force, from this statement there would be zero force in bar c. The component force table would then be as follows:

F_x	F_y	F_z	F
1	0	0	F_a
0	0	1	F_b
0	0	0	P

This proves that the force in bars a and b must be zero. Even if bars a and b did not form a right angle, the Z component of bar b (still calling the X axis parallel to bar a) would be unopposed and therefore equal to zero. If any component of a force is equal to zero, of course, the force has to be equal to zero. This leads to the second rule.

RULE 2. If it is known that all the bars intersecting at a joint except two noncollinear bars have no force in them and there is no external load applied to the joint, the forces in these two bars must also be zero.

8.2 Determination of Bar Forces

As with planar structures the first step is to determine the unknown reaction components if possible. The space frame of Fig. 8.4 is the simplest type of structure. It has six reaction components as shown, one horizontal and one vertical at each lower joint. The heavy lines at joints a, b, and c, indicate the only direction in which the reactions can act. In applying Eq. 8.4 to this structure

$$r = 6 \qquad b = 6 \qquad t = 0 \qquad n = 4$$
$$b + r = 2t + 3n$$
$$6 + 6 = 2 \times 0 + 3 \times 4$$
$$12 = 12$$

The structure is statically determinate. The external force acting at joint d is best broken into its x, y, and z components. These are

$$P_x = 20 \times \tfrac{4}{5} = 16 \text{ kips} \qquad P_y = 20 \times \tfrac{3}{5} = 12 \text{ kips} \qquad P_z = 0$$

Before proceeding further, it is best to restate the rules with regard to the relationship of moments about an axis.

1. A force parallel to an axis causes zero moment about that axis.
2. A force that intersects an axis causes zero moment about that axis.

The reactions are determined by taking moments about axes passing through the reaction joints. A proper choice of axis can eliminate all but

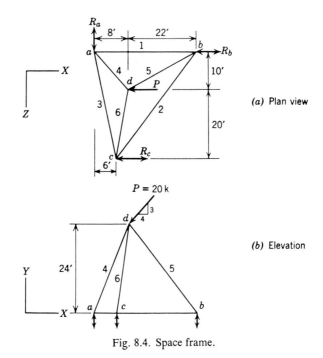

(a) Plan view

(b) Elevation

Fig. 8.4. Space frame.

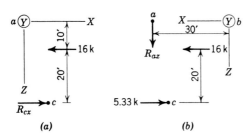

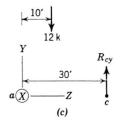

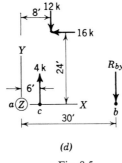

(d)

Fig. 8.5

one unknown. For instance, consider the moments about a Y axis passing through joint a as shown in Fig. 8.5a. The only unknown in the equation for moments about the Y axis is R_{cx}.

$$\sum M_y = 0 \qquad R_{cx} = \frac{16 \times 10}{30} = 5.33 \text{ kips} \rightarrow$$

The reactions at joints a and b are either parallel or intersect the Y axis passing through a. The next step is to pass a Y axis through joint b as shown in Fig. 8.5b and write the equation for moments about this axis.

$$\sum M_y = 0 \qquad 16 \times 10 - 5.33 \times 30 - 30R_{az} = 0 \qquad R_{az} = 0$$

Since one reaction at b is in the X direction, it can be found by taking

$$\sum F_x = 0 \qquad R_{bx} = 16 - 5.33 = 10.67 \text{ kips} \rightarrow$$

The next step is to find the three reactions in the Y direction at each joint. R_{cy} can be determined by taking moments about an X axis passing through joint a as shown in Fig. 8.5c.

$$\sum M_x = 0 \qquad R_{cy} = \frac{120}{30} = 4 \text{ kips} \uparrow$$

The value of R_{by} is found in a similar manner by taking moments about a Z axis passing through a as shown in Fig. 8.5d. It is noted that in taking moments about this axis both components of the applied force enter into the moment equation.

$$16 \times 24 + 6 \times 4 - 12 \times 8 - 30R_{by} = 0$$

$$R_{by} = \frac{312}{30} = 10.4 \text{ kips} \downarrow$$

The last reaction R_{ay} is simply found by summing the forces in the Y direction.

$$\sum F_y = 0 \qquad R_{ay} = 12 + 10.4 - 4 = 18.4 \text{ kips} \uparrow$$

Once all the reactions are determined, the bar force analysis can begin. Since the axes of the bars meet at a common point, the bar forces are concurrent. Three bar forces are the maximum number of unknowns at any point that can be handled by the equations of statics. The frame of Fig. 8.4 lies within this limitation, so the analysis can start at any joint. However, since bar 1 is parallel to the X axis, this bar will have no component forces in the Y or Z direction. There is a small saving in time in analyzing this joint first. It is best to proceed with the analysis of any joint by the use of a table and procedure as given in Section 2.1. The unknown force in the bar is carried in the table as an unknown quantity. The first step as shown by Table 8.1 is to determine the projections of each member in the X, Y, and Z directions. The true lengths of the members are then

Table 8.1   Three-Dimensional Analysis of a Joint

Bar	X	Y	Z	d	F_x	F_y	F_z	F
Joint a								
1	+30	0	0	30.00	+1.00	0	0	F_1
3	+6	0	+30	30.59	+0.196	0	+0.981	F_3
4	+8	+24	+10	27.20	+0.294	+0.882	+0.368	F_4
External Force						+18.4 kips		P
Joint b								
1	−30	0	0	30.0	−1.00	0	0	F_1
2	−24	0	+30	38.42	−0.625	0	+0.781	F_2
5	−22	+24	+10	34.05	−0.646	+0.705	+0.294	F_5
External Force					+10.67 kips	−10.4 kips		P
Joint c								
2	+24	0	−30	38.42	+0.625	0	−0.781	F_2
3	−6	0	−30	30.59	−0.196	0	−0.981	F_3
6	+2	+24	−20	31.31	+0.064	+0.767	−0.639	F_6
External Force					+5.33 kips	+4.00 kips		P

determined. The components of the forces in each of the orthogonal directions are proportional to the lengths of the member projected on the orthogonal axes. This can be stated in equation form as

$$\frac{F_x}{x} = \frac{F_y}{y} = \frac{F_z}{z} = \frac{F}{d} \tag{8.5}$$

The components in the X, Y, and Z directions are determined at each joint. Table 8.1 is used to determine these component forces at each joint. The summation of forces at each joint is as follows:

Joint a

$$\sum F_x = 0 \qquad 1.00F_1 + 0.196F_3 + 0.294F_4 = 0$$
$$\sum F_y = 0 \qquad\qquad\qquad + 0.882F_4 + 18.4 = 0$$
$$\sum F_z = 0 \qquad\qquad + 0.981F_3 + 0.368F_4 = 0$$

$$F_4 = \frac{-18.4}{0.882} = -20.86 \text{ kips}$$

$$F_3 = \frac{-0.368(20.86)}{0.981} = +7.83 \text{ kips}$$

$$F_1 = -1.53 + 6.13 = +4.60 \text{ kips}$$

At joint a, F_4 being minus means that the force is down toward the joint. This then means that F_4 is compression. F_3 and F_1 being positive are away from the joint and are thus tension.

The sign of the X, Y, and Z distances in Table 8.1 was based upon considering the joint as the origin. For instance, member 4 had its origin at joint a when the forces at joint a were analyzed. If the distance ad is the vector representation of the force in bar 4, the vector is away from the joint. This is so at all the joints; the vectors were all assumed to be acting away from the joint. A plus sign for the force in a member means that the force is in the direction assumed—away from the joint. A minus sign then means that the force is toward the joint (compression). The sign gives the direction of the force. From this direction the sense of the force in the member (tension or compression) is determined.

The solution of the forces at joint b is

$$\sum F_x = 0 \qquad -1.00F_1 - 0.625F_2 - 0.646F_5 + 10.67 = 0$$

$$\sum F_y = 0 \qquad \qquad \qquad + 0.705F_5 - 10.40 = 0$$

$$\sum F_z = 0 \qquad \qquad \qquad + 0.781F_2 + 0.294F_5 = 0$$

$$F_5 = \frac{10.40}{0.705} = +14.75 \text{ kips} \quad \text{(tension)}$$

$$F_2 = \frac{-0.294 \times 14.75}{0.781} = -5.55 \text{ kips}$$

Joint c

$$\sum F_y = 0 \qquad + 0.767F_6 + 4.00 = 0$$

$$F_6 = \frac{-4.00}{0.767} = -5.22 \text{ kips}$$

Since all other bar forces at joint c were known by the time the analysis reached this joint, it was only necessary to write one equation to find F_6.

A table such as Table 8.1 is definitely recommended when handling space frame problems. Confusion in signs and equations can easily result if an attempt is made to determine the bar forces without its use.

A structure that is quite common in engineering practice is the four-legged tower of Fig. 8.6. The possible reaction components are shown by the arrows at joints I, J, F, and L. Since there are three reaction components at each support, the total number of reactions is twelve. At first one might conclude that the structure is statically indeterminate. However, this conclusion should not be drawn without further investigation. Equation 8.4 provides a means for checking statical determinacy. The number of bars in the structure is twenty-four. The number of joints where all the intersecting bars or forces lie in a common plane is zero. The number of

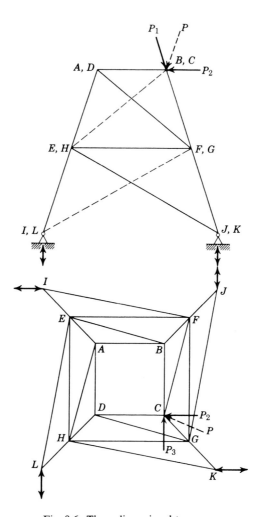

Fig. 8.6 Three-dimensional tower.

joints in which all of the intersecting bars do not lie in a common plane is twelve. Applying Eq. 8.4,

$$b + r = 2t + 3n$$

$$24 + 12 = 2(0) + 3(12)$$

$$36 = 36$$

Since the equation is satisfied, the structure is stable and statically determinate.

If the tower had horizontal bars connecting the panel points (I, J, K, L), there would then be a total of twenty-eight bars. Equation 8.4 would now read

$$b + r = 2t + 3n$$
$$28 + 12 = 3(12)$$
$$40 = 36$$

The structure with the four horizontal bars connecting points I, J, K, and L is statically indeterminate to the fourth degree. If the supports were changed so that the reactions were capable of resisting horizontal loads in only one direction, the number of reactions would then be eight, and the structure would again be statically determinate. However, the structure with the four horizontal bars connecting rigid supports is in reality statically determinate. Upon close inspection it is seen that the horizontal bars connecting the supports will have zero stress. If the supports cannot move, the bars cannot be strained. There is no necessity for horizontal bars connecting rigid supports. Many space frames appear statically indeterminate because the supports can resist forces in all directions. However, zero force members may result in a statically determinate structure.

The first step in the analysis of the tower is to break any load applied to a panel point into three components as shown in Fig. 8.6. One of the components is made parallel to the leg, and the other two components are parallel to the horizontal members meeting at the joint in question (members CE and CB of the illustration). The component P_1 lies in the plane $CDLK$. Since the bar CG is parallel to the force component P_1, this bar will carry the full force of P_1 and bar CD will not be stressed because of the action of P_1.

The second step is to consider the action of the force component P_2. By Rule 1 previously stated in this chapter, only the bars lying in the same plane as the force will be affected by this force. Therefore, as far as force P_2 is concerned the analysis can consist of analyzing the plane truss $CDLK$ only. The bar forces in this plane truss can be determined by any of the methods of truss analysis presented in Chapter 5.

The third step in the analysis is to determine the bar forces produced by the force component P_3. This force component lies in the plane $CBJK$. Therefore, the only bars affected by force component P_3 are those lying in the planar truss $CBJK$. The analysis of this truss can proceed by the same method as used in the second step.

Determination of the final total bar forces is just an algebraic summation of the forces resulting from the three independent analyses. This is permissible as long as the stresses are within the limits of the validity of the

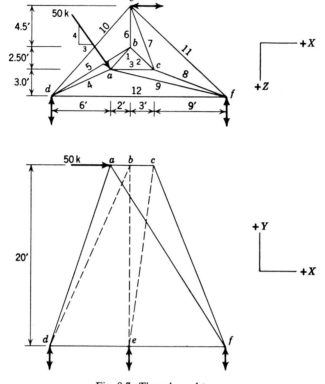

Fig. 8.7 Three-legged tower.

law of superposition. It will be noted that the only bars that will be stressed under all three components are the leg bars *CG* and *GK.*

The analysis for a condition of loading at any panel point can proceed in the manner just outlined.

An interesting problem for analysis is the three-legged tower of Fig. 8.7. The arrows at the supports show the possible direction of the reactions. There are six possible reaction components. The number of bars is twelve, and the number of joints is six. Applying Eq. 8.4 the result is

$$b + r = 2t + 3n$$
$$12 + 6 = 2(0) + 3(6)$$
$$18 = 18$$

The structure is stable and statically determinate.

The first step in the analysis is to resolve the 50-kip force into components parallel to the *X* and *Z* axes. These component forces are 30 kips and 40 kips in the $+X$ and $+Z$ direction, respectively. The analysis of each joint could proceed from joint to joint similarly to the problem of Fig. 8.4. However, the reactions are determined first. Since the reaction at e is the only reaction acting in the X direction, the value of this reaction is equal to the X component of the force applied at a. This value is 30 kips. The value of R_{ex} is then equal to -30 kips. The minus sign indicates the reaction is in the minus X direction. The reaction at f in the Z direction is found by taking moments about a Y axis passing through joint d. This equation is

$$\sum M_{dy} = 0 = 30 \times 3 + 40 \times 6 - 30 \times 10 + R_{fz}(20)$$

$$R_{fz} = -\frac{30}{20} = -1.50 \text{ kip}$$

The value of R_{dz} is now easily determined.

$$\sum F_z = 0 = 40 - 1.50 + R_{dz}$$

$$R_{dz} = -38.5 \text{ kips}$$

The vertical reactions are found by first taking moments about an X axis passing through d and f. This equation is

$$\sum M_{dx} = 0 = 40 \times 20 + R_{ey}(10)$$

$$R_{ey} = -\frac{800}{10} = -80 \text{ kips}$$

The vertical reaction at d is determined by taking moments about a Z axis passing through f.

$$\sum M_{fz} = 0 = 30 \times 20 - 80 \times 12 + R_{dy}(20)$$

$$R_{dy} = \frac{360}{20} = 18 \text{ kips}$$

$$R_{fy} = 80 - 18 = 62 \text{ kips}$$

Proceeding with the analysis of bar forces, it is observed that at each joint there are four apparent unknown bar forces and only three equations of static equilibrium. This does not, however, mean that the structure is statically indeterminate. The proof of determinancy has already been established. It means that at each joint the forces in all the members meeting at a joint could be written in terms of the unknown bar force in one of the members. This unknown bar force could be carried throughout all the joints of the structure and then evaluated at one of the supports where there are only three unknown forces. However, in this tower several short cuts can be utilized. An analysis of joint b shows that all bars meeting at this joint lie in a common plane except bar 2. Since there is no external force applied to joint b, then by Rule 1 of this chapter the force in bar 2 must be zero. Likewise at joint c bars 2, 7, and 8 lie in a common plane, but bar 3 does not lie in this plane. Since no load is applied at joint c, the force in bar 3 must be zero. When we consider joint c again and apply Rule 2, the force in bars 7 and 8 must be zero since bars 2 and 3 are zero. If the force in bar 3 is zero, the analysis can start at joint a as there is now only three unknown bar forces at this joint. The analysis of this joint follows the tabular form as shown.

Joint a

Bar	X	Y	Z	d	F_x	F_y	F_z	F
1	+2	0	−2.50	3.20	+0.625	0	−0.781	F_1
4	−6	−20	+3.00	21.10	−0.284	−0.948	+0.142	F_4
9	+14	−20	+3.00	24.60	+0.569	−0.813	+0.122	F_9
External force					+30.0k		+40.0k	P

The three equations of equilibrium for joint a yield

$$\sum F_y = 0 = -0.948F_4 - 0.813F_9$$

$$F_4 = -\left(\frac{0.813}{0.948}\right)F_9 = -0.858F_9$$

$$\sum F_x = 0 = +0.625F_1 - 0.284F_4 + 0.569F_9 + 30.0$$

$$+0.625F_1 - 0.284(-0.858F_9) + 0.569F_9 = -30.0$$

$$F_1 = -1.30F_9 - 48.0$$

$$\sum F_z = 0 = -0.781F_1 + 0.142F_4 + 0.122F_9 + 40.0$$

$$-0.781(-1.30F_9 - 48.0) + 0.142(-0.858F_9) + 0.122F_9 = 40$$

$$+1.02F_9 + 37.5 - 0.122F_9 + 0.122F_9 = -40.0$$

$$F_9 = -\frac{77.5}{1.02} = -76.3 \text{ kips}$$

$$F_4 = -0.858(-76.3) = +65.5 \text{ kips}$$

$$F_1 = -1.30(-76.3) - 48.0$$

$$= +51.2 \text{ kips}$$

The negative sign for F_9 indicates that the force is towards the joint and thus is in compression.

The process of analysis can now move to joint b. It has already been established that bar 2 has no force in it, so the only unknowns are bar forces 5 and 6. The analysis of this joint is as follows:

Joint b

Bar	X	Y	Z	d	F_x	F_y	F_z	F
1					$-32.0k$	0	$+40.0k$	F_1
5	-8	-20	$+5.50$	22.23	-0.360	-0.900	$+0.247$	F_5
6	0	-20	-4.50	20.50	0	-0.976	-0.220	F_6

$$\Sigma F_x = 0 = -0.360F_5 - 32.0$$

$$F_5 = -\frac{32.0}{0.360} = -89.0 \text{ kips}$$

$$\Sigma F_y = 0 = -0.900F_5 - 0.976F_6$$

$$F_6 = \frac{80.1}{0.976} = +82.0 \text{ kips}$$

Since there are only two unknowns, writing the F_z column in the table is not necessary. However, this last equation can be used as a check.

The remaining bar forces to be found are the bars 10, 11, and 12. The force in bar 10 and 12 can be found easily by writing the equations:

$$\Sigma F_z = 0 \quad \text{and} \quad \Sigma F_x = 0 \quad \text{at joint } d.$$

Joint d

Bar	X	Y	Z	d	F_x	F_y	F_z	F
4					$+18.6k$	$62.0k$	$-9.3k$	F_4
5					$-32.0k$	$-80.0k$	$+22.0k$	F_5
10	$+8$	0	-10.0	12.8	$+0.625$	0	-0.781	F_{10}
12	$+20$	0	0	20.0	1.00	0	0	F_{12}
R_d						$+18.0k$	$-38.5k$	R_d

Bar 10 is first found by taking $\Sigma F_z = 0$.

$$\Sigma F_z = 0 = -0.781F_{10} + 22.0 - 9.3 - 38.5 = 0$$

$$F_{10} = -\frac{25.8}{0.781} = -33.0 \text{ kips}$$

The force in bar 12 can now be determined by writing the equation

$$\Sigma F_x = 0 = F_{12} + 0.625(-33.0) - 32.0 + 18.6$$

$$F_{12} = +34.0 \text{ kips}$$

Here again the column headed F_y in the table is not needed but was used to show a check on forces in bars 4 and 5.

The only remaining bar force that has not been analyzed is that for bar 11. This force is determined by the equation $\Sigma F_z = 0$ at joint f.

$$\Sigma F_z = 0 = -\frac{10}{15.62} F_{11} + 76.3(0.122) - 1.50$$

$$F_{11} = \frac{7.8}{0.64} = 12.2 \text{ kips}$$

It should be noted that in the writing of all the equations for summation of forces at the joints the direction of the unknown bar force was assumed to be away from the joint. If this is done, a resulting negative sign indicates compression and a positive sign the usual tension.

Problems

1. Prove that the structure of Fig. 8.8 is statically determinate and determine the bar forces due to a 15-kip load at d as shown. Possible directions of the reactions are as shown.

Ans. $ab = -20.4, ac = -2.4, bc = +5.4, ad = -28.2, be = 0$
$cf = -22.4, af = +28.0, bd = +23.5, ce = 0, df = -16.8,$
$de = 0, ef = 0.$

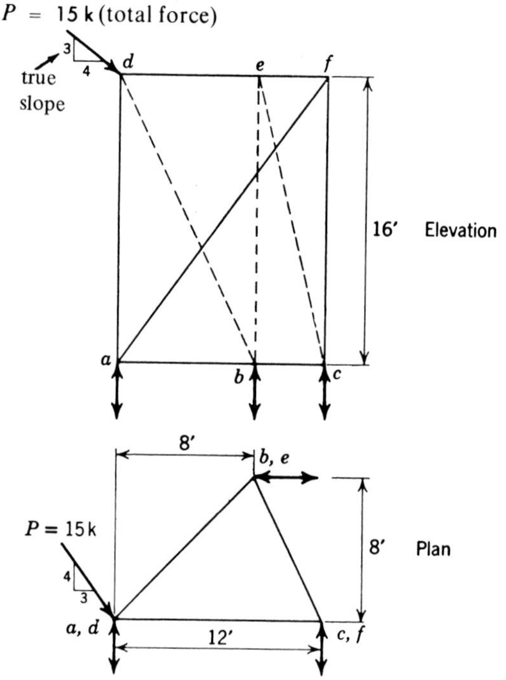

Fig. 8.8

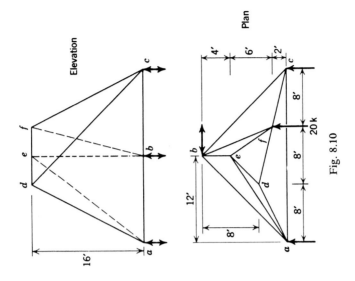

Elevation

P = 12 k

60°

10'

Plan

5'

3'

10'

P = 12 k

Fig. 8.9

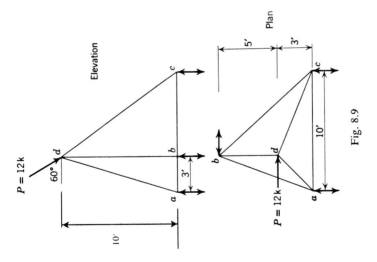

Elevation

16'

Plan

4'

6'

2'

8'

8'

8'

20 k

12'

8'

Fig. 8.10

225

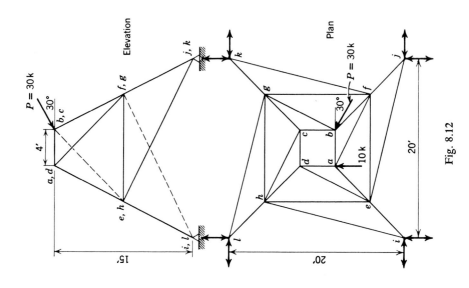

P = 30k

30°

b, c

4'

a, d

f, g

Elevation

j, k

15'

e, h

i, l

P = 30k

30°

Plan

k

g

c

b

10k

d

a

f

h

l

e

i

20'

20'

Fig. 8.12

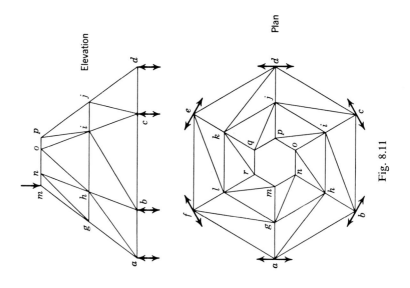

Elevation

d

j

n o p

m

i

c

h

g

b

a

Plan

e

d

k

j

c

f

l

q

p

r

i

m

n

o

h

g

b

a

Fig. 8.11

2. Find the bar forces for the frame of Fig. 8.9.

3. Determine the bar forces in the space frame of Fig. 8.10 produced by the 20-kip horizontal force. Bar *ab* is parallel to *de*.

4. By applying Rule 1 and Rule 2, indicate the bars of zero stress for the Schwedler dome of Fig. 8.11.

5. Determine the bar forces for the tower of Fig. 8.12. The 10-kip load is horizontal.

6. The frame as shown below has forces in the x, y, and z directions at joint *e*. The reaction at support *c* is 12.4 kips. Determine the values of the other reactions as shown and the force in each bar.

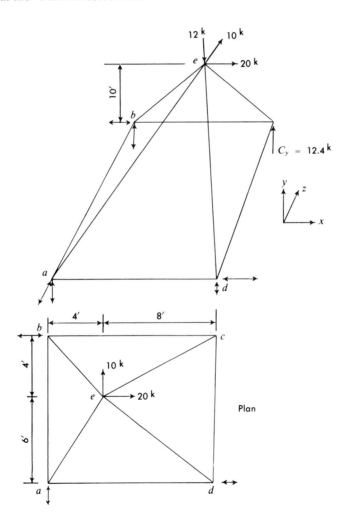

Plan

CABLE SYSTEMS

9.1 Introduction

One of the oldest forms of structural systems, as attested by the ancient vine and bamboo bridges of Asia, has had a new revival in structural engineering. This system is marked by the cable as the predominant load-carrying member. The cable has been used extensively in modern-day engineering for suspension bridges, tramways, and construction systems, but only recently has it found favor with architects and engineers for permanent buildings. Three striking examples are the Pan American Terminal at Idlewild Airport, the Olympic Arena at Squaw Valley, and the Health and Physical Education Building of the Central Washington College of Education, Ellensburg, Washington.

The cable is somewhat unique among structural members in that it is resistant to tension forces only. A cable will not carry compression forces or bending moments. When loads are placed on a cable, it assumes a configuration so that there are only tension forces in the cable. Since a cable can be fabricated from high-strength wires which are relatively low in cost with respect to strength, it has the lowest cost-strength ratio of any structural element. This cost factor as well as easy handling of long lengths make the cable ideal for many structural applications. Cables are made in a wide assortment of types and sizes, and the structural engineer should select carefully the proper cable for the particular application.

9.2 Single Concentrated Load Systems

If the shape of the suspended cable can be ascertained, it is usually relatively easy to determine the force in the cable. Loads on a cable can be placed into three categories, as shown in Fig. 9.1. Part (*a*) shows a concentrated load applied to the cable. Neglecting the weight of the cable, the cable forms two straight lines between the load and the supports. Part (*b*) shows a uniform load per unit length of span. The shape of the cable is a parabola for this type of loading. The third type of loading, as shown in (*c*), is a uniform load per unit length of cable. This type of loading causes the cable to take the form of a catenary.

If the weight of the cable can be neglected, the cable and the applied loads under a concentrated loading system form an equilibrium diagram. Figure 9.2 shows a cable in which the supports are at the same level.

The determination of the reactions and the tension in the cable can be made using the equations of static equilibrium.

(1) $\sum F_y = 0$ $V_1 + V_2 = P$

(2) $\sum F_x = 0$ $H_1 + H_2 = 0$

(3) $\sum M_A = 0$ $V_2 \cdot L = Px$

(4) $\sum M_B = 0$ $V_1 \cdot x = H_1 y$

When we solve the four equations for the four reaction components, the results are

$$V_1 = P\left(1 - \frac{x}{L}\right) \qquad V_2 = \frac{Px}{L}$$

$$H_1 = H_2 = \frac{Px}{y}\left(1 - \frac{x}{L}\right) \tag{9.1}$$

$$T = \sqrt{V^2 + H^2}$$

It is seen from Eqs. 9.1 that the tension in the cable is different from *A* to *B* than from *B* to *C*. As the maximum tension is the item of primary

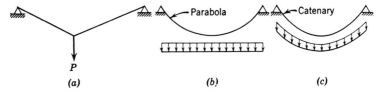

(*a*)   (*b*)   (*c*)

Fig. 9.1 Cable shapes. (*a*) Concentrated load. (*b*) Uniform load. (*c*) Uniform load per cable length.

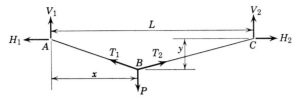

Fig. 9.2 Cable with concentrated load.

interest, the maximum value of V is substituted into the equation for T. The horizontal component of the cable tension is equal throughout the length of the cable. The smaller the value of cable sag y, the larger is the tension in the cable. If the concentrated load is placed midway between supports, Eqs. 9.1 simplify to

$$V_1 = V_2 = \frac{P}{2} \qquad (9.2)$$

$$H_1 = H_2 = \frac{PL}{4y}$$

It can be seen that for small values of y with respect to values of L extremely large cable tensions can occur even for small loads. At this point it should be observed that the deflection of the cable can greatly influence the stress analysis of the suspension system. For all previous structures covered in this book the effect on the structural analysis due to deflections of the structure can usually be ignored. The change in geometry for rigid structures is negligible. However, this may not be so for a cable suspension system. For instance, if a cable is supported at points 100 ft apart and a concentrated load of 12 kips is supported at midspan with a final sag of 4 ft, the tension in the cable is

$$H = \frac{12 \times 100}{4 \times 4} = 75 \text{ kips}$$

$$T = \sqrt{(75)^2 + (6)^2} = 75.25 \text{ kips}$$

Using a factor of safety of two, a $1\frac{1}{8}$ in. diameter bridge strand would be required (see Appendix B). If an additional load of 6 kips were to be placed on the cable at the same position as the previous load, allowing, of course, a reduction in the safety factor, the force in the cable based on the same sag of 4 ft is

$$T = 75.25 \times \tfrac{18}{12} = 113 \text{ kips}$$

The difference between the theoretical 113 kips and 75.25 kips produces an increase in the length of the cable. The length of the cable before the application of the 6 kips load was

$$S = 2\sqrt{(50)^2 + (4)^2} = 100.32 \text{ ft.}$$

The increase in the length will be the unit strain multiplied by the length

$$\epsilon = \frac{\Delta T}{AE} = \frac{(113 - 75.25)}{(0.776)\,24{,}000} = 2.03 \times 10^{-3}$$

$$\Delta S = 2.03 \times 10^{-3} \times 100.32 = 0.204 \text{ ft.}$$

The new cable length $= 100.32 + 0.20 = 100.52$ ft. The new value of

$$y = \left[\frac{(100.52)^2}{4} - (50)^2 \right]^{\frac{1}{2}} = 5.11 \text{ ft.}$$

Since this calculated sag of 5.11 ft is considerably different (percentage basis) than the estimated sag of 4 ft, the calculated value of $T = 113$ kips is not correct.

A new value of T could be calculated using $y = 5.11$ but reasoning will prove that this would not result in a calculated sag of 5.11 ft because this value of sag was for a T equal to 113 kips. It will save time to make a new estimate of sag somewhat less than 5.11 ft since calculations based on a sag greater than four feet will give $T < 113$ kips and thus a sag less than 5.11 ft.

For a second trial: $y = 4.6$ ft will be chosen

$$H = \frac{18 \times 100}{4 \times 4.6} = 97.8 \text{ kips}$$

$$T = \sqrt{81 + 9570} = 98.2 \text{ kips}$$

For

$$y = 4.6$$

$$S = 100.42 \text{ ft.}$$

$$\Delta S = \frac{(97.8 - 75.25)\,100.42}{(0.776)\,24{,}000} = 0.12 \text{ ft.}$$

New cable length = 100.32 + 0.12 = 100.44 ft.

$$y = \sqrt{(50.22)^2 - (50)^2} = 4.7 \text{ ft.}$$

Since there is a small difference between a sag of 4.6 and 4.7 feet, the final length of sag can be taken as the average between the initial and final sag.

$$H = \frac{18 \times 100}{4 \times 4.65} = 96.8 \text{ kips}$$

The correct value of cable tension for the total load of 18 kips is

$$T = \sqrt{81 + 9370} = 97.2 \text{ kips}$$

This value of cable tension is 15.8 kips less than that calculated ignoring the stretch of the cable due to the additional 6 kip load. To have a factor of safety of two a larger diameter cable would have to be used, which would change the computations again as the total strain would be reduced slightly and subsequently the sag and cable tensions. If the problem condition had a much greater sag-to-span ratio under the initial load, the cable strain would have had a smaller effect on the cable tension. This example shows that in cable systems the elastic deflections may have to be considered in the analysis. This is true in long-span suspension bridges.

In a cable system where a single concentrated load is acting, and where the supports are at different elevations, the procedure for determining the reactions is the same as for the previous example. Figure 9.3 shows this system.

The resultant equations are

$$H_1 = H_2 = \frac{Px(L - x)}{y_1 L - x(y_1 - y_2)}$$

$$V_1 = \frac{P(L - x)}{L - x\left(1 - \dfrac{y_2}{y_1}\right)}$$ (9.3)

$$V_2 = P\left[1 - \frac{L - x}{L - x\left(1 - \dfrac{y_2}{y_1}\right)}\right]$$

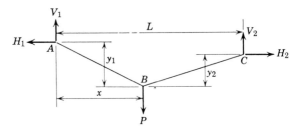

Fig. 9.3 Cable with supports at different elevations.

9.3 Multiple Concentrated Load Systems

When several concentrated loads are supported by a cable, it is necessary to know the elevation of the cable at one of the load points with respect to the elevation of the supports. If this cable sag is known, as in Fig. 9.4, the elevation at the other points of load as well as the cable tensions between loads can be readily determined.

Since the moment at all points along the cable must be zero, the cable assumes a shape that results in this condition. The first step in the solution of the cable problem of Fig. 9.4 is to determine the vertical components of the reactions at the supports. Since both supports are at the same level, the determination of the vertical components of the reactions is the same as the determination of the vertical end reactions of a simple beam. By taking moments about point f and neglecting the weight of the cable,

$$\sum M_f = 0 \quad V_a \times 150 - 1.2 \times 120 - 1.8 \times 80 - 1.2 \times 40 - 1.0 \times 15 = 0$$

$$V_a = \frac{351}{150} = 2.34 \text{ kips}$$

$$\sum F_y = 0 \quad V_f = 1.2 + 1.8 + 1.2 + 1.0 - 2.34 = 2.86 \text{ kips}$$

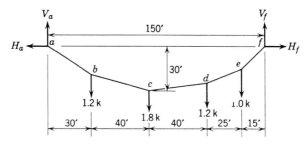

Fig. 9.4 Cable with multiple concentrated loads.

By taking moments about c, the point where the cable sag is known, the value of H_a is found.

$$\sum M_c = 0 \qquad 2.34 \times 70 - 1.2 \times 40 - H_a \times 30 = 0$$

$$H_a = \frac{115.8}{30} = 3.86 \text{ kips}$$

$$\sum F_x = 0 \qquad H_a = H_f$$

From the previous section it was shown that the horizontal component of tension in a cable is the same value anywhere along the cable.

The distance from the cable cord to the points of load along the cable can now be determined.

$$\sum M_b = 0 \qquad V_a \times 30 - y_b \times H = 0$$

$$y_b = \frac{30 \times 2.34}{3.86} = 18.19 \text{ ft}$$

$$\sum M_d = 0 \qquad 2.86 \times 40 - 3.86 \times y_d - 1.0 \times 25 = 0$$

$$y_d = \frac{89.4}{3.86} = 23.16 \text{ ft}$$

$$\sum M_e = 0 \qquad 2.86 \times 15 - 3.86 \times y_e = 0$$

$$y_e = \frac{42.9}{3.86} = 11.11 \text{ ft}$$

The total length of the cable can now be calculated

$$S = [(30)^2 + (18.19)^2]^{1/2} + [(40)^2 + (11.81)^2]^{1/2} + [(40)^2]^{1/2}$$

$$+ [(25)^2 + (12.05)^2]^{1/2} + [(15)^2 + (11.11)^2]^{1/2}$$

$$= 35.08 + 41.71 + 40.58 + 27.75 + 18.67$$

$$= 163.79 \text{ ft.}$$

The procedure is the same irrespective of the number of concentrated loads, provided the weight of the cable can be neglected. With the supports at different elevations the same reasoning applies. Of course, at all times a solution is possible only if the elevations of the supports, as well as the elevation of one point along the cable, are known. It is suggested that the student use deductive reasoning in applying the equations of equilibrium instead of mechanically applying some developed formula or graphical procedure for the solution of cable problems.

9.4   Uniformly Loaded Cable with Respect to Span Length

Supports at Same Elevation.   It has already been stated that a cable with a uniform load per unit length of span results in a parabolic configuration of the cable. This can be proven simply by writing the equation of static equilibrium for a portion of the cable. By taking a cable system as shown in Fig. 9.5a and cutting a section at any point on the cable, the equation for moment at the point of cut on the cable, Fig. 9.5b, is

$$\frac{wL}{2}x - Hy - \frac{wx^2}{2} = 0$$

This equation has to be zero as a cable has negligible bending resistance, and thus has to assume a shape that produces no bending moments anywhere along the cable. Solving for y from the preceding equation gives

$$y = \frac{wx}{2H}(L - x) \tag{9.4}$$

The value of y is positive downward and all values of x are positive as the origin is at point A, the left support. Equation 9.4 is of the form of a second degree parabola.

The horizontal component of the tension in the cable has to be constant throughout the entire length of the cable. Proof for this is the equilibrium equation of $\Sigma F_x = 0$ for any portion of the cable. If the value of y at the

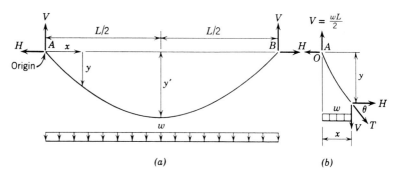

(a)   (b)

Fig. 9.5   Cable with uniform load.

lowest point on the cable is called y', the evaluation of H is easily made by substituting into Eq. 9.4 the terms y' and $L/2$ for y and x respectively. This gives

$$H = \frac{wL^2}{8y'} \qquad (9.5)$$

The tension at any point on the cable can be found by determining H and V at that point and applying the Pythagorean theorem. Another approach is to determine the slope of the cable at the requisite point. The slope is found by taking the derivative of Eq. 9.4.

$$\frac{dy}{dx} = \tan \theta = \frac{w}{2H} (L - 2x)$$

$$T = \frac{H}{\cos \theta}$$

The value of T will be greatest where the angle θ is the greatest. The angle θ will be the greatest at the support, and it will be the least at the low point on the cable. At the support

$$V = \frac{wL}{2}$$

The maximum cable tension for a uniformly loaded cable with both supports at the same elevation is

$$T^2 = H^2 + V^2 = \frac{w^2 L^4}{64 y'^2} + \frac{w^2 L^2}{4}$$

$$= \frac{w^2 L^2}{4} \left(\frac{L^2}{16 y'^2} + 1 \right)$$

$$T = \frac{wL}{2} \sqrt{1 + (L^2/16 y'^2)} \qquad (9.6)$$

From Eq. 9.6 it is seen that the smaller the sag with relationship to the span the greater is the tension in the cable. The tension in the cable approaches infinity as y' approaches zero as a limit.

It is no doubt apparent that it is nearly impossible to load a cable with a load that is uniform with respect to the span. This condition will be approached by supporting a uniform load from the cable with closely spaced hangers. If the hangers are closely spaced the error is negligible.

The length of the cable can be determined by integrating the expression $ds = \sqrt{dx^2 + dy^2}$ over the span:

$$S = 2 \int_0^{L/2} \sqrt{dx^2 + dy^2} = 2 \int_0^{L/2} \left[1 + \left(\frac{dy}{dx} \right)^2 \right]^{1/2} dx$$

The easiest solution of this equation is to expand the expression by the binominal theorem where the first four terms are

$$(a + b)^n = a^n + na^{n-1}b + \frac{n(n-1)}{2!} a^{n-2}b^2 + \frac{n(n-1)(n-2)}{3!} a^{n-3}b^3$$

By taking $a = 1$ and $b = (dy/dx)^2 = [(w/2H)(L - 2x)]^2$

$$S = 2\int_0^{L/2} \left(1 + \frac{1}{2}\left[\frac{w}{2H}(L - 2x)\right]^2 - \frac{1}{8}\left[\frac{w}{2H}(L - 2x)\right]^4 + \cdots\right) dx$$

Only the first three terms are of any significance.

$$S = L\left[1 + \frac{8}{3}\left(\frac{y'}{L}\right)^2 - \frac{32}{5}\left(\frac{y'}{L}\right)^4\right] \tag{9.7}$$

Only for large values of y'/L will the third term have any appreciable value.

EXAMPLE PROBLEM 9.1   Determine the size of cable with a safety factor of two required to support a uniform load of 1.2 kips per foot over a span of 120 ft. The sag is 20 ft.

$$H = \frac{1.2(120)^2}{8 \times 20} = 108 \text{ kips}$$

$$V_{max} = 1.2 \times \frac{120}{2} = 72 \text{ kips}$$

$$T = \sqrt{(108)^2 + (72)^2} = 130 \text{ kips}$$

The cable must have a breaking strength of at least 130 tons. A $1\frac{1}{2}$-in.-diameter bridge cable is satisfactory for this load (Appendix B). The required length of cable is

$$S = 120\left[1 + \frac{8}{3}\left(\frac{20}{120}\right)^2 - \frac{32}{5}\left(\frac{20}{120}\right)^4\right]$$

$$= 120(1 + 0.074 - 0.005) = 120 \times 1.069 = 128.3 \text{ ft}$$

Supports at Different Elevations.   A cable with the supports at different elevations as shown in Fig. 9.6 can be treated in a manner similar to the previous case of supports at the same elevation. In this case the mathematics is less cumbersome if the origin of coordinates is taken at the low point of the cable. By taking moments at any point on the cable a distance x from the origin, the expression for y can be found (Fig. 9.6b).

$$Hy = \frac{wx^2}{2}$$

$$y = \frac{wx^2}{2H} \tag{9.8}$$

If Eq. 9.8 is written for both the left and right portion of the cable,

$$y_a = \frac{wx_a{}^2}{2H}$$

$$y_b = \frac{wx_b{}^2}{2H}$$

Since H is the same in both equations, the two equations can be set equal to each other.

$$\frac{x_a{}^2}{y_a} = \frac{x_b{}^2}{y_b}$$

$$x_b = L - x_a$$

$$x_a{}^2 = \frac{y_a}{y_b}(L^2 - 2Lx_a + x_a{}^2)$$

$$x_a{}^2\left(1 - \frac{y_a}{y_b}\right) + \frac{2y_a}{y_b}Lx_a - \frac{y_a}{y_b}L^2 = 0$$

making

$$\frac{y_a}{y_b} = k$$

$$x_a{}^2(1 - k) + 2Lkx_a - kL^2 = 0 \tag{9.9}$$

If k and L are known then the location of the low point on the cable can be determined from Eq. 9.9. With the location of the low point on the

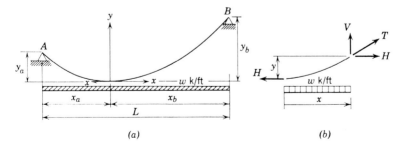

(a)   (b)

Fig. 9.6

cable known, the tension in the cable can be determined by procedures already presented.

To determine the length of the cable the expression

$$S = \int_0^x \sqrt{dx^2 + dy^2}$$

can be written for both portions of the cable. Using the method of employing the binominal theorem, the expression for length of the portion of the cable from the low point to the support is

$$S_a = \int_0^{x_a} \sqrt{1 + w^2 x_a{}^2/H^2} \, dx$$

$$= x_a\left(1 + \frac{w^2 x_a{}^2}{6H^2} - \frac{w^4 x_a{}^4}{40H^4} + \cdots\right)$$

$$= x_a\left[1 + \frac{2}{3}\left(\frac{y_a}{x_a}\right)^2 - \frac{2}{5}\left(\frac{y_a}{x_a}\right)^4 + \cdots\right]$$

and likewise

$$S_b = x_b\left[1 + \frac{2}{3}\left(\frac{y_b}{x_b}\right)^2 - \frac{2}{5}\left(\frac{y_b}{x_b}\right)^4 + \cdots\right]$$

The total length of the cable is then (considering only the first three terms)

$$S = x_a\left[1 + \frac{2}{3}\left(\frac{y_a}{x_a}\right)^2 - \frac{2}{5}\left(\frac{y_a}{x_a}\right)^4\right] + x_b\left[1 + \frac{2}{3}\left(\frac{y_b}{x_b}\right)^2 - \frac{2}{5}\left(\frac{y_b}{x_b}\right)^4\right] \quad (9.10)$$

EXAMPLE PROBLEM 9.2   A cable is supported and loaded as shown in Fig. 9.7. Determine the maximum tension in the cable and the required length of the cable.

$$y_a = 8 \text{ ft} \qquad y_b = 20 \text{ ft} \qquad L = 100 \text{ ft} \qquad k = \tfrac{8}{20} = 0.4$$
$$x_a{}^2(1 - 0.4) + 2 \times 100 \times 0.4x_a - 0.4(100)^2 = 0$$
$$0.6x_a{}^2 + 80x_a = 4000$$
$$x_a{}^2 + 133.3x_a = 6667$$

The solution to this equation is $x_a = 38.75$ ft and therefore, $x_b = 100 - 38.75 = 61.25$ ft.

The maximum tension will be at support B, the point of maximum slope.

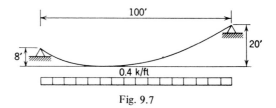

100′

20′

8′

0.4 k/ft

Fig. 9.7

$$H = \frac{0.4(38.75)^2}{2 \times 8} = 37.53 \text{ kips}$$

$$V_B = wx_b = 0.4(61.25) = 24.5 \text{ kips}$$

$$T_{max} = \sqrt{(37.53)^2 + (24.5)^2} = 44.8 \text{ kips}$$

The required length of the cable is

$$S = 38.75\left[1 + \frac{2}{3}\left(\frac{8.0}{38.75}\right)^2 - \frac{2}{5}\left(\frac{8.0}{38.75}\right)^4\right]$$

$$+ 61.25\left[1 + \frac{2}{3}\left(\frac{20.0}{61.25}\right)^2 - \frac{2}{5}\left(\frac{20.0}{61.25}\right)^4\right]$$

$$= 38.75 \times 1.028 + 61.25 \times 1.067 = 105.15 \text{ ft}$$

9.5 Uniformly Loaded Cable with Respect to Cable Length

A cable that has a uniformly distributed load along the length of the cable takes the shape of a catenary. A cable hanging under its own weight is an example of this type of loading.

The equation of a catenary is $y = (a/2)(e^{x/a} + e^{-x/a})$, where a is the y distance from the origin to the low point of the catenary. Since the relationships among hyperbolic functions gives

$$\cosh u = \tfrac{1}{2}(e^u + e^{-u})$$

the equation for a catenary can be written in the hyperbolic function form as

$$y = a \cosh \frac{x}{a}$$

A diagram of a cable carrying its own weight is shown in Fig. 9.8a and a free body portion of the cable is shown in (b). The load per unit length

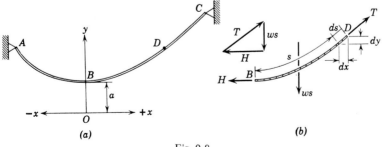

(a)   (b)

Fig. 9.8

of the cable is denoted by w. The total load applied to a portion of cable of s length is ws. The tension in the cable at D is then

$$T = \sqrt{H^2 + (ws)^2}$$

The following computations can be simplified by choosing the origin such that the constant a bears the relationship $H = wa$. Substituting this expression into the equation for T, the result is

$$T = w\sqrt{a^2 + s^2} \tag{9.10}$$

The value of T at any point along the cable cannot be found until the location of the low point of the cable is known and the resulting expression for s is also determined.

For a ds length of cable the relationship $dx = ds \cos \theta$ can be written. Since $\cos \theta = H/T$, it can be written

$$dx = \frac{H}{T} ds = \frac{wa\ ds}{w\sqrt{a^2 + s^2}} \quad \text{or} \quad dx = \frac{ds}{\sqrt{1 + s^2/a^2}}$$

If this expression is integrated between the limits from $B(0,a)$ to $D(x,y)$, the result is

$$x = \int_0^s \frac{ds}{\sqrt{1 + s^2/a^2}} = a\left(\sinh^{-1} \frac{s}{a}\right)\Big/_0^s = a \sinh^{-1} \frac{s}{a} \tag{9.11}$$

This result can be written in the form $s = a \sinh (x/a)$.

This expression relates the distance along the cable to the horizontal distance x. The relation between x and y can be written as

$$dy = dx \tan \theta = \frac{ws}{H} dx = \frac{s}{a} dx = \sinh \frac{x}{a} dx$$

Integrating again from $B(0,a)$ to $D(x,y)$, the result is

$$y - a = \int_0^x \sinh \frac{x}{a} dx = a\left(\cosh \frac{x}{a}\right)\Big/_0^x = a\left(\cosh \frac{x}{a} - 1\right)$$

$$y = a \cosh \frac{x}{a} \tag{9.12}$$

This is the equation of a catenary with vertical axis and the origin at a distance a below the lowest point of the cable.

Since

$$\cosh^2 u - \sinh^2 u = 1$$

$$s^2 = y^2 - a^2 \tag{9.13}$$

By substituting this expression for s^2 into Eq. 9.10, the cable tension is

$$T = wy \tag{9.14}$$

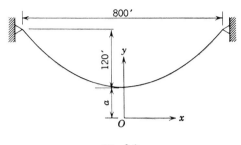

Fig. 9.9

The maximum cable tension T is at the point of maximum value of y. The value of y must include the catenary parameter a. In order to determine the tension in a cable with a known sag, the value of a has to be determined from the expression for the cable curve (Eq. 9.12).

When the cable supports A and C are not at the same elevation, the position of the lowest point of the cable is unknown. The additional relationships $x_B - x_A = L$ and $y_B - y_A = r$, where L is the horizontal distance and r is the vertical distance between the supports, can be utilized. The determination of the constant a in Eq. 9.12 requires a trial and error solution.

EXAMPLE PROBLEM 9.3   A uniform cable weighing 6 lb per foot is to be suspended between two points at equal elevations and 800 ft apart as shown in Fig. 9.9. If the sag of the cable is to be 120 ft, (a) what is the maximum cable tension due to the weight of the cable? (b) What is the length of the cable between supports?

The first step in the solution is the determination of the constant a.

$$y = a \cosh \frac{x}{a}$$

when $\qquad\qquad x = 400 \qquad y = 120 + a$

$$\frac{120 + a}{a} = \cosh \frac{400}{a}$$

The value of a is found by successive trials.

a	$\dfrac{120}{a}$	$\dfrac{120}{a} + 1$	$\dfrac{400}{a}$	$\cosh \dfrac{400}{a}$
680	0.176	1.176	0.588	1.178
690	0.174	1.174	0.580	1.173
685	0.175	1.175	0.584	1.175

With $a = 685$ ft

$$y_c = 120 + 685 = 805 \text{ ft}$$

$$T_{max} = wy = 6 \times 805 = 4830 \text{ lb}$$

The solution for the length of the cable is

$$S^2 = y^2 - a^2$$

$$S^2 = (805)^2 - (685)^2 = 178{,}800$$

Total length of the cable $= 2\sqrt{178{,}800} = 845.7$ ft.

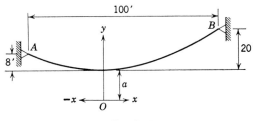

Fig. 9.10

EXAMPLE PROBLEM 9.4   Determine the maximum cable tension for the suspended system of Fig. 9.10 if the load is 0.4 kip per foot length of the cable. Also determine the length of the cable.

$$y_a = 8 + a \qquad y_b = 20 + a$$

$$-x_a = 100 - x_b$$

From the equation $y = a \cosh x/a$

(1) $$8 + a = a \cosh \frac{100 - x_b}{a}$$

(2) $$20 + a = a \cosh \frac{x_b}{a}$$

or

(1a) $$\frac{100 - x_b}{a} = \cosh^{-1} \frac{8 + a}{a}$$

(2a) $$\frac{x_b}{a} = \cosh^{-1} \frac{20 + a}{a}$$

Adding equations (1a) and (2a)

$$\frac{100}{a} = \cosh^{-1}\frac{8+a}{a} + \cosh^{-1}\frac{20+a}{a}$$

The solution of this equation by successive trials is $a = 96.2$ ft. Substituting $a = 96.2$ ft into Eq. 1

$$1.083 = \cosh\frac{100 - x_b}{96.2}$$

$$x_b = 61.03 \text{ ft}$$

$$x_a = 100 - 61.03 = 39.97 \text{ ft}$$

$$y_b = 20 + 96.2 = 116.2 \text{ ft}$$

$$y_a = 8 + 96.2 = 104.2 \text{ ft}$$

$$T_{max} = w(y_b) = 0.4(116.2) = 46.48 \text{ kips}$$

The length of cable is

$$S = \sqrt{(116.2)^2 - (96.2)^2} + \sqrt{(104.2)^2 - (96.2)^2}$$
$$= \sqrt{4248} + \sqrt{1603} = 65.18 + 40.04 = 105.2 \text{ ft}$$

A comparison of the results of Example Problem 9.2 with 9.4 shows that for low sag ratios the differences are small between the catenary solution and the parabolic solution.

Problems

1. Determine the tension in the cable of the suspended systems of Fig. 9.11. The cable sags are at rest values.

Ans. (*a*) 9.36 kips, (*b*) 19.0 kips, (*c*) 19.2 kips, (*d*) 26.6 kips.

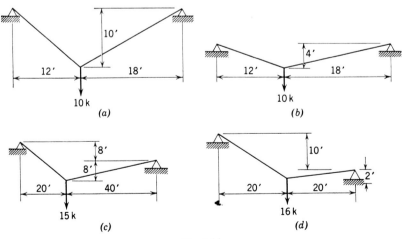

Fig. 9.11

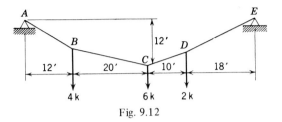

Fig. 9.12

2. Find the cable tension between each force and determine the required length of the cable for the system of Fig. 9.12.   *Ans.* Length of cable = 65.12 ft.

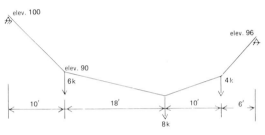

Fig. 9.13

3. Determine the maximum cable tension and length of cable for the load system shown in Fig. 9.13.

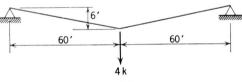

Fig. 9.14

4. A 1-in. diameter cable supports a central load of 4 kips with an at rest sag of 6 ft as shown in Fig. 9.14. An additional central load of 12 kips is added later. What is the final tension in the cable?   *Ans.* 63.7 kips.

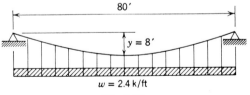

Fig. 9.15

5. Find the required size and length of cable with a safety factor of two for the system of Fig. 9.15.

6. If the sag for the cable of Fig. 9.15 is increased to 16 ft, what is the required length and diameter of the cable?

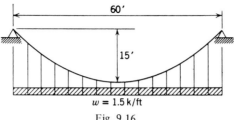

$$w = 1.5 \, k/ft$$

Fig. 9.16

7. What is the safety factor against breaking for the cable of Fig. 9.16? Diameter of cable equals $\frac{7}{8}$ in.

8. Determine the force in the cable of Fig. 9.16 if the load is changed to 1.5 kips per foot length of cable. What are the horizontal and vertical components of the cable tension acting on the support?

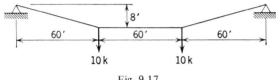

Fig. 9.17

9. A steel cable of $1\frac{1}{4}$ in. in diameter supports the loads as shown in Fig. 9.17. The sag as shown is the at rest sag at 80°F. What would be the sag and cable tension if the temperature dropped to −10°F?

$$\varepsilon = 6.5 \times 10^{-6} \, in./in./F°$$

Ans. $y = 7.59$ ft   Tension = 79.6 kips.

10. What would be the change in sag for the steel cable of Fig. 9.16 due to a decrease in temperature of 120°F? What change in tension would this change in sag produce?

Ans. Sag change = −0.06 ft   Tension change = +0.14 kip.

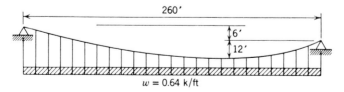

$$w = 0.64 \, k/ft$$

Fig. 9.18

11. Determine the maximum cable tension and required length of cable for the loading condition of Fig. 9.18.

12. A 2½ in. diameter cable is to be suspended between two supports at the same elevation. The supports are 600 ft. apart. The sag of the cable is 60 ft. What is the maximum tension force in the cable from the weight of the cable only. What are the values of the reactions and the length of cable between supports?

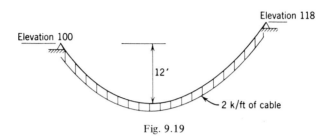

Fig. 9.19

13. Determine the maximum cable tension and the required length of cable for the loading condition of Fig. 9.19. Horizontal span of cable is equal to 50 ft.

14. Compare the maximum cable tension and length of cable between a catenary and a parabolic hung cable system. The spans are 180 ft and the sag for each system is 36 ft. Each system carries the same total load of 90 kips. Both loads are uniformly distributed (one with respect to cable and the other with respect to span).

DEFLECTIONS

10.1 Introduction

In addition to the two basic requirements that a structure support the imposed loads without exceeding permissive stresses and that it is economical, it is also necessary that a structure does not undergo excessive deflections during service. Most structures are designed in accordance with an established design specification. Nearly all structural design specifications contain limits of deflection for particular types of structures. Deflection of ceiling beams may be limited by the cracking of a plaster ceiling, the deflection of a bridge by the possible resulting vibration, or the limit of deflection of an airplane wing by the aerodynamic characteristics. Furthermore, it is a condition for most structures that an appearance of large deflections be avoided. A floor of a building that had a deflection that could be noted by the unaided eye, regardless of its ability to carry safely the imposed loads, would not be considered satisfactory. Similarly, a bridge that had a noticeable sag would be shunned by the motorist. Bridges have an inherent upward deflection fabricated into the main structure so that under dead-load conditions the net deflection will be zero or still slightly upward. In order to determine the amount of upward deflection (termed camber) to be fabricated into the structure, it will, of course, be necessary to compute the dead-load deflections.

Not only are deflections in a completed structure important but also deflections during erection are the concern of the engineer. Erection procedures will many times depend on an exact knowledge of the deflections to be expected during the erection. This is especially true of bridges. A close check of deflection during erection will indicate any possibility of failure. Excessive deflection of the Quebec cantilever bridge during its first erection in 1907 was noted a short time before it collapsed.

The third reason for a knowledge of the methods of computing deflections being so important to the engineer is that these methods form the foundation for analysis of statically indeterminate structures. When the equations of static equilibrium are not sufficient in number to determine all the unknown forces acting on a structure, it is necessary to employ deflection or slope conditions as the additional equations.

In the design of machinery, deflections come into important consideration. Clearance of moving parts has to be carefully controlled. Distortions beyond certain limits will cause malfunction of the equipment. The rolls of a steel plate rolling mill are a good example of necessary control of deflection. In order to reduce the center deflection of a roll and thus keep the variation of plate thickness within limits, double rolls are used on top and bottom. The outside rolls provide stiffness to the inner rolls.

Many different methods can be used to compute deflections. No one method is best for all applications. The engineer should be familiar with many methods and understand which method is best for each particular application. For some applications there will be little to choose from among two or three methods. The choice will usually result in using that method with which the engineer is most familiar. This chapter includes four of the most versatile of the several methods for computing deflections for beams and trusses. A treatment of the other methods can be found in texts covering the analysis of statically indeterminate structures.[1, 2, 3, 4]

10.2   Moment-Area Method

This well-known method is usually covered in texts dealing with mechanics of materials. It was first introduced in 1873 by Professor C. E. Greene of the University of Michigan. It is applicable to determining the slope and deflection of beams and frames. In Fig. 10.1 a small length dx of a cantilever beam is subjected to a bending moment M. The stress on any fiber is usually expressed by the well-known flexure formula $\sigma = My/I$. This equation is valid if the following conditions exist:

1. The maximum fiber stress in the beam does not exceed the proportional limit.

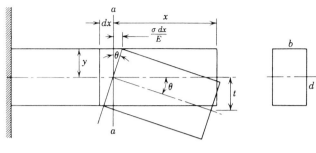

Fig. 10.1

2. There is no sudden change in cross section of the beam.

3. The material of the beam has the same modulus of elasticity in tension as in compression.

4. The beam is straight.

5. The loads act in the plane that contains an axis of symmetry of the beam.

6. The beam is long in proportion to its depth so that the effect of shearing strain on the axial strain is negligible.

The unit strain on any fiber is the stress divided by the modulus of elasticity, or $e = \sigma/E$. In a length of dx the total strain will be equal to $\sigma\, dx/E$.

The plane section a-a of Fig. 10.1 rotates through an angle θ due to a change in length of the fibers in the dx distance. Since the angle θ is very small, $\theta = \tan \theta$ and

$$\theta = \frac{\sigma\, dx}{Ey} = \frac{My}{I} \cdot \frac{dx}{Ey} = \frac{M\, dx}{EI}$$

Since θ is also the change in slope of the tangent to the elastic line of the beam in the dx distance, the change in slope of the tangent to the elastic line between any two points A and B along a beam will be the summation of the changes in the dx lengths along the beam. This can be expressed in equation form as

$$\Delta\theta = \int_A^B \frac{M\, dx}{EI} \tag{10.1}$$

If the moment diagram throughout the length of a beam is divided by the value of EI at all points along the beam, the resulting diagram is called a M/EI diagram. Since the base of the diagram is the distance along the beam, the right-hand side of Eq. 10.1 is the area of the M/EI diagram between points A and B along the beam. This concept is termed the first moment-area principle and can be stated thus:

The change in slope of the tangent to the elastic line between any two points A and B on the beam is equal to the area of the M/EI diagram between points A and B.

It should be observed that this principle defines the change in slope. Since Eq. 10.1 results in a dimensionless quantity, the resultant change of slope is expressed in radians.

Referring again to Fig. 10.1 it is noted that the deflection t of the end of the beam from the tangent to the elastic line at section a-a is equal to the angle θ times the distance x, providing the angle θ is sufficiently small so that the angle θ as expressed in radians is equal (within normal engineering practice) to the tangent of θ. This is true for angles under one degree. Changes in elastic slopes in most all beams are less than one degree.

The deflection of point A from the tangent at point B is the summation of all the effects of the dx lengths between A and B. This is expressed in equation form as

$$t_{AB} = \int_{A}^{B} \frac{Mx\,dx}{EI} \tag{10.2}$$

If $M\,dx/EI$ is the area of the M/EI diagram in the dx length, this area multiplied by the distance x is equal to the area multiplied by the lever arm x or specifically is the moment of the M/EI area taken about the point where the deflection is required. This concept is termed the second moment-area principle and can be stated thus:

The deflection of a point A on the elastic line from the tangent to the elastic line at point B is equal to the moment taken about A of the area of the M/EI diagram between A and B.

The second moment-area principle deals with the deflection, normal to the original axis of the beam, of one point on the beam *measured from the tangent to the elastic line at a second point.* It does not necessarily give the true deflection from the original position.

In applying the moment-area principles the determination of the direction of slope and deflection is usually obvious. However, sometimes this may not be so. If the areas of the M/EI diagrams are given the same sign as the moment diagrams, then if the sum of the areas between any two points A and B is positive, the tangent *on the right* is rotated counterclockwise with respect to the tangent on the left. Similarly, if the sign of the area is negative, the tangent *on the right* will be clockwise with respect to the other tangent. To determine the direction of the point on the elastic

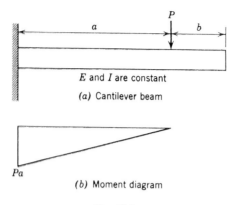

E and *I* are constant

(a) Cantilever beam

(b) Moment diagram

Fig. 10.2

line at *A* with respect to the tangent at *B*, the areas of the *M/EI* diagram are given the same sign as the moment diagram, and the lever arms to the *M/EI* areas are always taken as positive. Then a positive value for the tangential deviation, as determined by the second moment-area principle, indicates that the point on the elastic line (point *A*) is *above* the tangent drawn at point *B*.

The moment-area principles are readily adapted to determining the slopes and deflections of beams which have moment diagrams whose shapes are simple figures. Beams with constant section are more easily handled by this method.

The simplest application of the moment-area principles is the cantilever beam. It is required to find the deflection at the end of the cantilever beam of Fig. 10.2. If the support does not rotate, which is the usual case, the tangent to the elastic line at the support remains horizontal, and the deflection of the end of the beam from the tangent at the support is the true vertical deflection from the original position.

The deflection can then be determined by taking the moment about the end of the beam of the *M/EI* diagram between the end and the fixed support. The value of the vertical deflection at the end of the beam is simply

$$\Delta = \frac{Pa}{EI} \times \frac{a}{2}\left(\frac{2}{3}a + b\right) = \frac{Pa^2}{EI}\left(\frac{a}{3} + \frac{b}{2}\right)$$

The application of the moment-area method to a simply supported beam requires more detailed thought. A simple beam with a single concentrated load is shown in Fig. 10.3. The requirement is to find the true deflection under the point of load. *E* and *I* are considered constant throughout the length of the beam. Here, unlike the cantilever beam, it is not known where

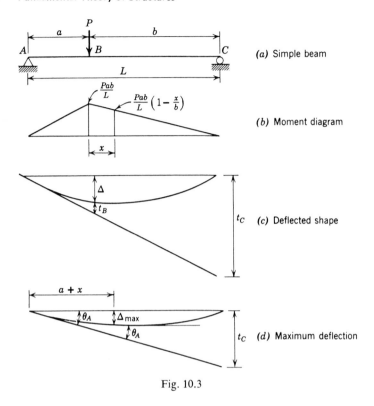

Fig. 10.3

the tangent to the elastic line is horizontal. The true deflection (Δ) under the load cannot be found directly. However, the deflection of point C from the tangent at A can be found readily. This value is t_C as shown in Fig. 10.3c and is equal to the moment about point C of the area of the M/EI diagram between A and C. The equation is

$$EIt_C = \frac{Pab}{L}\left[\frac{a}{2}\left(b + \frac{a}{3}\right) + \frac{b}{2}\left(\frac{2b}{3}\right)\right]$$

$$= \frac{Pab}{6L}(a^2 + 3ab + 2b^2)$$

Since E and I are constant, the moment diagram can be used instead of the M/EI diagram. The value of EIt_B can be evaluated in a similar manner by taking moments about B of the area of the moment diagram between A and B.

$$EIt_B = \frac{Pab}{L}\left(\frac{a}{2} \times \frac{a}{3}\right) = \frac{Pa^3b}{6L}$$

The summation of Δ and t_B is found by a proportionality of similar triangles.

$$\Delta + t_B = \frac{a}{L} t_C = \frac{Pa^2b}{6EIL^2} (a^2 + 3ab + 2b^2)$$

Since t_B has already been evaluated,

$$\Delta = \frac{a}{L} t_C - t_B = \frac{Pa^2b^2}{3EIL}$$

In order to determine the maximum deflection of a beam loaded as shown in Fig. 10.3a, it is first necessary to locate the point on the beam where the maximum deflection occurs. This can be determined from a consideration of the change in slope along the beam. The tangent to the elastic line at the point of maximum moment is, of course, horizontal. The slope of the elastic line at A is equal to t_C/L. This is also the value of the angle formed by the tangents to the elastic line at A and the point of maximum deflection. Using the first moment-area principle,

$$\frac{t_C}{L} = \text{area of the } M/EI \text{ diagram between } A \text{ and the point}$$
of maximum deflection

Assuming the point of maximum deflection is a distance x to the right of the point of load, the following expression can be written:

$$\frac{t_C}{L} = \frac{Pab}{EIL}\frac{a}{2} + \frac{Pab}{EIL}\left(x - \frac{x^2}{b}\right) + \frac{Pab}{EIL}\frac{x^2}{2b}$$

$$t_C = \frac{Pab}{EI}\left(\frac{a}{2} + x - \frac{x^2}{2b}\right)$$

$$x\left(1 - \frac{x}{2b}\right) = \frac{EIt_C}{Pab} - \frac{a}{2}$$

Solving for x gives the point of maximum moment. Once this point has been determined, the actual value of maximum deflection can proceed in the same manner as for determining the deflection at the point of load.

In applying the moment-area method to simple or continuous beams a particular manner of drawing the bending moment diagram will prove expedient in the determination of deflections. Instead of drawing the complete bending moment diagram as was shown in Fig. 10.3b, the moment diagram is drawn in parts. As an example, Fig. 10.4 shows a simple beam loaded with several loads. A separate moment diagram is drawn for each particular load. The moments are computed by moving along the beam from the right reaction to the left. Of course, the moment diagrams could also be developed by moving from the left reaction to the right. The

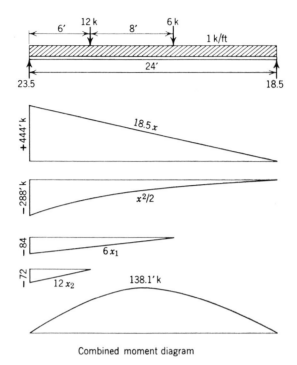

Combined moment diagram

Fig. 10.4 Moment diagram by parts.

determination of the areas and the moments of those areas is much easier when the moment diagram is drawn by parts. This is easily understood when a comparison is made with the combined moment diagram. Each moment diagram due to each separate load is in the shape of a simple right triangle or a parabola. Refer to Table 10.1 for areas and centroids of simple figures.

It is noted that in using the moment-area methods the slope determined is the *change* in slope of the tangents to the elastic line between two points on the beam, and the deflection determined is the deflection at one point on the elastic line from the *tangent to the elastic line at a second point*. However, in engineering practice the desired quantities are the slope measured from the unloaded position and the deflection of a point on the elastic line from its original position before loading.

The moment-area methods can be extended to develop procedures that are easier to use for some problems than the basic moment-area method just presented. This is especially true for simple spans unsymmetrically loaded or for continuous beams. These methods are developed in the next two sections.

Table 10.1

Shape		Area	$\bar{x}$	$\bar{y}$
Triangular		$\dfrac{bh}{2}$		$\dfrac{h}{3}$
Trapezoid		$\dfrac{b}{2}(h_1 + h_2)$	$\dfrac{h_1 + 2h_2}{3(h_1 + h_2)}$	$\dfrac{(h_1^2 + h_1 h_2 + h_2^2)}{3(h_1 + h_2)}$
Parabolic		$\dfrac{2bh}{3}$	$\dfrac{b}{2}$ *	$\dfrac{2h}{5}$
Segment of parabola		$\dfrac{2}{3}bh$	$\dfrac{b}{2}$	
General spandrel		$\dfrac{bh}{n+1}$	$\left(\dfrac{n+1}{n+2}\right)b$	$\left(\dfrac{n+1}{4n+2}\right)h$

* For a half parabola x is $3/16\ b$ from the centerline.

10.3   Elastic Load Method

The procedures developed in this section make it possible to determine the true slopes (true slope meaning rotation of the tangent from the unloaded position) and true deflections directly (true deflection meaning the deflection measured from the unloaded position). The deflections determined will be the deflections measured perpendicular to the axis of the beam and from the original unloaded position. Of course, incremental deflections due to incremental loads can also be found.

In Fig. 10.5a the elastic line of a beam supported at A and B is shown with the tangents to the elastic line at A and B drawn. The change in slope between A and B is $\theta_A + \theta_B$, and by the first moment-area principle this value is equal to the area of the M/EI diagram shown in part (b). The deflection of point B from the tangent at point A is the angle $\theta_A + \theta_B$

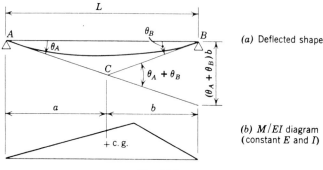

(a) Deflected shape

(b) M/EI diagram
(constant E and I)

Fig. 10.5

multiplied by the distance b. It should be remembered that the angle $\theta_A + \theta_B$ is so small that the intercept and the arc are considered equal. Since, by the second moment-area principle this tangential deflection is also equal to the moment of the M/EI diagram between A and B taken about point B, the distance b must be the distance from B to the centroid of the M/EI diagram. The angle θ_A, measured from the undeflected position of the beam, is equal to $(\theta_A + \theta_B)b/L$. However, as just stated $(\theta_A + \theta_B)$ is the area of the M/EI diagram between A and B. Therefore,

$$\theta_A = \left(\text{area } \frac{M}{EI} \text{ diagram}\right)_A^B \cdot \frac{b}{L}$$

or

$$\theta_A = \int_A^B \frac{M\,dx}{EI} \cdot \frac{b}{L}$$

If b is the distance from B to the centroid of the M/EI diagram, θ_A will be the numerical value of the reaction at A of a beam simply supported at A and B and loaded with the M/EI diagram. Similarly,

$$\theta_B = \left(\text{area } \frac{M}{EI} \text{ diagram}\right)_B^A \cdot \frac{a}{L}$$

or

$$\theta_B = \int_B^A \frac{M\,dx}{EI} \cdot \frac{a}{L}$$

and the true slope of the tangent at the right end is numerically equal to the right reaction of the simple beam loaded with the M/EI diagram. The value of the reaction is also the shear at the supports for the beam with the M/EI diagram as the loading. The M/EI diagram when used as a load is called an "elastic weight."

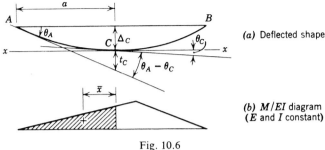

(a) Deflected shape

(b) M/EI diagram
(E and I constant)

Fig. 10.6

Figure 10.6 is a beam similar to the one of Fig. 10.5. It is a requirement to determine the slope θ_C of the tangent at point C. It can be written that $\theta_C = \theta_A - (\theta_A - \theta_C)$. It was previously proven that θ_A is equal to the reaction of the elastically loaded beam. The value $\theta_A - \theta_C$ is the change in slope between A and C, which is equal to the area of the M/EI diagram between A and C. If this area of the M/EI diagram is subtracted from the reaction of the elastically loaded beam, the result is the value of shear at point C of the elastically loaded beam. This then proves the following rule:

In order to determine the true slope of the elastic line at any point along a beam simply supported or continuous at both ends, the beam is considered simply supported and loaded with the M/EI diagram. The true slope is then numerically equal to the shear at the point in question of the elastically loaded beam.

Reference is again made to Fig. 10.6 for the next requirement is to determine the deflection of point C (Δ_C) from the unloaded position. The value of $\Delta_C + t_C$ is equal to the slope θ_A multiplied by the distance a. From moment-area principles the value of t_C is equal to the moment about point C of the area of the M/EI diagram between A and C. The value of Δ_C can then be written as

$$\Delta_C = \theta_A \cdot a - t_C = \theta_A \cdot a - (\text{area of } M/EI \text{ diagram})_A^C \cdot \bar{x}$$

where $\bar{x}$ is the distance from point C to the centroid of the area of the M/EI diagram between A and C.

In the preceding equation θ_A is the reaction at A of the elastically loaded beam. If this value is multiplied by the distance a, the first term in the equation is the moment about point C of the reaction of the elastically loaded beam. The second term is equal to the moment of the elastic load between A and C. The value of Δ_C is therefore equal to the bending moment at point C of the beam loaded with the M/EI diagram. The following rule can be stated with regard to the *true* deflection, where *true* deflection is defined as the deflection, normal to the axis of the beam, measured from the original unloaded position.

The true deflection at any point along the elastic line of a beam simply supported or continuous at both ends is numerically equal to the value of the bending moment of a simply supported beam loaded with the M/EI diagram.

As the basic given data at the start of the development were only the elastic line from *A* to *B*, and as this line could be representative of either a simple or continuous beam, these relationships with regard to slope and deflection are applicable for either a continuous or simple-span beam. Only two qualifications are in the two elastic-weight principles. The first is that the beam contains no hinges or pins between the supports. The second is that the supports do not deflect. If the supports deflect, the slopes and deflections must be measured from the final chord line connecting the points of support.

The advantage of the elastic-load method over the basic moment-area method is very apparent for continuous beams. Example Problem 10.1 illustrates the use of the elastic load method for a two-span continuous beam. The moment at the interior support is determined by the three-moment equation. The reactions of the beam produced by the real loading are then found. The elastic load is determined by drawing the moment diagram in parts (a separate diagram for the end reactions and each load). The reactions for the elastic loading are then found. Since *E* and *I* are constant, the elastic loading can in this case consist of only the moment diagram. Since the maximum deflection will be at a point of zero slope, the point of maximum deflection will be at the point of zero shear for the beam loaded with the elastic load. A simple algebraic expression can be written to locate the point of zero shear and thus the point of maximum deflection. Once the location of maximum deflection is known the value of the deflection is determined by computing the bending moment at that point for the beam loaded with the elastic loading. The advantage of the elastic-load method is that the continuous beam can be broken into a series of simple beams when the elastic load is applied. The computations give the maximum deflection in each span. There would also be an upward deflection in the second span, but this would be a very small value. A positive bending moment in accordance with the beam sign convention results in a downward elastic load, and a negative sign for the bending moment results in an upward elastic load. The sign convention for shear and bending moment due to the elastic load follows the usual beam convention. The resulting positive value of slope for a positive value of shear means a slope of the tangent that is down and to the right. A negative slope would be up and to the right. Positive values of bending moment due to the elastic load mean deflections down, and negative bending moments mean the deflections are up from the position of the chord connecting the ends of the beam.

EXAMPLE PROBLEM 10.1   Determine the maximum deflection in each span of the following beam. I is constant.

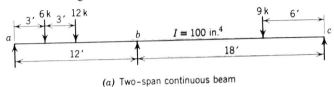

(a) Two-span continuous beam

$$2M_b(12 + 18) = -6 \times 12^2(0.25 - 0.25^3) - 12 \times 12^2(0.5 - 0.5^3)$$
$$-9 \times 18^2(0.33 - 0.33^3)$$
$$60M_b = -203 - 648 - 863$$
$$M_b = -28.56 \text{ ft-kip}$$

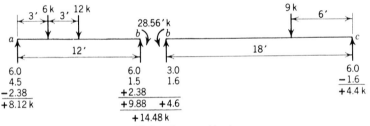

(b) Reactions - real loads

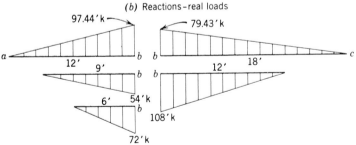

(c) Elastic loads (moment diagrams)

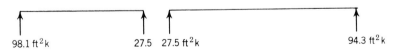

(d) Reactions from elastic loads

SPAN *ab*

$$V = 0 = 98.13 - \frac{97.44}{24} x^2 + \frac{54}{18} (x - 3)^2$$

$$x^2 + 16.98x = 118.05$$

$$x = 5.3 \text{ ft}$$

$$EI \, \Delta_{max} = 98.13 \times 5.3 + (2.3)^3 - \frac{8.12}{6} (5.3)^3$$

$$\Delta_{max} = \frac{330.8 \times 12^3}{30 \times 10^3 \times 100} = 0.19 \text{ in.}$$

SPAN *bc*

$$V = 0 = 94.3 - \frac{79.43}{36} x^2 + \frac{108}{24} (x - 6)^2$$

$$x^2 - 23.53x = -111.7$$

$$x = 6.59 \text{ ft or } 16.93 \text{ ft (max. at 6.59 ft)}$$

$$EI \, \Delta_{max} = 94.30 \times 6.59 - \frac{4.41}{6} (6.59)^3 + \frac{9}{6} (0.59)^3$$

$$\Delta_{max} = \frac{416.95 \times 12^3}{30 \times 10^3 \times 100} = 0.24 \text{ in.}$$

When the beam does not have a constant moment of inertia, such as the beam of Fig. 10.7, the moment diagram has to be divided by I to obtain the elastic load. E can be held in abeyance until the final step.

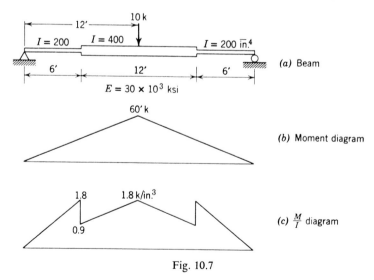

(a) Beam

(b) Moment diagram

(c) $\frac{M}{I}$ diagram

Fig. 10.7

Since the loading is symmetrical, the point of maximum deflection is at midspan. The reactions for the beam subjected to the elastic load is equal to one-half the area of the M/I diagram. This value is

$$R = 1.8 \times 3 + 0.9 \times 6 + 0.9 \times 3 = 13.5 \text{ ft-kips/in.}^3$$

The bending moment at midspan of the elastically loaded beam is

$$M = 13.5 \times 12 - 5.4 \times 8 - 5.4 \times 3 - 2.7 \times 2 = 97.2 \frac{\text{ft}^2\text{-kips}}{\text{in.}^3}$$

The deflection is then

$$\Delta = \frac{97.2 \times 12^2}{30 \times 10^3} = 0.466 \text{ in.}$$

10.4   Conjugate Beam Method

When beams have a hinge between supports, have cantilever ends, or have fixed supports, the elastic load method cannot be used without ramifications. A method has been developed whereby only the operations of computing shear and bending moment need be used to determine slopes and deflections. This method is called the conjugate beam method, *conjugate* meaning interchangeable. In this method a substitute beam is interchanged for the real beam after the bending moments in the real beam have been determined. The substitute beam has for its load the elastic load (M/EI diagram) as used in the elastic-load method. Its support conditions are sometimes changed with respect to the real beam. With the conjugate beam loaded and supported in the required manner, the following principle applies:

The shear in the conjugate beam at any location is equal to the slope of the real beam at the corresponding location, and the bending moment in the conjugate beam is equal to the corresponding deflection of the real beam.

The slopes and deflections are measured from the original unloaded position of the real beam. The span lengths between supports or hinges remain the same for both the conjugate and the real beam.

The support condition of the conjugate beam can be deduced by applying the conjugate beam principle to a real beam with any particular support condition. As a first example a simple end supported beam is used. At the support of the real beam the elastic line has a finite value of slope but no deflection. The conjugate beam must have shear but no bending moment at the corresponding location. In order to achieve this finite shear and zero moment condition, the conjugate beam must also have a simple support. Using a beam with one end connected to a rigid support as a second example, the real beam has no slope or deflection at this rigid support. Therefore the conjugate beam has both zero shear and bending moment at this end of the beam. The only way to achieve no shear or bending moment is to have a free end.

At an interior support for a continuous beam there must be only one value of slope directly above the support, and the deflection at this point must be zero. The shear in the conjugate beam at an infinitesimal distance both to the left and right of this corresponding point must be equal. The bending moment in the conjugate beam at the point of interior support in the real beam must be zero. These two requirements can only be met by a hinge (no support) in the conjugate beam. Other relationships between the supports of real and conjugate beams can be derived similarly. Table 10.2 can be used by the student as a check on his reasoning ability.

It should be observed that the conjugate beam is always statically determinate and at first sometimes appears, to be unstable. For instance, a real beam fixed at both ends would have as its conjugate beam a beam

Table 10.2 Real Beams and Corresponding Conjugate Beam

Real Beam	Conjugate Beam
Simply supported	Simply supported
Continuous	Hinges $\left(\begin{array}{c}\text{replace}\\\text{supports}\end{array}\right)$
Fixed and simply supported	Free and simply supported
Fixed and free	Free and fixed
Hinge	Interior support

with no supports. This beam would be in equilibrium under its elastic load only: Moments and shears for the conjugate beam are determined for the beam without specific supports.

The sign convention for slopes and deflections for the conjugate beam is the same as that given for the elastic load method; that is, plus shears result in slopes down and to the right, and plus bending moments result in downward deflections. Elastic loads (M/EI diagram) are downward if the bending moments in the real beam are positive.

A three-span continuous beam of constant section, with a load in the center span only, is used to show the application of the conjugate beam method. This application is shown in Example Problem 10.2. The positive and negative portions of the moment diagram are treated as two separate loadings and the results combined. It should be noted that the maximum upward deflections in the end spans occur at the same distance from the extreme reactions. The student should satisfy himself why this is so.

EXAMPLE PROBLEM 10.2   Determine the maximum deflection in each span of the three-span continuous beam of Fig. 10.8.

The loading and the reactions for the conjugate beam are shown in Fig. 10.8c. Since the maximum deflection is at the point of zero shear for the conjugate beam, the point of zero shear is first located.

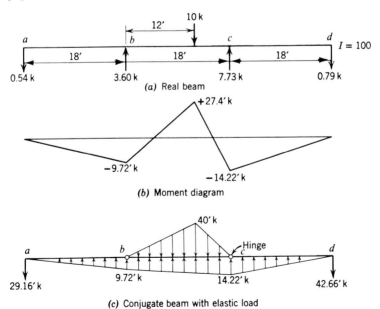

(a) Real beam

(b) Moment diagram

(c) Conjugate beam with elastic load

Fig. 10.8

DEFLECTION IN SPAN ab

$$V = 0 = 29.16 - \frac{9.72}{18} \cdot \frac{x^2}{2}$$

$$x^2 = 108$$

$$x = 10.4 \text{ ft to the right of } A$$

$$M = -29.16(10.4) + \frac{9.72}{108}(10.4)^3 = -202 \text{ ft}^3\text{-kips}$$

$$\Delta = -\frac{202 \times 12^3}{30 \times 10^3 \times 100} = -0.116 \text{ in. (up)}$$

DEFLECTION IN SPAN bc

$$V = 0 = 58.32 - \frac{3.33}{2}x^2 + 9.72x + \frac{4.5}{36}x^2$$

$$x^2 - 6.30x = 37.82$$

$$x = 10.06 \text{ ft to the right of } B$$

$$M = 58.32(10.06) - \frac{3.33}{6}(10.06)^3 + \frac{9.72}{2}(10.06)^2 + \frac{0.25}{6}(10.06)^3$$

$$= 555.3 \text{ ft}^3\text{-kips}$$

$$\Delta = \frac{555.3 \times 12^3}{30 \times 10^3 \times 100} = 0.32 \text{ in. (down)}$$

DEFLECTION IN SPAN cd

$$V = 0 = 42.66 - \frac{14.22}{36}x^2$$

$$x = 10.4 \text{ ft to the left of } D$$

$$M = -42.66(10.4) + \frac{14.22}{108}(10.4)^3 = -295.6 \text{ ft}^3\text{-kips}$$

$$\Delta = -\frac{295.6 \times 12^3}{30 \times 10^3 \times 100} = -0.17 \text{ in. (up)}$$

It would not have been necessary to determine the point of maximum moment in span CD since it had already been determined for the first span.

EXAMPLE PROBLEM 10.3 Determine the deflection and slope at the hinge and also the maximum deflection in the beam of Fig. 10.9.

The beam shown is statically indeterminate, and a solution by a method of analysis of statically indeterminate structures produced a shear at hinge

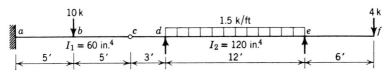

Fig. 10.9

c of 3.18 kips. The computation for deflections can then be simplified by dividing the beam into two separate parts.

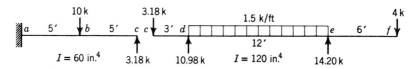

The conjugate beam of beam *ac* is a beam with a free end at *a* and a fixed end at *c*. The computations for slope and deflection in this beam are as follows:

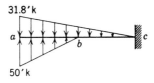

Conjugate beam *ac*

Slope at *c* is equal to the shear at *c* for the conjugate beam.

$$EI\theta_c = -31.8 \times 5 + 50 \times 2.5 = -34.0 \text{ ft}^2\text{-kips}$$

$$\theta_c = \frac{-34 \times 12^2}{30 \times 10^3 \times 60} = -0.0027 \text{ radian}$$

Slope of the tangent is up and to the right.

$$EI \Delta_c = -31.8 \times 5 \times 6.67 + 50 \times 2.5 \times 8.33 = -18.4 \text{ ft}^3\text{-kips}$$

$$\Delta_c = \frac{-18.4 \times 12^3}{30 \times 10^3 \times 60} = -0.0176 \text{ in. (up)}$$

The maximum deflection between *a* and *c* will occur where the shear on the conjugate beam is zero. It is assumed that the point of maximum deflection will be between *b* and *c*.

$$\text{Shear} = 34 - 3.18\left(\frac{x^2}{2}\right) = 0$$

$$x = 4.62 \text{ ft to left of } c$$

Moment at 4.62 ft to the left of c is

$$M = 18.4 + 34x - 3.18\left(\frac{x^3}{6}\right) = 86.4 \text{ ft}^3\text{-kips}$$

$$\Delta_{max} = \frac{86.4 \times 12^3}{30 \times 10^3 \times 60} = 0.083 \text{ in. (down)}$$

The conjugate beam for the *cf* portion of the beam consists of fixed supports at *c* and *f* and hinges at *d* and *e*. Since the loading between *d* and *e* is a uniform load, it is more convenient to draw the moment diagram by parts starting from *c* and moving to *f*. The elastic load then consists of four separate moment diagrams. These four loadings are shown below the conjugate beam. It should be remembered that a positive moment (by the standard beam sign convention) is a downward load and a negative moment an upward load.

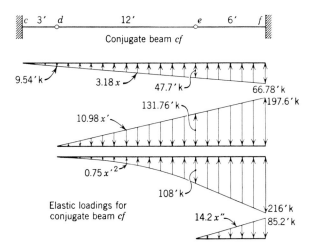

Shear at *d* and *e* of portion *de*

$$12V_d = -9.54 \times 12 \times 6 - 38.16 \times 6 \times 4 + 131.76 \times 6 \times 4$$

$$- 108 \times 4 \times 3$$

$$V_d = \frac{263.5}{12} = +21.96 \text{ ft}^2\text{-kips}$$

$$V_e = +6.8 \text{ ft}^2\text{-kips}$$

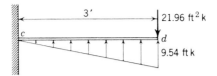

Shear at $c = 21.96 - 9.54 \times 1.5 = +7.65$ ft^2 kip

$$\theta_c = \frac{7.65 \times 12^2}{30 \times 10^3 \times 120} = 0.31 \times 10^{-3} \text{ radian}$$

It should be observed that the slope just to the left and right in the hinge at c is not the same.

Moment at $c = -21.96 \times 3 + 9.54 \times 1.5 \times 2 = -37.3$ ft^3-kips

$$\Delta_c = \frac{-37.3 \times 12^3}{30 \times 10^3 \times 120} = -0.018 \text{ in. (up)}$$

This checks the value of Δ_c as found from the left-hand portion of the beam.

Moment at $f = +6.8 \times 6 + 47.7 \times 6 \times 3 + 19.1 \times 3 \times 2$

$$- 131.8 \times 6 \times 3 - 65.8 \times 3 \times 2 + 108 \times 6 \times 3$$

$$+ 108 \times 3 \times 2 - 85.2 \times 3 \times 2 = 328 \text{ ft}^3\text{-kips}$$

$$\Delta f = \frac{328 \times 12^3}{30 \times 10^3 \times 120} = 0.16 \text{ in.}$$

The point of zero shear between d and e is found as follows:

$$V_x = 21.96 + 9.54x + 3.18\left(\frac{x^2}{2}\right) - 10.98\left(\frac{x^2}{2}\right) + 0.25x^3 = 0$$

$$x = 5.32 \text{ ft}$$

Moment at 5.32 ft. to right of point d is

$$M = 21.96 \times 5.32 + 4.77(5.32)^2 + 0.53(5.32)^3$$

$$- 1.83(5.32)^3 + 0.0625(5.32)^4 = 106.2 \text{ ft}^3\text{-kips}$$

$$\Delta_{\max} = \frac{106.2 \times 12^3}{30 \times 10^3 \times 120} = 0.051 \text{ in. (down)}$$

The maximum deflection is at point f.

10.5 Energy Methods

The energy concepts are very useful methods for determining the deflections of structures. In 1833 Clapeyron established the equality between the external work done by the loads deflecting a structure and the total strain energy produced in the structure.

External work = internal work (strain energy)

This relationship is the basis for several procedures of stress analysis that are presented in the following sections of this book.

When loads are applied to a structure, these loads move through a distance which is the deflection of the structure at the point of loading. If the reactions do not move then the total external work is the summation of the average magnitude of each of the loads multiplied by the respective deflection at the point of application of each load. A simple example will be the beam of Fig. 10.10 subject to a single concentrated load P. The relationship of the magnitude

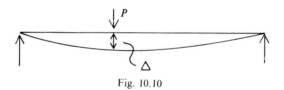

Fig. 10.10

of the load and the deflection Δ is shown by the following graph.

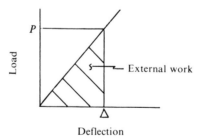

Deflection

Since the load-deflection relationship is a straight line, if all stresses in the structure are below the proportional limit, the external work is the area under the graph between the limits of P and $\triangle$. In equation form this is

$$\text{External work} = \frac{P\Delta}{2}$$

The internal work is the strain energy stored in the material of the structure when it deflects. This strain energy is recoverable when stresses do not exceed the elastic limit. Upon removing the loads the structure returns to its original unloaded position.

The evaluation of the strain energy, or internal work, for a beam can be determined by considering a small segment of the beam whose length is dx. In the length dx the

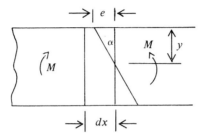

top fiber will shorten a distance e and the bottom fiber will lengthen this same amount (symmetrical section). The value of e will be the unit stress in the top fiber multiplied by the length dx and divided by E, the modulus of elasticity of the beam material.

$$e = \frac{\sigma}{E} dx = \frac{My}{EI} dx$$

As the beam is strained by bending, M will rotate through the angle α. This relationship of M and α is similar to P and $\triangle$.

Internal work $= \dfrac{M\alpha}{2}$

This internal work has to be the strain energy in the total length of beam L.

$$\text{Internal work} = \int_0^L \frac{M}{2} \left(\frac{Mdx}{EI} \right) = \frac{1}{2} \int_0^L \frac{M^2 dx}{EI}$$

If the energy relationship is now used

External work = internal work

$$\frac{P\Delta}{2} = \frac{1}{2}\int_0^L \frac{M^2 dx}{EI}$$

then the deflection Δ is

$$\Delta = \frac{1}{P}\int_0^L \frac{M^2 dx}{EI} \tag{10.3}$$

It should be remembered that equation 10.3 is only the deflection due to bending strain. There will be an additional deflection due to shear strain, but for most actual beams this shear strain deflection is very small and can be neglected. The internal work for shear is

$$\text{Internal work} = K\int_0^L \frac{V^2 dx}{2AG}$$

where K is a constant depending on the shape of the cross-section.

$$K = 1.2 \text{ for rectangular sections}$$

$$= 1.0 \text{ for } I \text{ shaped sections}$$

$$= 1.1 \text{ for circular sections}$$

V is the beam shear at any location along the beam, A is the cross-section area, and G is the shear modulus.

This method will now be used to determine the deflection of a simple beam with a concentrated load at the mid-span.

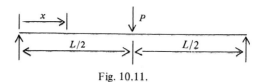

Fig. 10.11.

The internal work can be found when the equation for the bending moment at all points along the beam is written. For this case

$$M = \frac{P}{2}(x)$$

then

$$\Delta = \frac{2}{PEI}\int_0^{L/2}\left(\frac{P}{2}x\right)^2 dx = \frac{P}{2EI}\int_0^{L/2} x^2 dx = \frac{P}{2EI}\left[\frac{x^3}{3}\right]_0^{L/2} = \frac{PL^3}{48EI}$$

for a steel beam with $I = 200$; $L = 20$ ft; and $P = 6$ kips

$$\Delta = \frac{6(20)^3(12)^3}{48(3 \times 10^3)(200)} = 0.29 \text{ in.}$$

If it is desired to check the deflection due to shear strain the calculation for a beam with an I-shaped cross-section of 8.0 square inches is

$$V = \frac{P}{2}$$

$$\Delta = \frac{2}{PAG}\int_0^{L/2}\left(\frac{P}{2}\right)^2 dx = \frac{P}{2AG}\int_0^{L/2} dx = \frac{P}{2AG}[x]_0^{L/2}$$

$$= \frac{PL}{4AG} = \frac{6(20)(12)}{4(8)(12 \times 10^3)} = 0.004 \text{ in.}$$

The deflection due to shear strain is less than two percent of the total deflection for this particular beam.

The method just presented is very useable and straightforward for the case where only one concentrated load is applied and the desired deflection is at the point of application of the load. For multiple loads or required deflections at locations other than the point of application of the loads then the equation for external work will contain more than one unknown and the single energy equation is not sufficient.

The method just presented can be used also for truss type structures. The internal strain energy can easily be proven to be

$$\text{Internal work} = \sum \frac{S^2 L}{2AE}$$

where S is the force in any bar due to the load P and L/A are the dimensions of the member.

In 1879 the noted engineer Alberto Castigliano published a book containing two theorems pertaining to structural analysis. His second theorem will be presented here. His first theorem is not developed in this text because of its limited usefulness. It can be found in more comprehensive texts (1).

The second theorem provides a method of determining the elastic deflection of structures. It is stated as follows:

The displacement component of the point of application of an action on a structure, in the direction of that action, can be found by evaluating the first partial derivative of the total strain energy of the structure with respect to the applied action.

The action referred to in the theorem can be either a force or a couple and the displacement component will be a deflection or a rotation, respectively.

This theorem can be applied to any kind of a structure. The deflection due to axial forces, shear forces, bending moment, or torsional moments can be determined. The expressions for strain energy (internal work) as developed previously in this section are used in this method. The method may appear to have limitations in that the displacements cannot be determined at points on the structure where an action is not applied. However, the method can be modified to determine deflections at points where a load is not applied. The result is the use of a dummy load similar to the virtual work method that is to follow. (Section 10.6.)

The mathematical proof of the second theorem of Castigliano is developed by considering a beam acted upon by two concentrated loads, Fig. 10.12. These two loads produce deflections Δ_1 and Δ_2 at the point of application of these loads. If the loads are applied gradually the external work can be expressed as the average magnitude of the load times its displacement

$$\text{External work} = \frac{P_1\Delta_1}{2} + \frac{P_2\Delta_2}{2} \tag{10.4}$$

Now consider that a small additional load dP_1 is added to the load P_1. There will be small additional deflection produced at the point of application of the loads. This second condition is shown in Fig. 10.13.

There will now be an additional increment of strain energy due to the additional load dP_1. This additional increment of strain energy will be equal to the increment of external work. The increase in external work will be the sum of the products of the average load acting through the increased deflection. This can be stated mathematically.

$$dw = P_1 d\Delta_1 + \frac{dP_1}{2}(d\Delta_1) + P_2 d\Delta_2$$

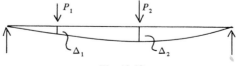

Fig. 10.12

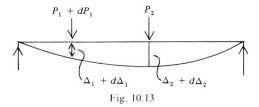

Fig. 10.13

the second term is a product of two differentials and can be neglected. The equation for incremental work can now be

$$dw = P_1 d\Delta_1 + P_2 d\Delta_2. \tag{10.5}$$

It can be stated mathematically

$$dw = \frac{\partial w}{\partial P_1}(dP_1)$$

and

$$d\Delta_1 = \frac{\partial \Delta_1}{\partial P_1}(dP_1); \qquad d\Delta_2 = \frac{\partial \Delta_2}{\partial P_1}(dP_1)$$

If these partial derivative expressions are substituted into equation 10.5, the resulting equation is obtained.

$$\frac{\partial w}{\partial P_1} \cdot dP_1 = P_1\left(\frac{\partial \Delta_1}{\partial P_1}\right)dP_1 + P_2\left(\frac{\partial \Delta_2}{\partial P_1}\right)dP_1$$

if the term dP_1 is divided out then

$$\frac{\partial w}{\partial P_1} = P_1\left(\frac{\partial \Delta_1}{\partial P_1}\right) + P_2\left(\frac{\partial \Delta_2}{\partial P_1}\right) \tag{10.6}$$

The first equation, 10.4, equating the total external work to the total internal strain energy, was

$$w = \frac{P_1\Delta_1}{2} + \frac{P_2\Delta_2}{2}$$

If this equation is differentiated with respect to P_1 then

$$2\left(\frac{\partial w}{\partial P_1}\right) = \Delta_1 + P_1\left(\frac{\partial \Delta_1}{\partial P_1}\right) + P\left(\frac{\partial \Delta_2}{\partial P_1}\right) \tag{10.7}$$

If equation 10.6 is subtracted from equation 10.7 then the resulting equation is

$$\frac{\partial w}{\partial P_1} = \Delta_1 \qquad (10.8)$$

Since w can be the internal strain energy, Equation 10.8 states that the deflection at load point 1 due to any set of loads on a structure is equal to the partial derivative of the strain energy with respect to the load applied at point 1. This is Castigliano's second theorem.

A simple example of this method is the cantilever beam of Fig. 10.14.

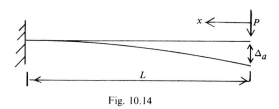

Fig. 10.14

The bending strain energy in a beam is

$$w = \int_0^L \frac{M^2\, dx}{2EI}$$

The value of M anywhere along the beam is Px, therefore

$$w = \frac{1}{2EI} \int_0^L (Px)^2\, dx = \frac{P^2 L^3}{6EI}$$

$$\Delta_a = \frac{\partial w}{\partial P} = \frac{PL^3}{3EI}$$

In the above procedure the integration was performed before the differentiation. When there are multiple loads on a structure it may be better to perform the differentiation under the integral. The resulting mathematical expression of Castigliano's second theorem is then

$$\Delta = \int \frac{M \dfrac{\partial M}{\partial P}}{EI}\, dx$$

The following example problem illustrates this procedure.

EXAMPLE PROBLEM 10.4 Determine the deflection at the 4 kip load for the beam shown below. *EI* is constant and equals 30×10^3 *k-in*^{2}

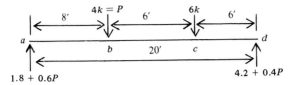

The first step is to write the mathematical expressions for bending moment at all locations along the beam. If deflection due to shear strains were required, the equations for shear would also be necessary.

The 4 kip load will be designated as *P*. The reaction at *a* will then be $1.8 + 0.6P$ and the reaction at *d* will be $4.2 + 0.4P$.

Section	$x = 0$ at	M	$\partial M/\partial P$
ab	*a*	$(1.8 + 0.6P)x$	$+0.6x$
bc	*a*	$(1.8 + 0.6P)x - P(x - 8)$	$+0.6x - x + 8$
dc	*d*	$(4.2 + 0.4P)x$	$+0.4x$

The value of $P = 4k$ can now be substituted

$$EI\Delta_b = \int M\left(\frac{\partial M}{\partial P}\right)dx = \int_0^8 (4.2x)(0.6x)dx + \int_8^{14}(32 + 0.2x)(0.4x + 8)dx$$

$$+ \int_0^6 (5.8x)(0.4x)dx$$

$$= \int_0^8 2.52x^2 dx + \int_8^{14}(-0.08x^2 - 11.2x + 256)dx + \int_0^6 2.32x^2 dx$$

$$= [0.84x^3]_0^8 + [-0.0267x^3 - 5.6x^2 + 256]_8^{14} + [0.773x^3]_0^6$$

$$= 1334.3 \text{ ft}^3\text{-k}$$

$$\Delta_b = \frac{1334.3(12)^3}{30 \times 10^5} = 0.77 \text{ inches}$$

If there were no load at point *b*, yet it was necessary to calculate the deflection at *b*, the same procedure could be followed through the above table. Instead of substituting the value of 4 kips for P the value of P would be zero. The deflection at b would then be;

$$EI\Delta_b = \int_0^8 (1.8x)(0.6x)dx + \int_8^{14} (1.8x)(-0.4x + 8)dx + \int_0^6 (4.2x)(0.4x)dx$$

$$= [0.36x^3]_0^8 + [-0.24x^3 + 7.2x^2]_8^{14} + [0.56x^3]_0^6$$

$$= 720 \text{ ft}^3\text{-k}$$

$$\Delta_b = \frac{720(12)^3}{30 \times 10^5} = 0.41 \text{ inches}$$

10.6   The Method of Virtual Work

In 1717 Johann Bernoulli presented the principle of virtual velocities. This led to the principle of virtual work and the development of a very useful method of determining deflections in structures. This method can be applied to trussed structures, beams, or frames. Either linear or angular deflections can be determined as a result of loads in any direction or resulting from temperature changes, shrinkage, or generally to any measurable cause.

The principle of virtual work can be stated:

If a deformable structure in equilibrium under a given known load or set of loads is subject to a virtual displacement as the result of an additional action, the resulting external work of the given load or loads due to the virtual displacement is equal to the resulting internal work of the stresses caused by the given load or set of loads.

The term *virtual displacement* means that the action producing the displacement is independent of the system of loads acting on the structure just before the virtual displacement takes place. It should be remembered that the original given loads remain on the structure during the virtual displacement, and the external and internal virtual work is *only* the work produced by the original given loads and forces. In application, the original given load system is most always a unit load applied to the structure at the required point of deflection and acting in the direction of the required deflection. Because of the use of the unit load as the original given force acting on the structure, the virtual work method is sometimes referred to as the "unit load method" or the "dummy load method."

The method can be best introduced by referring to the pin-jointed trussed structure of Fig. 10.15. In (a) a unit load is applied to the middle lower panel point. This is the original given load system. This unit load will induce forces in all the members of the truss. The force in each member due to this unit load will be designated by the letter u. Figure 10.15 shows the same structure with the additional loads P_1 and P_2 applied. The dashed lines show the deflected position of the structure, due to loads P_1 and P_2, from the original position when only the unit load was acting. Panel point c has deflected a distance Δ_c in the direction of the unit load. The application of the loads P_1 and P_2 is the action causing the virtual displacement of the structure. The external work due to the virtual displacement is the unit load multiplied by the deflection (Δ_c).

In order to evaluate the internal virtual work, attention is given to any member of the truss.

Figure 10.16 shows one of the members of the truss. With the original unit load applied to the truss the force in the member is u. With the additional application of the loads P_1 and P_2 each member of the truss will be subjected to an elongation or shortening. This change in length of the member is SL/AE, where S is the force in the member due to the loads P_1 and P_2. The value L is the original length of the member, and A and E are the area and modulus of elasticity respectively. If the member changes length an amount SL/AE, the work done by the original force of u acting on the member will be equal to $SL(u)/AE$. This quantity is the internal virtual work of the particular member. The total internal virtual work will

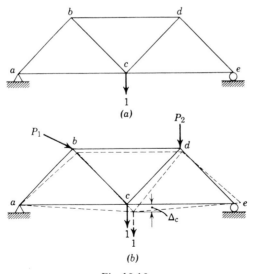

Fig. 10.15

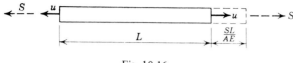

Fig. 10.16

be $\Sigma(SuL/AE)$ for the entire truss. The virtual work principle can then be written

External virtual work = internal virtual work.

$$1 \times \Delta = \frac{SuL}{AE}$$

The equation for deflection at any point on a truss is then

$$\Delta = \Sigma \frac{SuL}{AE} \tag{10.9}$$

where S = force in each member due to the applied loads.

      u = force in each member due to a unit load applied at and in the direction of the required deflection.

      L = length of each member.

      A = area of each member.

      E = modulus of elasticity of each member.

Equation 10.3 is easy to apply and is best handled by tabulating the value of each term for each member of the truss.

A cantilever truss is shown in Fig. 10.17. It is required to find the vertical deflection of the end of the truss produced by the given loads. The first step is to determine the u force in each member due to a unit load acting vertically at the required point of deflection. These u forces are then tabulated for each member. The next step is to determine each bar force

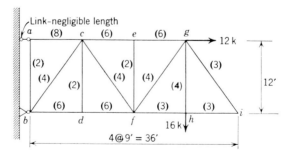

Areas are shown in ( ) sq. in.

$E = 10.6 \times 10^3$ ksi for all members

Fig. 10.17

due to the given set of loads. This value is also tabulated. A listing of the length of each member and its cross-sectional area is then made. If E is constant for all members, there is a saving of time by first performing the arithmetical operation of SuL/A. Then, after obtaining the sum of this quantity for all the members, divide the resultant sum by E. The tabular solution for the truss of Fig. 10.17 is shown in Table 10.3.

The application of the virtual work method to members in which the deflection is caused primarily by bending can also be made. The external virtual work is again equated against the internal virtual work. The external virtual work is the same as it was for the truss; that is, the unit load multiplied by the deflection caused by the system of applied loads. The internal virtual work can be evaluated by considering that the moment at any point on the member caused by the unit load at the required point of deflection is m. At any section of a member subject to pure bending, the stress in the extreme fiber due to the applied loads can be readily determined by the flexure formula, $\sigma = My/I$. The elongation of the extreme fiber of length dx is $\sigma\,dx/E$ or $My\,dx/EI$. A plane section perpendicular to the axis of the member rotates through the angle $M\,dx/EI$ upon application of a bending moment M (see Fig. 10.1). The work done by a moment is equal to the magnitude of the moment multiplied by the angle through which the moment moves. The value of the internal work due to the virtual

Table 10.3

Member	u	S	L	A	$\dfrac{SuL}{A}$
			Feet	Inches^2	
ab	0	0	12	2	0
ac	+3.00	+48.00	9	8	+162.0
cg	+1.50	+24.00	18	6	+108.0
bf	−2.25	−24.00	18	6	+162.0
fi	−0.75	0	18	3	0
bc	−1.25	−20.00	15	4	+93.75
cd	0	0	12	2	0
cf	+1.25	+20.00	15	4	+93.75
ef	0	0	12	2	0
fg	−1.25	−20.00	15	4	+93.75
gh	0	+16.0	12	4	0
gi	+1.25	0	15	3	0

$$= 713.25 \text{ kip-ft/in}^2.$$

$$\Delta_i = \frac{713.25 \times 12}{10.6 \times 10^3} = 0.807 \text{ in.}$$

rotation of the section of the beam of dx length is $mM\,dx/EI$. Therefore, the relationship for the entire member is

$$1 \times \Delta = \int_0^L \frac{mM\,dx}{EI}$$

The deflection due to a bending moment is then

$$\Delta = \int_0^L \frac{mM\,dx}{EI} \tag{10.10}$$

Equation 10.10 is applied in like manner as Eq. 10.9. A unit load is placed on the member at the point and in the direction of the required deflection. An equation or equations of moment due to the unit load and the applied loads are written for all points along the length of the member. Similarly, equations are written for the bending moments throughout the entire length of the member for the applied loads. The integration as noted in Eq. 10.10 is then performed.

It is required to find the maximum deflection in a uniformly loaded beam of constant E and I and of length L. If the intensity of the load is w pounds per unit length, the equation for moment at any point x along the beam is

$$M = \frac{wL}{2}x - \frac{wx^2}{2} = \frac{w}{2}(Lx - x^2) \qquad \text{(see Fig. 10.18)}$$

Because of symmetry the maximum deflection is obviously at the center of the span. Therefore, the unit load is placed at the midspan. The expression for moment from the end of the span to the midpoint due to the unit load is $m = x/2$. The maximum deflection is determined as follows:

$$EI\,\Delta = \int_0^L Mm\,dx = 2\int_0^{L/2} \frac{w}{2}(Lx - x^2)\frac{x}{2}\,dx$$

$$EI\,\Delta = \frac{w}{2}\int_0^{L/2}(Lx^2 - x^3)\,dx = \frac{w}{2}\left(\frac{Lx^3}{3} - \frac{x^4}{4}\right)_0^{L/2}$$

$$= \frac{w}{2}\left(\frac{5L^4}{192}\right) = \frac{5wL^4}{384}$$

$$\Delta = \frac{5wL^4}{384EI}$$

The student should verify this equation for deflection by applying the moment-area or elastic weight principles.

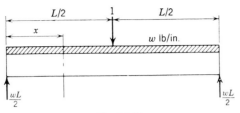

Fig. 10.18

Figure 10.19 shows a beam with two concentrated loads. The deflection at midspan is required. Since a continuous expression for the moment from one end of the beam to the other cannot be written, it is necessary to write the expressions for moments between the loads. The unit load, of course, is placed at the midspan.

$x = 0$ to $x = 7$	$x = 7$ to $x = 15$	$x = 15$ to $x = 20$	$x = 20$ to $x = 30$
$m = \dfrac{x}{2}$	$m = \dfrac{x}{2}$	$m = 15 - \dfrac{x}{2}$	$m = 15 - \dfrac{x}{2}$
$M = 6.3x$	$M = 21.0 + 3.3x$	$M = 21.0 + 3.3x$	$M = 261 - 8.7x$

$$EI\Delta = \int_0^7 3.15x^2\,dx + \int_7^{15}(10.5x + 1.65x^2)\,dx + \int_{15}^{20}(315 + 39x - 1.65x^2)\,dx$$

$$+ \int_{20}^{30}(3915 - 261x + 4.35x^2)\,dx$$

$$EI\,\Delta = \left(1.05x^3\right)_0^7 + \left(5.25x^2 + 0.55x^3\right)_7^{15} + \left(315x + 19.5x^2 - 0.55x^3\right)_{15}^{20}$$

$$+ \left(3915x - 130.5x^2 + 1.45x^3\right)_{20}^{30}$$

$$= 6846 \text{ ft}^3 \text{ kips}$$

$$\Delta = \frac{6846 \times 1728}{30 \times 10^3 \times 500} = 0.79 \text{ in.}$$

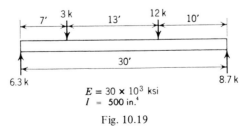

$$E = 30 \times 10^3 \text{ ksi}$$
$$I = 500 \text{ in.}^4$$

Fig. 10.19

In these deflection computations the unit load was placed at the midspan. However, this position would not be the point of maximum deflection for the unsymmetrical loading. The virtual work method is not as readily adaptable for determining the maximum deflection for unsymmetrically loaded beams as the conjugate beam method.

As previously stated, the virtual work method can be used to determine deflections caused by shearing stress, torsional stresses, or temperature. Then, the solution involves the determination of the internal virtual work due to these causes. In the determination of deflections due to shearing strains the internal virtual work is the shearing force at any section, designated v, due to the unit load applied at the required point of deflection, multiplied by the shearing strain in the direction of v produced by the applied loads. This resulting internal work can be proven[1] to be equal to

$$\text{Internal virtual work} = K \int_0^L \frac{vV}{AG}\,dx$$

and the deflection is

$$\Delta = K \int_0^L \frac{vV}{AG}\,dx \tag{10.11}$$

where K is a form factor equal to 1.2 for rectangular cross sections and 1.0 for WF or I beams. V is the shear due to the applied loads. The term A is the cross-sectional area of the member, except that for WF or I beams it is the area of the web. G is the shear modulus.

It is of interest to compare the deflection of a beam caused by moment to that caused by shear. As an example a simply supported beam of 20-ft span carries a uniform load of 4 kips per foot. If the beam is a $W14 \times 87$ steel section the deflection due to moment is

$$\Delta m = \frac{5wL^4}{384EI} = \frac{5 \times 4 \times (20)^4 \times (12)^3}{384 \times 30 \times 10^3 \times 967} = 0.498 \text{ in.}$$

The deflection due to shear is

$$\Delta_v = 2 \int_0^{L/2} \frac{vV}{AG}\,dx$$

$$v = \tfrac{1}{2} \qquad V = \frac{wL}{2} - wx$$

$$\Delta_v = \frac{1}{AG} \int_0^{L/2} w\left(\frac{L}{2} - x\right) dx = \frac{wL^2}{8AG}$$

$A = 5.88$ sq in. $G = 12 \times 10^6$ lb per square inch

$$\Delta_v = \frac{4 \times (20)^2 \times 12}{8 \times 5.88 \times 12 \times 10^3} = 0.034 \text{ in.}$$

It is seen that the ratio of shearing strain deflection to flexural strain deflection is 0.068. The smaller the span to depth ratio the larger will be the ratio of shear deflection to bending deflection. However, usually the ratio is small and the deflection due to shearing strains can be neglected.

The virtual work method is easily adapted to determining deflections of beams due to a uniform difference in temperature between the top and bottom of a beam. If there is a uniform change in temperature from top to bottom, there will be a difference in elongation of fibers due to the thermal expansion or contraction. This difference in a distance dx along the beam is

$$\Delta_e = \varepsilon \, \Delta t \, dx$$

where ε is the coefficient of thermal expansion and Δt is the temperature difference between the top and bottom of the beam.

The rotation of a plane section due to this temperature difference is the value Δ_e divided by the depth of the beam. The internal virtual work is then the moment due to the unit load multiplied by the angle this moment rotates through as a result of the temperature difference.

$$\text{Internal virtual work} = \int_0^L \frac{m \varepsilon \, \Delta t}{d} \, dx$$

and the deflection is

$$\Delta = \varepsilon \int_0^L \frac{m \, \Delta t}{d} \, dx \tag{10.12}$$

It is required to determine the deflection at midspan of a 16-ft simply supported steel beam of 15 in. in depth due to a temperature difference of $40°$ F between the top and bottom of the beam. The coefficient of thermal expansion for steel is 6.5×10^{-6} in. per inch per degree Fahrenheit.

With the unit load at midspan the moment between the end and the midspan due to the unit load is $m = x/2$.

The deflection is

$$\Delta = 2\varepsilon \int_0^{L/2} \frac{m \, \Delta t}{d} \, dx$$

$$= \frac{2 \times 6.5 \times 10^{-6} \times 40}{2 \times 15} \int_0^{96} x \, dx$$

$$= 8.67 \times 10^{-6} (96)^2 = 0.080 \text{ in.}$$

10.7   The Williot-Mohr Method

In 1877 the French engineer Williot developed a graphical method for determining the deflections of a structure where all the members are two force members. The original method as developed by Williot is quite limited, but when used with the supplement to the method as proposed in 1887 by the German, Otto Mohr, it becomes a very useful tool for determining the deflections of all the joints of a truss. If the deflection is only required for one joint, the method of virtual work is usually the most expedient. However, if the deflections of all the joints, or even several joints, are required, the Williot-Mohr method usually offers a time-saving advantage. Sufficient accuracy for almost all practical applications can be obtained with a moderate size diagram. This method, like all graphical methods, requires care in scaling and drawing lines.

When a bar is subjected to an axial load, the total strain in the bar is equal to SL/AE, where S is the force in the bar, L is the total length, and A and E are the cross-sectional area and modulus of elasticity of the bar respectively. The change in length of all the members of a truss, therefore, can easily be determined once all the bar forces are found. From these changes in length the deflected shape of the structure can be obtained. For instance, if a two-bar truss as shown in Fig. 10.20a is subjected to a concentrated load at the point of intersection of the bars, the change in length of each bar is determined by applying the equation SL/AE to each bar. Bar *AB* elongates and bar *BC* shortens. Figure 10.20b shows by the heavy line segments the exaggerated change in length of these two members. The change in length of the members of a truss is so small that they certainly would not be visible on a scaled drawing of the entire structure. For instance, a 60-ft steel member stressed to 20,000 psi would only change length by 0.4 in. By referring again to Fig. 10.20b, if the changes of length took place as shown, the members would rotate about their ends *A* and *C* until the point *B* reached the position of *B'*. If instead of striking arcs as shown in (b), perpendiculars were drawn from the stressed ends of

(a)

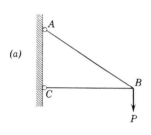

(b)

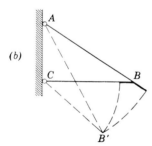

(c)

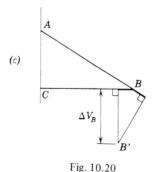

Fig. 10.20

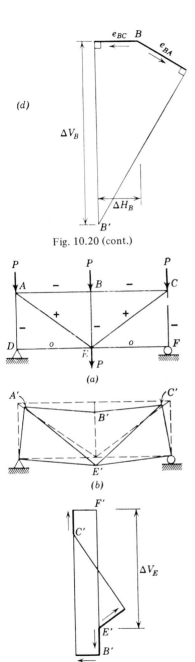

Fig. 10.20 (cont.)

(a)

(b)

(c)

Fig. 10.21

the members as shown in part (c), the new position of B' would be as shown. In Fig. 10.20b and c, the two locations of B' appear to be separated by a measurable quantity. However, in an actual structure the members AB and BC would not rotate through more than a few minutes of arc. Since the tangent of an angle and the measure of the angle itself are equal within five-place accuracy for angles up to one degree, the location of B' obtained by drawing perpendiculars is sufficiently accurate for all engineering purposes. Figure 10.20d shows the enlargement of the change in lengths of the members. This diagram is called the Williot diagram. Drawing perpendiculars from the ends of e_{BC} and e_{BA} gives the new location of B (point B'). Measuring the vertical and horizontal displacement of B' from B gives the respective deflections of point B of the actual structure. The Williot diagram as shown in (d) only utilizes the change in lengths of the members. It is obvious that it is not practical to draw a diagram that would combine the length of the member as well as its change in length drawn to the same scale.

The truss of Fig. 10.21 is considered symmetrical about the midspan with respect to shape, size of members, and loading. It is necessary, as will be shown, to start the Williot diagram with a member that does not rotate. Figure 10.21b shows the exaggerated deflected shape of the structure. If the structure is symmetrical, member BE does not rotate. The Williot diagram as shown in c starts with this nonrotating member. Since member BE is in compression, point B moves down with respect to point E. The point B' is then located directly below point E' by drawing a vertical line $E'B'$ representing, to some convenient scale, the change in length of member EB due to the applied loads. The next objective is to locate point C'. Member EC increases in length; hence, point C moves away from point E (up and to the right). The change in length of EC is then drawn in the Williot diagram to the same selected scale. A perpendicular is then drawn from the end of the graphical representation of the change in length of EC. Point C is also allied with point B through member BC. Since member BC is in compression, it shortens and point C moves to the left towards B. In the Williot diagram a line is drawn horizontally to the left from B' equal to the change in length of BC. From the end of this line a perpendicular is drawn. The intersection of this perpendicular and the previous one drawn from the line representing the total strain in EC locates the point C'. It is now only necessary to find the point F' to complete the diagram. Since there is no force in member EF, point F does not move laterally with respect to E, but it rotates (moves vertically). From E' a vertical line is then drawn in the Williot diagram. Joint F is connected to joint C by the bar CF. This bar is in compression; therefore, F moves vertically upward toward C. In actuality point C moves, not point F. This is immaterial in the Williot diagram as movements are relative. To locate point F' the total strain in CF is represented by a heavy vertical line from C' and a perpendicular is then drawn from the end of this line. The intersection of this perpendicular and the vertical line drawn from E' locates F'. In the actual structure panel point F does not move, so the deflections from the unloaded position, of panel points B, C, and E, can be determined by measuring from point F' in the Williot diagram. The vertical deflection of panel point E is the measured quantity ΔV_E as shown in Fig. 10.21c. Horizontal or vertical displacement components or the total displacement can readily be found from the Williot diagram. The procedure is simple, quick, and accurate.

If the truss in Fig. 10.21 were not symmetrical or had not been loaded symmetrically, then member BE would not have remained vertical. The amount of rotation would be indeterminate, and in drawing the displacement diagram the location of B' would have been unknown. It is apparent that if the line $E'B'$ were rotated about an axis through E', the resulting vertical and horizontal displacements would be changed from the previous values.

A symmetrical truss is shown in Fig. 10.22 with an unsymmetrical load. The vertical deflections of the panel points B, D, F, H, and J are required. Since the deflections at several points are required, the Williot-Mohr method is the best approach. The first step is to determine the bar forces

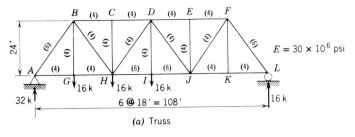

(a) Truss

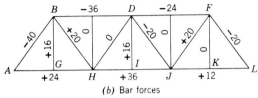

(b) Bar forces

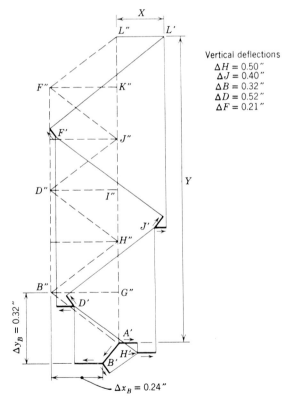

Vertical deflections
$\Delta H = 0.50''$
$\Delta J = 0.40''$
$\Delta B = 0.32''$
$\Delta D = 0.52''$
$\Delta F = 0.21''$

(c) Williot-Mohr diagram

Fig. 10.22

for all the members due to the given loads. These bar forces are shown in Fig. 10.22*b*. The next step is to calculate the total strain in each bar. Table 10.4 shows the necessary calculations.

It is obvious that all the members of the pin-connected truss will rotate to some extent about their pins. Since this is so, it is not possible at this time to locate the final position of the second point in the Williot diagram. The location of the first point is, of course, arbitrarily chosen. The procedure is to start the diagram at one of the points of support— this is usually at the end of the structure, but not necessarily so. The Williot diagram is then started by assuming that one of the members whose one end is common to the support point does not rotate. In the Williot diagram, Fig. 10.22*c*, the member *AB* is assumed not to rotate. The Williot diagram then proceeds in the usual manner. The movement of *B* with respect to *A* is indicated by the heavy line *A'B'*. Since member *AB* is in compression, joint *B* moves down and to the left. Therefore, point *B'* is located down and to the left a distance 0.080 in. multiplied by the magnification factor. Since member *AB* is assumed not to rotate, the line *A'B'* is drawn parallel to *AB*. The location of *H'* can now be determined by finding the movement of *H* with respect to *A* and *B*. This can be done since the movement of *H* with respect to *A* and *B* is determined by the change in lengths and rotations of members *AH* and *BH*. Member *AH* is in tension, so *H* will move to the right with respect to *A* an amount equal to the total strain in member *AH*. In the Williot diagram, the strain in *AH* is represented by the heavy line drawn to the right from *A'*. A perpendicular is then drawn from the end of this line.

Table 10.4

Bar	Force = S	L	A	SL/AE
	kips	feet	square inches	inches
AB	−40	30	6	−0.080
BD	−36	36	4	−0.130
DF	−24	36	4	−0.086
FL	−20	30	6	−0.040
AH	+24	36	4	+0.086
HJ	+36	36	6	+0.086
JL	+12	36	4	+0.043
GB	+16	24	4	+0.038
BH	+20	30	4	+0.060
DI	+16	24	4	+0.038
DJ	−20	30	4	−0.060
JF	+20	30	4	+0.060

The strain in *BH* is represented by the heavy line drawn down and to the right from *B'* (parallel to member *BH*). A perpendicular drawn from the end of this line intersects the previous perpendicular locating *H'*. The next step in the procedure is to locate point *D'* in the Williot diagram. This can be done by determining the movement of *D* with respect to *B* and *H*. From *B'* a line is drawn to the left (movement of *D* with respect to *B*), whose length is scaled to the total strain in *BD*, and a perpendicular is then drawn. Since there is no force in *HD*, there is no change in length of this member. Therefore, a line is drawn from *H'* that is perpendicular to the slope of member *HD*. In essence, this is the same procedure as before since the line in the Williot diagram representing the total strain in *HD* is of zero length. The intersection of the two perpendiculars locates *D'*. The process of completing the Williot diagram is a repeat of the past procedure for each joint. Point *J'* would be located next, then *F'*, and last, *L'*.

The distance from *A'* (the point on the structure that cannot move) to *L'* represents, by the Williot diagram, the movement of the right end of the truss with respect to the left end. This measurement indicates that *L* moves up an amount equal to *Y* and to the right an amount equal to *X*. It is apparent that it is physically impossible for *L* to move vertically with respect to *A* because panel point *L* rests on the right support. The Williot diagram must be in error. The exaggerated deflected shape of the truss, as determined from the Williot diagram, is given by the dashed lines of Fig. 10.23. Returning end panel point *L* to the support would place the deflected truss in its correct position. It is then necessary to rotate the truss until end *L* is restored to its correct vertical position. This means that the truss would be rotated through an angle $\alpha = Y/S$, where *S* is the span of the truss. If this were done, each panel point would move through the same angle and the vertical correction for each panel point would be equal to angle α times the horizontal distance of each respective panel point from point *A*. Since the angle α would be very small, the horizontal displacement of any of the lower chord points during the rotation would be infinitesimal.

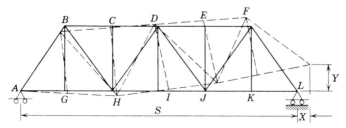

Fig. 10.23

The reference point for measuring the correct deflection of panel point
L cannot be point A' in the Williot diagram, but must be a point of intersection of lines drawn vertically from A' and horizontally from L'.
This point is designated L''. Since rotating the truss changes the point of
reference for panel point L, it also changes the point of reference for
measuring the deflections of the other panel points of the truss. In
rotating the truss through the angle α, the vertical movement of panel point
K will be equal to α times the distance from A to K. This vertical movement
is $\frac{5}{6} Y$. The correct reference point for measuring deflections of panel point
K must be $\frac{5}{6} Y$ above point A'. The correction points for all the lower chord
panel points then lie along the vertical line from A' and their position along
this line is in the same relationship as their position along the lower chord
of the truss. The correct displacement of any point is measured from the
double prime point to the single prime point. For example, the displacement of panel point H due to the loading shown in Fig. 10.22a is measured
from point H'' to point H' in Fig. 10.22c.

In the rotation of the truss through the angle α, the upper chord panel
points move down and also horizontally to the right. The horizontal
movement is equal to the angle α times the height of the truss. The
reference point on the correction diagram for an upper chord panel point
is then on a horizontal line to the left and passes through the respective
lower chord panel point. For instance, the correction point for panel point
B is on a line drawn horizontally through G''. The distance from G'' to
B'' has to be in the same proportion as the true distances on the truss itself.
This can be written in ratio form as

$$\frac{\text{Distance } G'' B''}{\text{Distance } A' G''} = \frac{24}{18}$$

This, then, means that the correction diagram is a scaled diagram of the
truss superimposed on the Williot diagram. This correction diagram
was first introduced by the German engineer, Otto Mohr, and it carries
his name. The Mohr diagram for the truss in Fig. 10.22 is shown by the
dashed lines in part (c). The vertical displacement between the two support points, as given in the Williot diagram, sets the scale for drawing
the Mohr correction diagram. It should be remembered that the correct
displacements are measured from the respective points on the Mohr
correction diagram to the related points on the Williot diagram. For
example, the correct displacement of panel point B is measured as vertically down 0.32 in. and horizontally to the right 0.24 in.

The following basic concepts should be remembered in drawing the
Mohr correction diagram. The initial points on both the Williot and
Mohr diagrams are common and represent the panel point at the fixed

support. The other established point on the Mohr diagram is the point related to the panel point at the second support (expansion support). The location of this second point is at the intersection of two lines. One of the lines is drawn through the point representing the fixed support and is drawn perpendicular to the allowable direction of movement at the expansion bearing. The second line is drawn through the point on the Williot diagram (single prime letter) representing the displacement of the panel point at the expansion support. This second line is drawn parallel to the allowable direction of movement at the expansion bearing. With these two points on the Mohr correction diagram established, the additional points relating to the other panel points can be obtained by a ratio of the dimensions of the structure superimposed on the two points already established. The Mohr correction diagram is then a scale drawing of the structure superimposed on the Williot diagram.

The values of deflection so obtained will be larger than actual deflections to be expected from a riveted, bolted, or welded truss. The effect of contributing stresses in the floor system and bracing will also reduce the actual deflections. This effect is difficult to evaluate. Limited tests have indicated that the true deflections may only be one-half to three-fourths of the theoretical deflections. Since the Williot-Mohr diagram gives the maximum possible deflection, these values will not be exceeded in the actual structure.

10.8   Law of Reciprocal Deflections

In Chapter 1, reference was made to Maxwell, the British engineer-physist, who developed the basic concepts of analysis of statically indeterminate structures. He also presented a theorem that bears his name—Maxwell's theorem of reciprocal deflections. This theorem can be stated:

A deflection at point A on a structure caused by a force at point B will be equal to the deflection at point B due to a load of the same magnitude (and in the same direction) applied at point A on the structure.

This law applies only to stable structures and to structures in which the stress-strain relationship at all points in the structure is linear. It is equally valid for statically determinate or indeterminate structures.

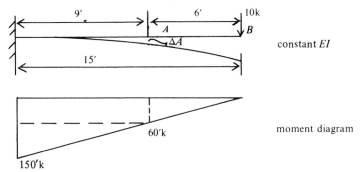

constant EI

moment diagram

A simple proof of this law can be shown. Consider the deflections at two points A and B on a cantilever beam with a 10 kip load acting alternately at B and then A.

By the moment-area method, the deflection at point A is

$$EI\Delta_A = 60(9)(4.5) + 90(4.5)(6) = 4860 \text{ ft}^3\text{-k.}$$

The 10 kip load is now placed at A and the deflection at B is determined.

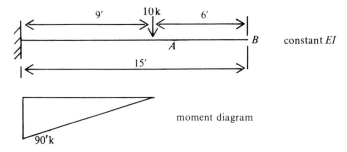

constant EI

moment diagram

$$EI\Delta_B = 90(4.5)(6 + 6) = 4860 \text{ ft}^3\text{-k.}$$

This theorem can easily be proven on other structures. Although the statement of the theorem spoke of deflections, it should be remembered that it is also true of rotations due to forces and/or couples. This can be proven also by a simple example.

Consider the case of a simple beam AB shown below with constant E and I.

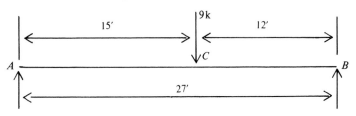

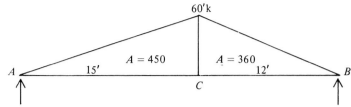

60'k

A = 450   A = 360

A   15'   12'   B

C

Moment diagram on conjugate beam.

The left reaction at point A and thus the shear at point A for the conjugate beam is

$$R_A = V_A = \frac{360(8) + 450(17)}{27} = 390 \text{ ft}^2\text{-k}$$

The rotation of the tangent to the elastic line at A is then:

$$EI\theta_a = 390 \text{ ft-k}$$

The deflection at C due to a couple of 9 ft-kips at point A will now be determined.

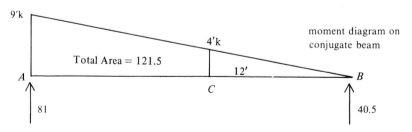

9'k

15'   12'

C

A   B

27'

-0.33k   0.33k

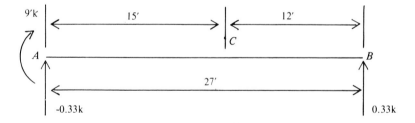

9'k

4'k

Total Area = 121.5

moment diagram on conjugate beam

A   12'   B

C

81   40.5

The deflection at point C is then the bending moment at point C for the conjugate beam

$$EI\Delta_C = M_C = 40.5(12) - 24(4) = 390 \text{ ft}^3\text{-k}.$$

In the previous example it is seen that

$$\theta_a = \Delta_C$$

however the units of θ_a would be in radians and Δ_C in a linear quantity (ft in the previous example). If the values were determined for a unit load at C and a unit couple at A then

$$EI\alpha_{AC} = 43.33 \text{ ft}^2\text{-k} \qquad \text{and} \qquad EI\delta_{CA} = 43.33 \text{ ft}^3\text{-k}$$

therefore,

$$\alpha_{AC} = \delta_{CA}$$

α_{AC} will be in units of radians per kip while δ_{CA} will be in ft. per ft. kip; therefore,

$$\frac{\text{radians}}{\text{kip}} = \frac{\text{ft.}}{\text{ft.-kip}}$$

$$\frac{1}{\text{kip}} = \frac{1}{\text{kip}}$$

A similar demonstration could be given to prove the validity of the theorem for equal couples applied at any two points on a structure and the resulting equality of rotations.

References

1. Carpenter, S. T., *Structural Mechanics;* Robert E. Krieger Pub., 1960 (reprint 1976).
2. White, R. N., P. Gergely, and R. G. Sexsmith, *Structural Engineering,* John Wiley & Sons, 1972.
3. Laursen, H. I., *Structural Analysis,* McGraw Hill, 1969.
4. Norris, C. H., J. B. Wilber, and S. Utku, *Elementary Structural Analysis,* McGraw-Hill, 1976.

Problems

1. Find the deflection at the end of a 12-ft cantilever beam subjected to a uniform load of 0.8 kip/ft. E is 30×10^3 ksi and I is equal to 400 in.[4]. Use the moment area method.   *Ans.* $= 0.3$ in.

2. For the cantilever beam of Fig.10.24 determine the maximum deflection by the moment-area method. (*Hint:* Draw separate moment diagrams for each loading.)

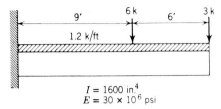

$$I = 1600 \text{ in.}^4$$
$$E = 30 \times 10^6 \text{ psi}$$

Fig. 10.24

3. What would be the deflection of the beam in Problem 2 if the moment of inertia was 2400 in.⁴ from the 6-kip load to the support?   *Ans.* = 0.314 in.

4. Determine the maximum deflection for the beam of Fig.10.25. $I = 500 \text{ in.}^4$, $E = 30 \times 10^6$ psi. (*Hint:* Draw separate moment diagrams for the end moment and the uniform load.)

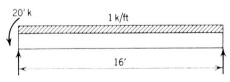

Fig. 10.25

5. Determine the deflection at the cantilever end of the beam of Fig.10.26 by conjugate beam. Also determine the maximum upward deflection in the beam. EI is constant and equals 12×10^6 k-in².

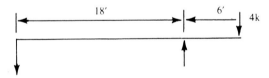

Fig. 10.26

6. Determine the maximum deflection between supports for the beam of Fig. 10.27 and the deflection at the end of the cantilever. The beam is timber with $E = 1.6 \times 10^3$ ksi and the I values are as shown. The increase in I between supports is obtained by gluing joists together.

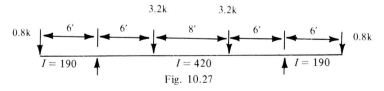

Fig. 10.27

7. Determine the maximum deflection in the beam of Fig. 10.28. Also determine the deflection at each hinge. Use conjugate beam method.

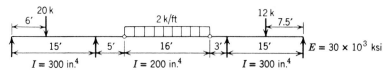

Fig. 10.28

8. (a) Work Problem 5 using the general energy method (Eq. 10.3). (b) Determine the deflection due to shear strains (use k = 1.0).

9. Using Castigliano's method determine the deflection under the 8 kip load and also at mid-span for the beam in Fig. 10.29. $E = 10.6$ x 10^3 ksi, $I = 420$ in⁴.

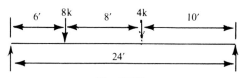

Fig. 10.29

10. For the beam shown in Fig. 10.30, find the deflection of point B due to bending by the method of virtual work. $I = 200$ in.⁴ $E = 10 \times 10^6$ psi.

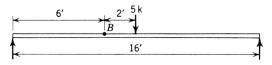

Fig. 10.30

11. Determine the horizontal deflection of point A of the structure shown in Fig. 10.31 by the method of virtual work. $I = 400$ in.⁴ and $E = 30 \times 10^6$ psi.
Ans. $\Delta_{Ah} = 1.65$ in.

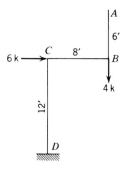

Fig. 10.31

12. It is required to reduce the deflection in the structure of Problem 11 to 1.0 in. by increasing the moment of inertia of the portion *CD* only. What is the new required moment of inertia of *CD*?

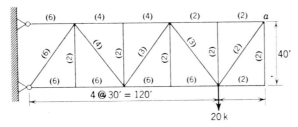

Fig. 10.32

13. Determine the vertical deflection at the end of the cantilever beam of Fig. 10.33. Use virtual work and consider bending strains only. *EI* is constant at 29 x 10^5 k-in².

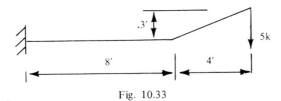

Fig. 10.33

14. An aluminum beam has a simply supported span of 20 ft and carries a total uniform load of 800 lb/ft. The rectangular cross section of the beam is 6 in. wide and 12 in. deep. The top of the beam is 50° F cooler than the bottom. What are the maximum deflections produced by bending, shear, and temperature? $Ea = 10.6 \times 10^6$ psi, $Ga = 4 \times 10^6$ psi, $\varepsilon_a = 13 \times 10^{-6}$ in./in./F°.

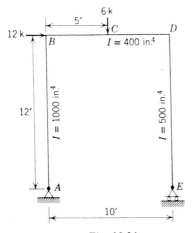

Fig. 10.34

15. (*a*) Determine the horizontal deflection of point *D* of the bent in Fig.10.34. (*b*) Determine the vertical deflection of point *C*. Use virtual work in both parts. $E = 30 \times 10^6$ psi.

16. Determine the vertical deflection at the load point for the steel truss shown in Fig. 10.35. The cross-sectional areas of the members are shown in ( ).

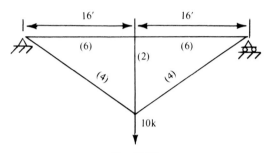

Fig. 10.35

17. Determine the horizontal movement of point *a* of the steel truss shown in Fig.10.32. The areas of the members in square inches are shown in parentheses. What would be the deflection if the truss were made of aluminum? Use the virtual work method. $E_s = 30 \times 10^6$ psi, $E_{al} = 10.6 \times 10^6$ psi.

18. Determine the deflection due to the truck load for the steel truss shown in Fig. 10.36. Cross-sectional areas are shown in ( ).

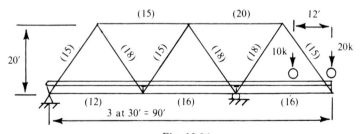

Fig. 10.36

19. Find the dead load deflections of the lower chord panel points for the bridge truss of Fig.10.37 by the Williot method. Areas of the members in square inches are shown in parentheses. $E = 30 \times 10^6$ psi.

20. Find the vertical deflections of the lower chord panel points by the Williot-Mohr method for the truss of Fig. 10.38. Areas of members in square inches are shown in parentheses. $E = 30 \times 10^6$ psi.

21. Find the vertical and horizontal deflection of joint *d* by the Williot-Mohr method for the structure of Fig. 10.39. The total strain in each member is shown.

22. It is required to obtain a camber at the center of the truss of 4 inches for the structure of Fig. 10.40. It is planned to do this by increasing the length of each top chord member and the end posts the same amount. What is the necessary increase in length. (*Hint:* Determine the camber for an arbitrary increase in length of each top chord and end post member.)

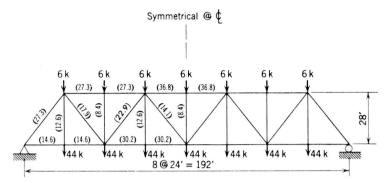

Fig. 10.37

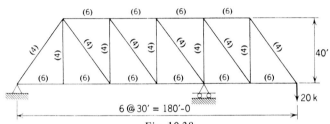

Fig. 10.38

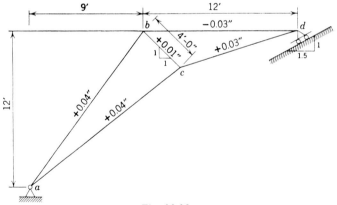

Fig. 10.39

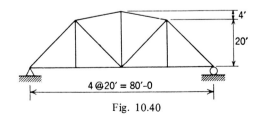

Fig. 10.40

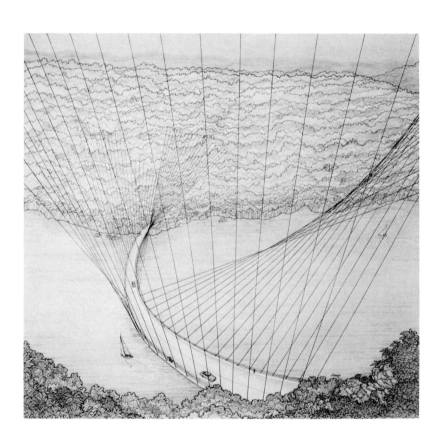

PHILOSOPHY OF STATICALLY INDETERMINATE STRUCTURES

11.1 Introduction

The analysis of structures requires the engineer to classify all structures into two general categories—*statically determinate* and *statically indeterminate*. In a previous chapter, it was pointed out that before an engineer can determine the shape and size of any element of a structure, he must be able to quantify four force actions to which the structural element can be subjected. These four actions are:

axial force (tension or compression)

shear force

bending moment

torsional moment

He must be able to ascertain the values of each of these. This is the entire objective of structural analysis. If he can determine the magnitude of these four actions by the use of the principles of static equilibrium, then the structure is classified as a statically determinate structure. This was the case for all the structures analyzed in the previous chapters with the exception of some frames in Chapter 7. However, in those cases the frames were analyzed as statically determinate structures with the use of some simplifying assumptions.

It then follows that the definition of a statically indeterminate structure is one that cannot be analyzed using the principles of static equilibrium only. Some engineers have not liked the designation of statically indeterminate and have tried to develop a new name for such structures, but with little success. The name *statically indeterminate* is descriptive and thus its usage remains.

The engineer responsible for the design of a structure can select which class of structure would be most suitable on the basis of conditions and economy. As a simple example, consider the design of a bridge. It may be possible to use a beam of one single span or to place a pier in between the ends of the bridge and make two spans. The two span bridge would require beams of smaller cross-section than a single span. However, the saving in cost of the beams would have to be equated against the cost of the pier. Assume that it has been decided that two spans may be better than one single span (erection of the bridge may also be a deciding factor). The next decision is whether to make each span a simple beam or to provide continuity between the two spans. This would be done by making the two span beams one single piece or from two or more pieces spliced together so the bending moments and shears can be carried by the splices.

The decision as to which class of structure to use will depend on costs, stiffness, erection requirements, and appearance. All of these factors should be taken into consideration in the design selection of any structure.

11.2 Advantages and Disadvantages of Statically Indeterminate Structures

There will be both advantages and disadvantages of using one type structure over the other, or all structures would either be statically determinate or indeterminate. If one single span beam was required, it would, of course, be obvious that the only way to make it indeterminate would be to restrain one or both ends from rotation. Unless this beam was part of a more complex structure, this would not likely be advisable.

There are several advantages of statically indeterminate structures. First, there is the matter of costs. Refer back to the two span bridge. The maximum bending moment in the bridge would be less for the continous structure than for the simple span. This would even be more pronounced if there were three or more continuous spans as contrasted with the same number of simple spans. For instance, the maximum bending moment in a simple beam with the uniformly distributed load of w and span L is $wL^2/8$. While the continuous beam of three equal spans of length L and same loading with simply supported ends, would have a maximum bending moment of $wL^2/10$. Generally speaking, a statically indeterminate structure will require less material than a statically determinate structure. There can be exceptions to this statement.

Another advantage of the statically indeterminate structure is in the reduction of deflections. A continuous structure will have less deflections and thus be less flexible than a simple span structure. Design codes all have limits on the magnitude of the maximum deflection. Sometimes a structural element may have to be enlarged over that required on a basis of stress limit in order to meet the deflection limitation. In most cases a smaller deflection will also mean less problems from vibration.

The next consideration is the matter of erection of a structure. It may be simpler to erect a continuous structure than a structure with a series of simple beams. This has proven to be the case on long span girder bridges. If a three span bridge was to be built, as shown in Figure 11.1, with long spans so that one piece of sufficient length to reach from support to support was too large to handle, intermediate temporary piers would have to be used in order to erect the bridge. These temporary piers would be required in all three spans if the spans were simple spans only. However, in the case of continuous spans, no temporary piers would be required to erect the center span. This span could be erected by cantilever construction where sections of the bridge could be added to the cantilever portions from both directions. The bridge would then be joined at the center. This cantilever erection would be most advantageous where the structure was over a deep ravine or deep water.

In all fairness it should be pointed out that the erection method shown in Figure 11.1 (b) can be used on a statically determinate structure. The two ends could be cantilevered over the permanent piers a certain distance as shown. Pins could then be inserted at the ends of the cantilevered ends and the gap closed by a simple span connecting at the pins. The student will recognize this as the structure of Figs. 4.9 and 4.44. This bridge type is given the name of cantilever bridge. The fabrication of the pins may be costly and the structure will be less stiff than the continuous bridge. Material requirements to carry the live loads will be greater for the cantilever bridge, although dead load moments could be equal for the same loading.

The last advantage of statically indeterminate structures to be discussed is that of appearance. A statically indeterminate bridge (continuous spans) can have a more aesthetically pleasing shape. The maximum bending moment in a simple span will occur at midspan, while in

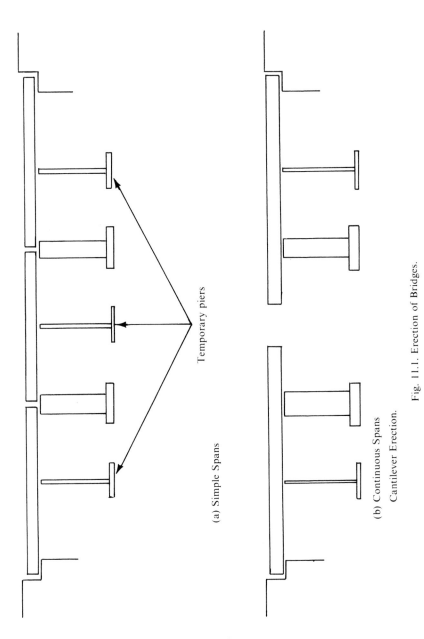

(a) Simple Spans

Temporary piers

(b) Continuous Spans
Cantilever Erection.

Fig. 11.1. Erection of Bridges.

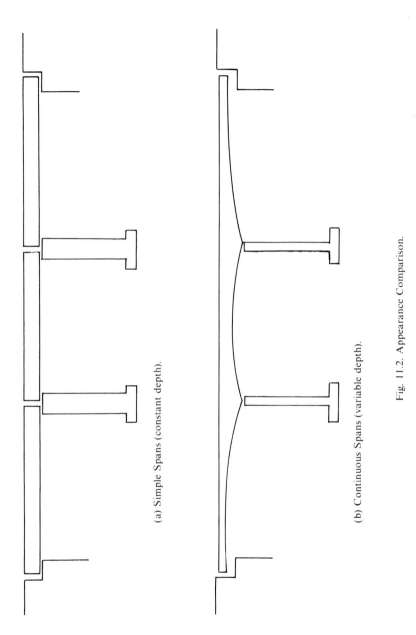

(a) Simple Spans (constant depth).

(b) Continuous Spans (variable depth).

Fig. 11.2. Appearance Comparison.

a continuous span, this maximum will be at the interior supports. Therefore, the girder (or truss) can be deeper at the interior supports than at midspan. Figure 11.2 shows this difference. Another factor with regard to appearance is that since only one set of bearing devices will be required at the interior piers, it may be possible to make the piers more narrow.

A rigid frame bridge may also be more pleasing in shape than a simple girder span (see Fig. 11.3).

It should not be thought that all the advantages are on the side of the statically indeterminate structure. In some cases, the erection of simple spans, where each span can be erected in one piece, will be faster and more economical than adding splices in order to achieve continuity. The rigid frame bridge will require a foundation that will have to resist horizontal reactions (and in some types of frames, also moments). This could result in larger foundations and footings.

A distinct difference between determinate and indeterminate structures is that stresses will be developed in statically indeterminate structures by support settlements, changes in temperature, and fabrication errors. Such stress development will be more fully explained in later sections of this book. If a support of a statically determinate structure moves (not a sufficient amount to change the geometry), the magnitude of the reactions do not change.

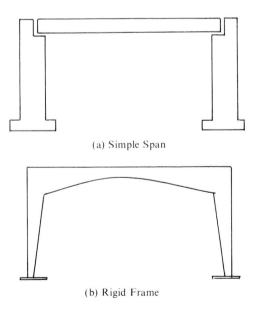

(a) Simple Span

(b) Rigid Frame

Fig. 11.3. Appearance Comparison.

11.3 Degree of Indeterminacy

Previous chapters (4, 5, and 7) developed procedures and equations for evaluating structures to determine if they were stable, statically determinate or indeterminate. Basically speaking, there are two approaches to this determination.

(1) Can the magnitude of all reactions, bar forces, and in the case of frames, all internal bending moments and shears be determined by only using the equations of statical equilibrium? If so, and the structure is stable, then it is statically determinate.

(2) Can reaction components be removed or internal members cut without making the structure unstable? If so, then the structure is statically indeterminate.

In the case of the second approach, those reactions that can be removed or those members that can be cut are called *redundants.* Redundant meaning that they are not necessary for the state of stability. The number of reactions that can be removed, and/or the number of members cut until a further removal or cut would make the structure unstable, is the *degree of indeterminacy.* As an example, the structures in Fig. 11.4 show the statically indeterminate structure. For comparison, the determinate structure is shown with the redundants of the original structure removed. The degree of indeterminacy (D) is thus indicated. The engineer has a choice of selection of several possible redundants for his analysis.

In the beam of Figure 11.4 (a) the structure had only one redundant. Additional spans would add one more redundant for each additional span. The truss (b) has two redundant bars. Either one of the crossed diagonals in each panel could be classified as the redundant. In the frame of (c) there are three reaction restraints at each support; restraint in horizontal and vertical translation and rotation. The determinate structure can be achieved by removing all three restraints at one support or two restraints at one and one restraint at the other support. A third method would be to remove the moment reaction at each support and add an internal pin.

In this third case, one of the redundants is the internal bending moment at the chosen location of the pin. A fourth possibility would be to leave the supports as was originally shown and to completely sever the frame at one location, thus removing the ability of the frame at the location to carry axial force, shear force, or bending moment. The proper selection of redundants will sometimes simplify the analysis or will reduce the mathematical operations. Experience is usually the best teacher on the proper selection of redundants.

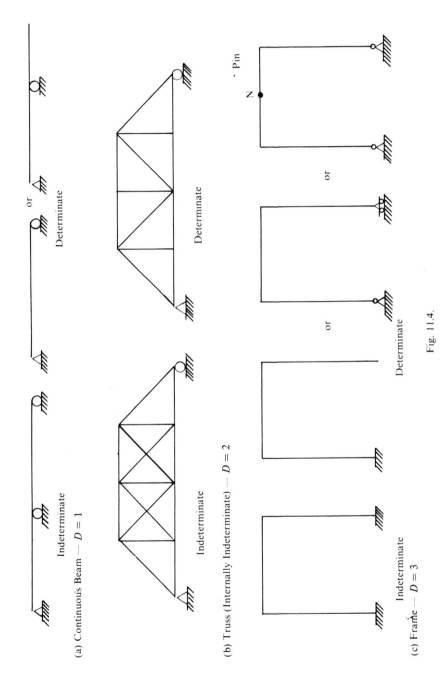

Fig. 11.4.

11.4   Factors Affecting Analysis

In order to analyze a statically determinate structure, it is only necessary to know the forces to which the structure is subjected and also the geometry of the structure. The simple proof of this is that in applying the equations of equilibrium, the only parameters are force and distance. This limitation of two parameters is not so in statically indeterminate structures. In the analysis of such structures, values of rotation and translation of the structure are determined, either directly or indirectly. In Chapter 10 it was shown that in order to determine rotations or translations at any point, on any structure, the numerical value of the modulus of elasticity of the material and the moment of inertia at all points on the structure must be known. For structures that have members carrying axial forces the cross-sectional areas of the members must be known.

At the beginning of the design of a structure, the area (A) and the moment of inertia (I) of the elements of the structure are not known. Estimates of these values have to be made first. The analysis and design is then completed. The values of A and I determined from the design are then checked against the original estimate. If these two sets of values do not agree, then the analysis and design has to be refined by another cycle using new values of A and I. The absolute values of A and I are not required for a correct analysis, only the relative values of each member of the structure. For instance, the reactions of a continuous beam on rigid supports will be the same for any size beam as long as the relative I's for each span remain the same and the span lengths and the loading are the same. This is not true if there is a settlement of the supports.

Most all methods of analysis of statically indeterminate structures are based on linearity of the material of the structure. That is, all materials should obey Hook's Law, and any settlements of the supports are proportional to the load on the supports, or the settlement is a specific known value regardless of the force the support is required to carry. This linearity relationship makes the law of superposition valid. This law states that all stresses in a structure are directly proportional to the load applied. This also means that if the stresses are determined for two given sets of loads independent of each other, the stresses resulting from the simultaneous application of both sets of loads can be determined by adding the stresses calculated from the two independent analyses. There are, of course, exceptions where the law of superposition is not valid even though the material has a linear stress-strain relationship. This is when the geometry of the structure makes an appreciable change when loaded. A case of this was the suspension structures in Chapter 9. Flexible arch structures may also have to be investigated for change in geometry.

11.5 Development of Methods of Analysis of Statically Indeterminate Structure

Analysis and design of statically indeterminate structures is a relatively latter-day development. Although James Clerk Maxwell published,[1] in 1864, the first significant paper on the analysis of statically indeterminate structures, it was many years later before engineers made any significant use of indeterminate analysis. Such methods of analysis were adopted in Europe before much use was made in America. Teaching of these more complex methods in engineering schools in America was given great emphasis by the publication in 1930[2] of the Hardy Cross method of moment distribution. In 1939, the writer was a student in the first class in the analysis of statically indeterminate structures given at his alma mater. At that date, such a course was only available at the more advanced engineering schools. During the 1930's, several textbooks were published in the U.S.A. treating the analysis of statically indeterminate structures. They treated the classical methods known at the time.

There are several basic methods and many adaptations of these basic methods available for the analysis of statically indeterminate structures. The early methods all approached the problem by the simultaneous solution of equations relating to the boundary conditions of the structure. The German, Otto Mohr, in 1874,[3] presented a paper on this approach to indeterminate analysis. Maxwell had also written on this subject a few years previous. Several different methods for calculating slopes and deflections of structures were developed in the latter half of the nineteenth century.

Castigliano published[4] his theorems of work in 1879. In the early twentieth century, the method of slope deflection came into use. Up to the time of Professor Cross' paper, the analysis of frame structures was a tedious and time consuming hand calculation. Complex structures could only be analyzed by approximate methods. Classical methods at that time depended on the solution of many simultaneous equations.

The Hardy Cross method has been thought of as ushering in the modern era in structural analysis. Even with this new and more rapid procedure, analysis of frames with several bays and several stories was still a very formidable task. For such large scale projects simplified versions of the Hardy Cross moment distribution method were developed. Since this method is an iterative process, time could be saved with little loss in accuracy by a reduced number of cycles of iteration.

With the development of the high speed digital computer after World War II, new procedures in structural analysis for highly redundant structures became possible. It was now mathematically feasible to solve large sets of simultaneous equations. The basic approach to the solution of redundant forces and reactions did not change. It still followed the concept of Maxwell and Mohr in developing the formulation of condition equations representing the geometric coherence of the structure. The simultaneous solution of these equations could now be performed in a fraction of the time that was possible before computers. In fact, it was now possible to solve large numbers of such equations where previously it was not humanly possible to do so.

Chapter 12 of this text will cover the basic concepts in developing condition equations for the solution of indeterminate structures. Examples will be presented of structures of a few redundants wherein results can be found by hand calculations. Importance of understanding basic concepts will be stressed. In Chapter 14 the formulation of equations that can lead to computer solutions will be treated. This procedure, called matrix analysis of structures, will be presented in fundamental form.

With the availability of computers and the development of computer programs, very complex structures can be analyzed. Another great advantage the present-day structural engineer has is that he can study several complex structures in less time than the engineer could previously investigate one structure. Optimum design can now be based on real numbers where before it had to be based on intuition and experience only.

References

1. Maxwell, J.C., "On the Calculation of the Equilibrium and Stiffness of Frames," *Philosophical Magazine,* Series 4, Vol. 27, 1864, pp. 294-299.
2. Cross, H., "Analysis of Continuous Frames by Distributing Fixed-End Moments," *Proc. ASCE,* Vol. 56, May 1930, p. 919.
3. Mohr, O., "Beitrag zur Theorie des Fachwerks," Zeitschrift des Architekten—und Ingenieer, Vereins zu Hannover, Vol. 20, no. 4, 1874, pp. 509-526.
4. Castigliano, A., Theorie de l'equilibre des Systemes Elastiques et ses Applications, published by A.F. Negro, Turin, 1879.

Problems

1. For the structures shown in Fig. 11.5 determine the degree of indeterminacy. Consider structures c, d, e and h have pin joints.

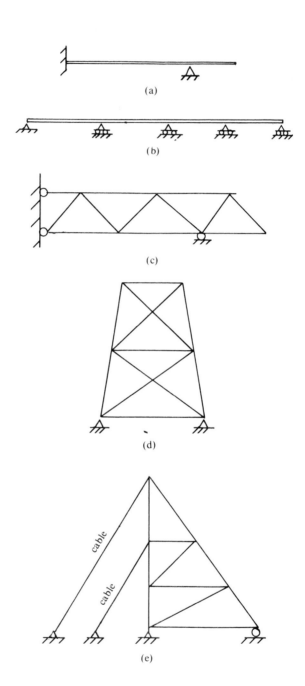

(a)

(b)

(c)

(d)

(e)

Fig. 11.5

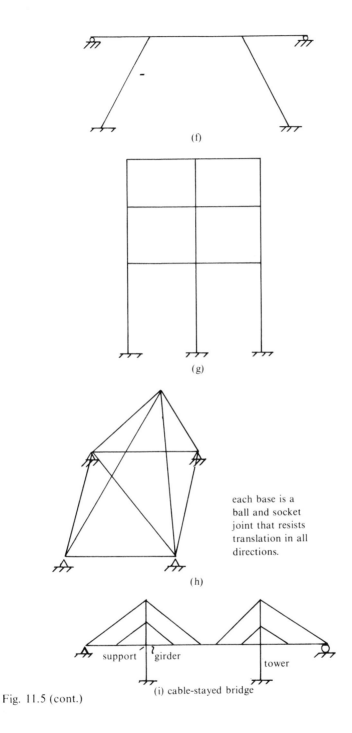

(f)

(g)

each base is a
ball and socket
joint that resists
translation in all
directions.

(h)

support girder

tower

(i) cable-stayed bridge

Fig. 11.5 (cont.)

317

METHODS OF CONSISTENT DISPLACEMENTS

12.1 Introduction

As was stated in the previous chapter, the first method of analysis of statically indeterminate structures consisted of formulating equations of displacement and then making those equations consistent with the displacements of the supports of the structure. Some authors have labeled this the *general method*. It is indeed general, since all methods are a special formulation or adaptation of this method. The so-called matrix methods are in essence based on this same approach. The matrix method consists of two approaches to the analysis of statically indeterminate structures. They are called the *force method* and the *displacement method* by some writers or the *flexibility method* and the *stiffness method* by others. Whatever the names, they both utilize relationships between the displacements and forces that exist in a structure.

It is apparent that reactions are just forces applied to a structure at a given location, but that the displacement (using the general term of displacement as meaning translation and rotation) at these locations is controlled by physical conditions. Before an analysis can be made, the physical conditions at the supports must be evaluated. The condition at a support must be no displacement, a specific value of displacement, or a displacement that is related to the magnitude of the reaction force at the support. This latter case is the condition of an elastic or elastic-plastic support.

In actuality, there is no such condition as a support that will have zero displacement. All support structures are made from material that has a finite value of modulus of elasticity so all supports are subject to some displacement. In many cases, the displacements are so small that they can be considered equal to zero with no appreciable loss in accuracy. The structural analysis is simplified if the displacement of the support can be treated as being zero. Most examples in textbooks (including this one) will have such support conditions. However, the structural engineer should beware that he does not just blindly make such assumptions. An accurate foundation analysis should be made of any indeterminate structure to determine the magnitude of the support displacements of the actual structure. In other words, the *mathematical model* should fit the *physical model* (actual structure as it will be built). Assumptions should never be blindly or superficially made. The paths to engineering disasters are paved with inaccurate assumptions.

The popular use of statically indeterminate structures gave birth to the development of the science of soil mechanics. Such structural types demanded a more precise method of foundation analysis that could predict the magnitude of the support displacements.

12.2 Analysis of Beams

(a) Flexibility Method

The objective of this method is to determine the magnitude of the redundant forces. As was explained in the previous chapter, there is usually a choice of which forces are redundant. The term forces as used here can be reactions, internal axial forces, internal shears, or internal moments. The proper choice of redundant can in some cases simplify the solution of the problem.

In order to apply this method it is necessary to determine displacements of the structure at the point of application of the redundants and in the direction of the redundants. These deflections are not determined for the real structure but for the real structure reduced to a *statically determinate* structure by removal of the redundants. This statically determinate structure will be called the "primary structure." It is the statically indeterminate structure reduced to a statically determinate one by the removal of the redundants. The calculation of the deflections of statically determinate structures has been covered in Chapter 10. Several methods were presented for determining the deflections of structures. Some methods are more applicable to specific types of structures than others. The method used should be the one that produces the most direct and simple solution. The student may want to go back and review the material in Chapter 10 before proceeding further in this chapter.

As a simple and classic example of this method, consider a single span beam with a simple support (constraint from translation in vertical direction only) at one end and a fixed support (constraint from translation in two directions as well as rotation) at the other end (Fig. 12.1). It is very likely the student has been exposed to this example in his course in mechanics of materials, but many years in the classroom have proven to the author that repetition is a great teacher.

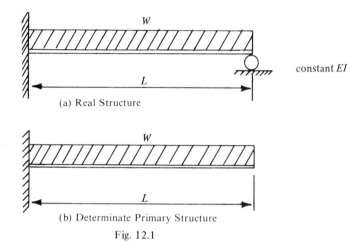

(a) Real Structure

(b) Determinate Primary Structure

Fig. 12.1

In Figure 12.1, it is observed that the real indeterminate structure (a) has been reduced to a determinate structure (b) by selecting as the redundant the vertical reaction at the right support in the cantilever (primary) beam. The moment area method is probably the simplest approach to determining this value. Example Problem 12.1 shows the calculation of Δ_b.

The next step is to place a unit force at the location of the redundant and in the direction of the redundant. If the correct direction of the redundant is unknown, then the direction can be assumed.

The third step is to calculate the magnitude of the deflection due to the unit load. This value is labeled δ_{bb} with lower case delta indicating deflection due to a unit load, the first subscript indicating location where the deflection is calculated and the second subscript where the unit load is located. If the deflection due to a unit load is δ_{bb} and if the law of superposition (and linearity) is valid, then the deflection at b due to a force equal to R_b must be equal to $(\delta_{bb})R_b$.

The final step is to equate the deflection at b due to the real loading to the deflection at b due to the redundant reaction R_b. If it can be safely assumed that the settlement of the support at b is so small as to be assumed equal to zero, then the sum of the two deflections can be made zero. The value of R_b will then be equal to $3/8(wL)$.

EXAMPLE PROBLEM 12.1. Determine magnitude of redundant reaction of beam in Fig. 12.2

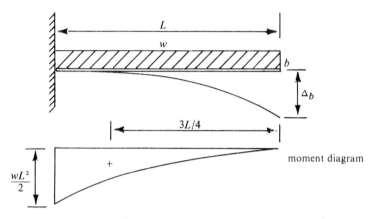

Fig. 12.2

$$\text{Area of moment diagram} = \frac{1}{3}\left(\frac{wL^2}{2}\right)(L) = \frac{wL^3}{6}$$

$$\text{Moment about } b \text{ of moment diagram} = \frac{wL^3}{6}\left(\frac{3L}{4}\right) = \frac{wL^4}{8}$$

$$\Delta_b = \frac{wL^4}{8EI}$$

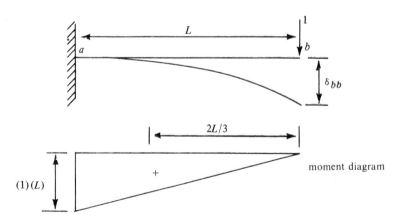

$$\text{Area of moment diagram} = \frac{L^2}{2}$$

$$\text{Moment about } b \text{ of moment diagram} = \frac{L^2}{2}\left(\frac{2L}{3}\right) = \frac{L^3}{3}$$

$$\delta_{bb} = \frac{L^3}{3EI}$$

If total deflection at b in real beam is zero, then by superposition

$$\Delta_b - \delta_{bb}(R_b) = 0$$

$$R_b = \frac{\Delta_b}{\delta_{bb}} = \frac{wL^4/8EI}{L^3/3EI} = \frac{3}{8}wL$$

If the support b did settle a value K and support a did not, then it can be seen that $\Delta_b - R_b(\delta_{bb}) = K$ and

$$R_b = \frac{\Delta_b - K}{\delta_{bb}}$$

in which R_b would be a lesser value than formerly. When the support at b deflects an amount equal to K the values of EI cannot be canceled and the units of E and I must be made consistent with the units of the deflection K.

Another point to remember is that if the value of I varies along the beam, then the M/I diagram must be used instead of just the moment diagram. Example Problem 12.2 treats this case.

The deflection can be determined by any other applicable method such as conjugate beam, virtual work, or Castigliano's method. For cantilever beams, the moment-area method is simple and direct for deflection calculations.

The redundant chosen in Problem 12.1 was the reaction at support b. The chosen redundant could have been the reaction moment at a as shown in Figure 12.3. If this had been the case, then the released restraint would have been the rotation of the tangent to the elastic line at point a. The calculated values would have been the rotation of the beams at $a(\theta_a)$ and the rotation of the beam at a due to a unit moment applied at $a(\alpha_{aa})$. Then the rotation of the beam at a due to the redundant moment M_a would be $(\alpha_{aa})M_a$, and the compatibility equation would be:

$$\theta_a - (\alpha_{aa})M_a = \text{rotation at } a \text{ in the real beam}$$

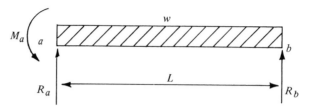

Fig. 12.3 Beam with Redundant M_a.

The detailed calculations are shown in Example Problem 12.2.

EXAMPLE PROBLEM 12.2.

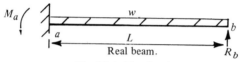

Real beam.

Fig. 12.4 Fixed end beam.

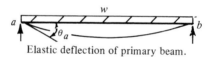

Elastic deflection of primary beam.

Conjugate beam with elastic load.

conjugate beam: $R_a = \dfrac{1}{2}\left(\dfrac{wL^2}{8}\right)\left(\dfrac{2L}{3}\right) = \dfrac{wL^3}{24}$

$$\theta_a = \frac{R_a}{EI} = \frac{wL^3}{24EI}$$

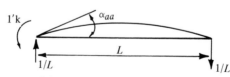

Primary beam with unit end moment.

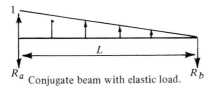

Conjugate beam with elastic load.

R_a R_b

$$R_a = \frac{2}{3}\left(\frac{L}{2}\right) = -\frac{L}{3}$$

$$\alpha_{aa} = \frac{R_a}{EI} = -\frac{L}{3EI}$$

$\theta_a + \alpha_{aa}(M_a) = 0$ (no support rotation in real beam)

$$M_a = \frac{wL^3/24EI}{L/3EI} = \frac{wL^2}{8}$$

For the structure that is a single-span beam fixed at one end, the choice of redundant is immaterial. The amount of effort is really the same in either Example Problem 12.1 or 12.2.

The same concept of indeterminate beam analysis will now be applied to a beam with more than one redundant—a three span continuous beam with simple supports at both ends and different values of I in each span (Example Problem 12.3).

EXAMPLE PROBLEM 12.3.   Determine the reactions in the continuous beam of Figure 12.5. The deflections Δ_b and Δ_c can be calculated by any one of several methods. The conjugate beam is very appropriate and will involve less mathematical work than the energy methods. First the primary structure, second the shear diagram, third the moment diagram, fourth, the M/I diagram, and last the conjugate beam acted on by the elastic loads are drawn in that order (Figure 12.6). The areas of the M/I diagram are calculated using

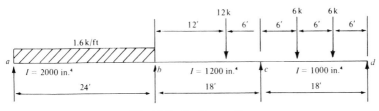

Fig. 12.5   Continuous Beam.

convenient geometrical shapes. The values of M/I are multiplied by 1,000 for convenience. It should also be noted that the values of I used are in in^4, not ft^4. The dimensional correction can be made later if true deflections are required. Relative deflections are only necessary in this method. The areas of the M/I diagram are then treated as concentrated loads placed on the end-supported conjugate beam at the locations of the centroids of the respective areas. Then by conjugate beam principles, the bending moment at any location along the conjugate beam is equal to the deflection in the primary beam at this location. The beam for which deflections are being calculated is the primary structure.

These relative values of deflection are:

$$EΔ_b = 115.697(10^{-3})$$

$$EΔ_c = 105.384(10^{-3})$$

It is now necessary to determine the values of $Eδ_{bb}$, $Eδ_{cb}$, and $Eδ_{cc}$. For the first two of these values, a unit load will be placed at point b and the deflections calculated at points b and c using the conjugate beam method in the same manner as previously (Figure 12.7). The relative values of $Eδ_{bb}$ and $Eδ_{cb}$ are:

$$Eδ_{bb} = 151.9(24) - 86.4(8) = 2.955(10)^3$$

$$Eδ_{cb} = 161.3(18) - 64.8(6) = 2.514(10)^3$$

The last deflection to be determined is the value of $Eδ_{cc}$. In order to do this, a unit load is placed at point c in the statically determinate beam, and by the same procedure (Fig. 12.8), the relative value of $Eδ_{cc}$ is:

$$Eδ_{cc} = 185.9(18) - 113.4(6) = 2.666(10)^3$$

The two simultaneous equations of consistent deflections can now be written

$$Δ_b + δ_{bb}R_b + δ_{bc}R_c = 0$$

$$Δ_c + δ_{cb}R_b + δ_{cc}R_c = 0$$

substituting

$$2.955R_b + 2.515R_c = -115.697$$

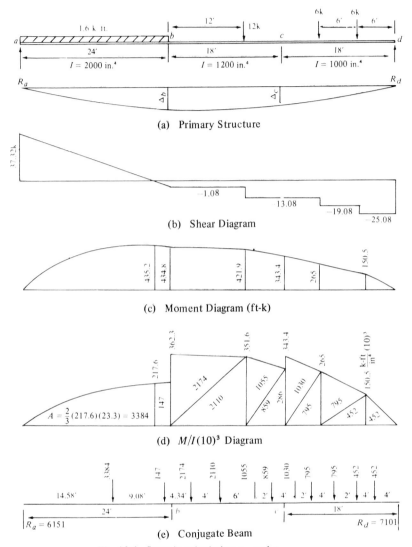

(a) Primary Structure

(b) Shear Diagram

(c) Moment Diagram (ft-k)

(d) $M/I(10)^3$ Diagram

(e) Conjugate Beam

Fig. 12.6   Steps in calculating Δ_b and Δ_c.

$$2.515 R_b + 2.666 R_c = -105.384$$

solving

$$R_b = -27.95 \text{ kips} \qquad R_c = -13.17^k$$

The negative sign for the reactions indicate that the reactions R_b and R_c

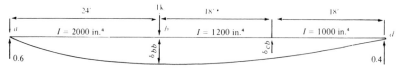

(a) Unit Load on Primary Structure

(b) Moment Diagram

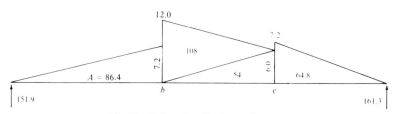

(c) Elastic Load on Conjugate Beam

Fig. 12.7. Steps in calculating δ_{bb} and δ_{cc}.

are in the opposite direction to the deflections Δ_b and Δ_c, respectively.

It should be noted that in the simultaneous equations, the values of E, 10^{-3}, and 12^4 would be present in all terms so they would cancel since the equations equal zero deflection. However, if the deflections at the supports in the real beam did not equal zero, then these terms would have to remain in order to be dimensionally correct. As an example, it is assumed that support b settled 1/2 inch and support c 1/4 inch. The simultaneous equations would now read with $E = 30 \times 10^3$ ksi.

$$115.967 + 2.955\,R_b + 2.515\,R_c = 0.5 \left(\frac{30 \times 10^3}{12} \right)$$

$$105.384 + 2.515\,R_b + 2.666\,R_c = 0.25 \left(\frac{30 \times 10^3}{12} \right)$$

rearranging

$$2.955\,R_b + 2.515\,R_c = -107.016$$

$$2.515\,R_b + 2.666\,R_c = -101.064$$

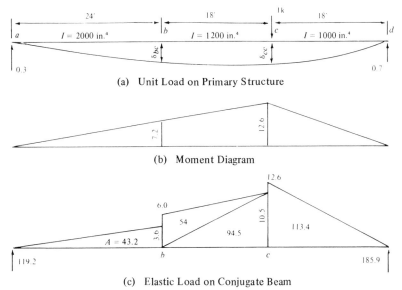

(a) Unit Load on Primary Structure

(b) Moment Diagram

(c) Elastic Load on Conjugate Beam

Fig. 12.8. Steps in calculating δ_{bc} and δ_{cc}.

solving

$$R_b = -20.04 \qquad R_c = -19.01$$

It is noted that the reaction R_b decreases in value while the reaction R_c increases. This is to be expected, since the support settlement at b is greater than at c. Support b tries to "get out from under the load" more than support c, therefore support b will carry less load and thus this reduction in load will be taken primarily by support c, but also by the end supports.

The student may want to plot the moment diagram for the continuous beam for both cases of no support settlement as well as for support settlements as given. This would indicate the sensitivity of bending moment with respect to support settlement for continuous beams.

It should be observed that the coefficients of R_b and R_c in the simultaneous equations have small differences in magnitude. This means that there will have to be several digits in each coefficient in order to have accuracy in the solution of the simultaneous equations. Numerical accuracy is generally improved when the off-diagonal terms of the δ matrix are smaller by an appreciable amount than the diagonal terms. Also, numerical accuracy is improved when the terms in the δ matrix are of considerable numerical difference. It may have been better to select other redundants than R_b and R_c in order to achieve a statically determinate structure. This might have been done by removing R_a and R_c. The numerical values of Δ_a and Δ_c and δ_{aa}, δ_{ac}, and δ_{cc} would have to be determined. The values of Δ_c and δ_{cc} would be different than those just calculated since the primary structure is changed. When these new deflections are determined, the simultaneous equations are (case of no support settlement):

$$8.208\,R_a - 1.728\,R_c = -113.51$$

$$-1.728\,R_a + 0.891\,R_c = +16.96$$

solving

$$R_a = -16.60^k R_c = -13.17^k$$

The proper selection of the determinate structure is important. This importance increases with an increase in redundants.

(b) Method of Three Moments

The principle of consistent displacements can be applied in another way to the analysis of continuous beams. Consider two adjacent spans of a continuous beam of several spans, as shown in Fig. 12.9, with any nature of loading. The deflected shape of this beam is drawn and the tangent to the elastic line over the support j is drawn. This tangent makes an angle θ with the original position (before loading) of the neutral axis of the beam. The slope at j is common to both spans. From the geometry of the tangent to the elastic line

$$\frac{y_i}{L_1} = \frac{y_k}{L_2}$$

From the second moment-area principle the value of y_i (displacement of elastic line at i from tangent at j) and y_k can be determined. To do this in general terms, the moment diagram for the two spans is drawn in parts. The

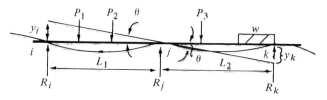

Fig. 12.9 Continuous beam.

simple beam moment diagram (representing a general loading on the beam) is drawn above the axis and the moment diagram representing the bending moments over the supports is drawn below the axis in Fig. 12.10.

The areas A_1 and A_2 are the areas of the simple beam diagrams for spans 1 and 2 respectively and the distances $\bar{x}_1$ and $\bar{x}_2$ are the distances to the centroids of these areas from i and k. A limitation that E is constant and the value of I is constant between any two supports is imposed. From the second moment area principle

$$EI_1 y_i = A_1 \bar{x}_1 + \frac{M_i L_1}{2}\left(\frac{L_1}{3}\right) + \frac{M_j L_1}{2}\left(\frac{2L_1}{3}\right)$$

$$EI_2 y_k = A_2 \bar{x}_2 + \frac{M_k L_2}{2}\left(\frac{L_2}{3}\right) + \frac{M_j L_2}{2}\left(\frac{2L_2}{3}\right)$$

since

$$\frac{y_i}{L_1} = \frac{y_k}{L_2}$$

$$\frac{A_1 \bar{x}_1}{I_1 L_1} + \frac{M_i L_1}{6 I_1} + \frac{M_j L_1}{3 I_1} = \frac{A_2 \bar{x}_2}{I_2 L_2} + \frac{M_k L_2}{6 I_2} + \frac{M_j L_2}{3 I_2}$$

The above equation rules out any translation of the supports. To include support settlement, as shown in Fig. 12.11, then the deviations of the elastic line at i and j from the tangent at k are

$$t_{ij} = \Delta_1 - y_i$$

$$t_{kj} = \Delta_2 + y_k$$

The signs in the above two equations are due to the fact that the tangent

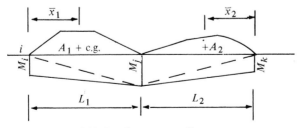

Fig. 12.10 Bending moment diagram.

is above the support i and below the support k in Fig. 12.9. The values Δ_1 and Δ_2 are (+) if support j moves downward from i and k respectively. The compatibility equations are now

$$\frac{E\Delta_1}{L_1} - \frac{A_1\bar{x}_1}{I_1L_1} - \frac{L_1}{6I_1}(M_i + 2M_j) = \frac{A_2\bar{x}_2}{I_2L_2} + \frac{L_2}{6I_2}(2M_j + M_k) - \frac{E\Delta_2}{L_2}$$

rearranging this equation in more useable form

$$\frac{M_iL_1}{I_1} + 2M_j\left(\frac{L_1}{I_1} + \frac{L_2}{I_2}\right) + \frac{M_kL_2}{I_2}$$

$$= -6\left(\frac{A_1\bar{x}_1}{I_1L_1} + \frac{A_2\bar{x}_2}{I_2L_2}\right) + 6E\left(\frac{\Delta_1}{L_1} + \frac{\Delta_2}{L_2}\right) \tag{12.1}$$

The above equation (12.1) is the well-known Three-Moment Equation first presented by the Frenchmen Bertot, Clapeyron, and Bresse in the mid-nineteenth century. It is an equation that is easy to apply. The equation can be written for any two spans of a multiple span continuous beam. The usual procedure is to start at the first two spans of a continuous beam and substitute the known values into the Three-Moment Equations. These known values will be E, I, and L for each span as well as the Δ values if there is support settlement. The $A\bar{x}$ terms are calculated from the simple beam moment diagrams considering each span as a simple beam span subjected to the given loadings. The unknowns in Equation 12.1 will be M_i, M_j, and M_k; the bending moments over the supports. If the beam is simply supported at the first support M_i will be zero in value.

After writing the first Three-Moment Equation a shift in one span is made and the second Three-Moment Equation is written. This repeating of the operation by moving to the next span is continued until all spans have been included in the process. There will then be n simultaneous equations containing n unknown bending moments. With the availability of the electronic hand calculator or digital computer the solution in the simultaneous equations is readily accomplished.

Figure 12.12 shows a three-span continuous beam with all three spans

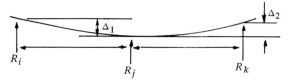

Fig. 12.11 Support settlement.

loaded. There are no settlements of the supports. The first step in the solution is to draw the simple beam moment diagrams, compute the areas of the moment diagrams, and locate the distance from the centroid to the support for each area. The second step is to write the three-moment equations, and the last step is to solve the equations and plot the final moment diagram.

EXAMPLE PROBLEM 12.3.   Draw the final bending moment diagram and determine the final reactions for the continuous beam shown in Fig. 12.12.

$$\frac{M_A L_1}{I_1} + 2M_B\left(\frac{L_1}{I_1} + \frac{L_2}{I_2}\right) + \frac{M_C L_2}{I_2} = -\frac{6A_1 \bar{X}_1}{I_1 L_1} - \frac{6A_2 \bar{X}_2}{I_2 L_2}$$

$\Delta_1 = \Delta_2 = 0$, so the last term in Eq. 12.1 has been dropped. As the ends are simply supported $M_A = 0$ for the first three-moment equation and $M_D = 0$ for the last equation.

(1) $2M_B\left(\dfrac{24}{20} + \dfrac{18}{12}\right) + \dfrac{M_C(18)}{12}$

$$= -\frac{6 \times 1843 \times 12}{20 \times 24} - \frac{6 \times 288 \times 10}{12 \times 18} - \frac{6 \times 144 \times 4}{12 \times 18}$$

(2) $\dfrac{M_B(18)}{12} + 2M_C\left(\dfrac{18}{12} + \dfrac{18}{10}\right)$

$$= -\frac{6 \times 288 \times 8}{12 \times 18} - \frac{6 \times 144 \times 14}{12 \times 18} - \frac{6 \times 432 \times 9}{10 \times 18}$$

(1) $5.4M_B + 1.5M_C = -276.4 - 80.0 - 16.0$

(2) $1.5M_B + 6.6M_C = -64.0 - 56.0 - 129.6$

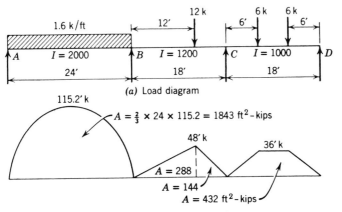

(a) Load diagram

$A = \frac{2}{3} \times 24 \times 115.2 = 1843$ ft^2-kips

$A = 288$

$A = 144$

$A = 432$ ft^2-kips

(b) Simple beam moments

Fig. 12.12

Multiplying Eq. 1 by the ratio 1.5/5.4 and subtracting 1*a* from 2*a* the resulting equations are

$$(2a) \quad 1.5M_B + 6.6 \quad M_C = -249.6$$

$$(1a) \quad \underline{1.5M_B + 0.417M_C = -103.4}$$

$$6.183M_C = \quad 146.2$$

$$M_C = - \quad 23.6 \text{ ft-kips}$$

$$M_B = \frac{-249.6 + 156.1}{1.5} = -62.3 \text{ ft-kips}$$

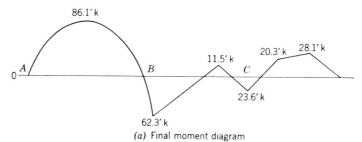

(a) Final moment diagram

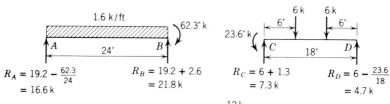

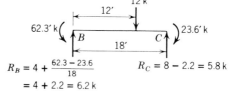

(b) Free-body diagrams

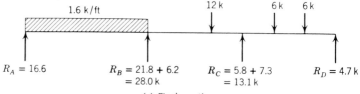

(c) Final reactions

Fig. 12.13

In the development of Eq. 12.1 all the moments were assumed positive. As the solution to this problem gives negative values for M_B and M_C it means there will be tension in the top of the beams over the supports.

The final moment diagram and the reactions are shown in Fig. 12.13.

Where the ends of the beam are fixed, the three-moment equations have to be written with an imaginary beam of zero length and a simple end support on each end of the actual beam. The procedure then follows as for the previous example. This is shown in Fig. 12.14.

EXAMPLE PROBLEM 12.4. Draw the shear and moment diagrams for the beam shown in Fig. 12.14a.

If an imaginary beam of zero length to the left of A is considered, the first three-moment equation is

$$2M_A(0 + 20) + M_B(20) = -\frac{6 \times 500 \times 10}{20}$$

(1) $$40M_A + 20M_B = -1500$$

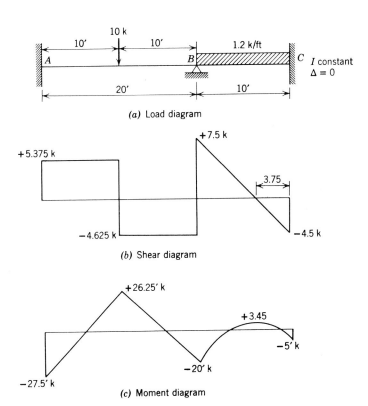

(a) Load diagram

(b) Shear diagram

(c) Moment diagram

Fig. 12.14

The second three-moment equation is

$$M_A(20) + 2M_B(20 + 10) + M_C(10) = -1500 - \frac{6 \times 100 \times 5}{10}$$

(2) $$20M_A + 60M_B + 10M_C = -1800$$

The third equation is written assuming a zero-length span to the right of C:

(3) $$M_B(10) + 2M_C(10 + 0) = -300$$

The three equations can be written

(1) $$20M_A + 10M_B \qquad\qquad = -750$$
(2) $$20M_A + 60M_B + 10M_C = -1800$$
(3) $$\qquad\qquad 10M_B + 20M_C = -300$$

Subtracting Eq. 1 from 2 and multiplying Eq. 3 by 5, we have

(1a) $$50M_B + 10M_C \ = -1050$$
(2b) $$50M_B + 100M_C = -1500$$

Subtracting (1a) from (2b)

$$90M_C = -450$$

$$M_C = -5 \text{ ft-kips}$$

$$M_B = \frac{1050 - 50}{50} = -20 \text{ ft-kips}$$

$$M_A = \frac{750 - 200}{20} = -27.5 \text{ ft-kips}$$

The shear and moment diagram are shown in Fig. 12.14b and c.

Settlement of supports can greatly alter the moments in a continuous beam. Suppose in the beam shown in Fig. 12.14a support B settled 0.05 ft and support C settled 0.04 ft, but support A did not settle. The three-moment equations will be the same as before, except they will contain the $6E\,\Delta I/L$ term. Since I is constant, it can be multiplied through and will appear only in the deflection term; E will be taken as 30×10^6 psi and I as 200 in^4. We must be careful to keep units straight, for all other terms are in kips and feet. The three-moment equations will now be

(1) $$40M_A + 20M_B = -1500 + \frac{6 \times 30 \times 10^3(-0.05)200 \times 12^2}{20(12)^4}$$

(2) $$20M_A + 60M_B + 10M_C = -1800 + \frac{6 \times 30 \times 10^3(+0.05)200 \times 12^2}{20(12)^4}$$

$$+ \frac{6 \times 30 \times 10^3(+0.01)200 \times 12}{10(12)^4}$$

(3) $\qquad 10M_B + 20M_C = -300 + \dfrac{6 \times 30 \times 10^3(-0.01)200 \times 12^2}{10(12)^4}$

(1a) $\qquad 20M_A + 10M_B = -750 - 312 = -1062$

(2a) $\qquad 20M_A + 60M_B + 10M_C = -1800 + 625 + 250 = -925$

(3a) $\qquad 10M_B + 20M_C = -300 - 250 = -550$

The solution of these three equations is

$$M_A = -57.7 \text{ ft-kips}$$
$$M_B = +9.2 \text{ ft-kips}$$
$$M_C = -32.1 \text{ ft-kips}$$

The new shear and moment diagrams are shown in Fig. 12.15. It is readily seen that small differential support settlements alter the reactions, shears, and moments considerably. The greater the I/L ratio for the spans, the greater is the effect of differential settlement. Foundation conditions

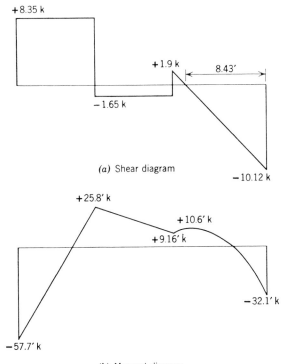

(a) Shear diagram

(b) Moment diagram

Fig. 12.15

have to be more accurately known for statically indeterminate structures than for those that are statically determinate.

As was stated previously, there are several other methods for analysis of continuous beams. In some instances these methods will prove more expedient than the theorem of three-moments. These other methods are treated in books dealing with the analysis of statically indeterminate structures.

(c) Continuous Beams–Maximum Shear and Moment Diagrams

It is common practice in continuous beam design to make the beam of variable moment of inertia along its length. For metal beams the moment of inertia is usually not changed gradually but is changed abruptly by adding plates to the flanges. In order to know the location of the changes in section when the beam is subjected to a moving load, it is usually necessary to plot *maximum* shear and *maximum* moment diagrams. This may also be so for a simple beam as well, where the cross section changes along the span.

To obtain the *maximum* shear diagram several specific points along the beam are chosen. The maximum shear at each point is calculated. The criterion developed in Section 4.8 is used to dictate the placement of the moving load for simple supported beams. After calculating the maximum total shear at the specific points, a maximum shear diagram can be plotted. Figure 12.16 shows a typical maximum shear diagram for a three-span continuous beam. It should be noted that on approximately one-half the span the maximum shear is positive and on the other portion it is negative.

The values of maximum moment at specific points along the beam are required in order to plot a maximum moment diagram. The moving load has to be placed in the position which will give the maximum possible live-load bending moment. From the typical maximum moment diagram of Fig. 12.37, it is seen that there are short portions of the beam where both

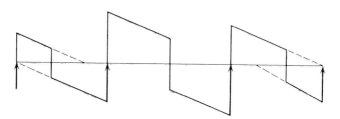

Fig. 12.16.   Maximum shear diagram.

Fig. 12.17.   Maximum moment diagram.

positive and negative moments can occur because of a moving load. Of course, these two values would occur with the live load in different positions.

(d) Continuous Beams on Elastic Supports

The majority of beams in engineering practice are considered to rest on nonyielding supports. However, with either continuous beams of short span or large moment of inertia, or both, a slight yielding of the supports will materially change the bending moments in the beam. This was illustrated in the beam of Fig. 12.15. All supports will, of course, yield to a certain extent when they are subjected to loads. The greater amount of the settlement will be elastic, especially for short duration loads. The effect of this elastic settlement of the support can be evaluated by using the three-moment equation (Eq. 12.1) derived previously. When we consider Fig. 12.17, where the spans are equal and have constant I and there is a concentrated load of P in the first span, the three-moment equation is

$$(M_A + 4M_B + M_C)L = -\sum PL^2(2r - 3r^2 + r^3) + \frac{6EI}{L}(\Delta_1 + \Delta_2)$$

If the supports are considered elastic,

$$\Delta_1 = k(R_B - R_A) \qquad \Delta_2 = k(R_B - R_C)$$

where k is the spring constant of the supporting medium in terms of inches per unit load. If P is taken as unity, the usual case for developing influence diagrams, the three-moment equation is then

$$(M_A + 4M_B + M_C)\frac{1}{L} = -(2r - 3r^2 + r^3) - \frac{6EI}{L^3}k(R_A - 2R_B + R_C)$$

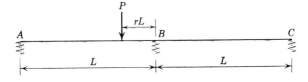

Fig. 12.18.   Continuous beam on elastic supports.

If the ends of the beam are considered simply supported and

$$\frac{6EIk}{L^3} = W \qquad (12.2)$$

$$WR_A - 2WR_B + (4 + W)R_C = -(2r - 3r^2 + r^3)$$

Writing the other two necessary equations

$$R_B + 2R_C = (1 - r)$$

$$R_A + R_B + R_C = 1$$

and solving these equations the values of the reactions are

$$R_A = \frac{2r + 3r^2 - r^3 + W(2 + 3r)}{4 + 6W}$$

$$R_B = \frac{4 - 6r^2 + 2r^3 + 2W}{4 + 6W}$$

$$R_C = \frac{-2r + 3r^2 - r^3 + W(2 - 3r)}{4 + 6W}$$

For $k = 0$ and therefore $W = 0$, the expressions for the reactions will be those values for which the supports are nonyielding.

EXAMPLE PROBLEM 12.5   Determine the maximum reactions for a three-span continuous steel beam of 12-ft span with a 1 kip moving load. Assume the supports rest on soil of which $k = 0.10$ in./kip and I is equal to 830 in.[4]

$$W = \frac{6 \times 30,000 \times 830 \times 0.10}{(12)^3 \times (12)^3} = \frac{14.9 \times 10^6}{2.98 \times 10^6} = 5.0$$

A comparison of the maximum values of the reactions where W equals 0, 5, and 10 is

	$W = 0$	$W = 5$	$W = 10$
R_A	1.0000	0.8529	0.8437
R_B	1.0000	0.4118	0.3750

It is seen that the value of the center reaction decreases considerably even for slight yielding of the supports. This yielding of the supports would greatly alter the moments in the beam.

For three equal spans the equations for the reactions are

LOAD IN FIRST SPAN

$$R_A = \frac{W[9 + 8W + r(31 + 6W + 18r - 6r^2)] + 7r + 12r^2 - 4r^3}{15 + 56W + 20W^2}$$

$$R_B = \frac{15 + W[32 + 6W - r(7 - 2W + 24r - 8r^2)] + 3r - 27r^2 + 9r^3}{15 + 56W + 20W^2}$$

$$R_C = \frac{W[21 + 4W - r(23 + 2W + 6r - 2r^2)] - 6(2r - 3r^2 + r^3)}{15 + 56W + 20W^2}$$

$$R_D = \frac{W[-6 + 2W - r(1 + 6W - 12r + 4r^2)] + 2r - 3r^2 + r^3}{15 + 56W + 20W^2}$$

where r is measured from the second reaction R_B.

LOAD IN SECOND SPAN

$$R_A = \frac{W[-6 + 2W + r(1 + 6W + 12r + 2r^2)] - r(2 + 3r - 5r^2)}{15 + 56W + 20W^2}$$

$$R_B = \frac{W[21 + 4W + r(23 + 2W - 6r - 6r^2)] + r(12 + 18r - 15r^2)}{15 + 56W + 20W^2}$$

$$R_C = \frac{15 + W[32 + 6W + r(7 - 2W - 24r + 6r^2)] - r(3 + 27r - 15r^2)}{15 + 56W + 20W^2}$$

$$R_D = \frac{W[9 + 8W - r(31 + 6W - 18r + 2r^2)] - r(7 - 12r + 5r^2)}{15 + 56W + 20W^2}$$

where r is measured from the third reaction R_C.

The maximum values of reaction for values of W of 0, 5, and 10 for the three-span beam are

	$W = 0$	$W = 5$	$W = 10$
R_A	1.0000	0.7862	0.7515
R_B	1.0000	0.4088	0.3631

Beyond a three-span beam the general expressions for reaction become too involved for general solution, and expressions for reactions at specific values of W is the most expedient procedure. Appendix A gives the influence lines for reaction for continuous beams up to and including four equal spans. Reference (2) gives tables and charts for influence lines for continuous beams up to twelve equal spans and up to four variable spans.

The yielding of the supports will, of course, change the bending moments in a continuous beam. For instance, if a three-span continuous beam, as shown by Fig. 12.19a is subjected to concentrated loads at midspan, the reactions for no yielding of the supports ($W = 0$) as obtained from the charts of Appendix A are

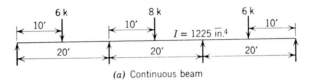

(a) Continuous beam

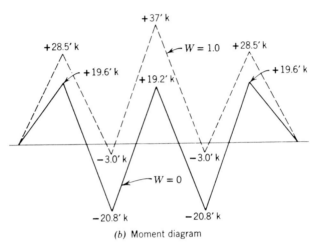

(b) Moment diagram

Fig. 12.19. Effect on bending moment of yielding supports.

$$R_a = 0.4 \times 6 - 0.073 \times 8 + 0.024 \times 6 = 1.96$$
$$R_b = 0.725 \times 6 + 0.576 \times 8 - 0.152 \times 6 = +8.04$$

The bending moment diagram for nonyielding supports is shown by the solid line of Fig. 12.19b. If the supports are resting on soil or relatively long columns, there will be an elastic deformation of the supporting structure. If investigations indicate that this yielding will be elastic and of the magnitude of 0.075 in. per kip, the value of W is

$$W = \frac{6EIK}{L^3} = \frac{6 \times 30 \times 10^6 \times 1225 \times 0.075 \times 10^{-3}}{(240)^3} = 1.0$$

The reactions for $W = 1.0$ are

$$R_a = 0.498 \times 6 + 0.019 \times 8 - 0.05 \times 6 = +2.85$$
$$R_b = 0.453 \times 6 + 0.48 \times 8 + 0.098 \times 6 = +7.15$$

The bending moment diagram for this condition is shown by the dashed line of Fig. 12.19b.

12.3 Analysis of Frames

The method of consistant displacement (flexibility method) will be used to analyze the bent shown in Figure 12.20. This bent has only one redundant since both legs are pinned at the base. Of the four reactions, any one can be chosen as the redundant. For this example, the horizontal reaction at d will be selected as the redundant reaction.

If there is no movement of support d, the basic equation can be written as

$$\Delta_d + \delta_{dd}(R_{dx}) = 0$$

The solution of Δ_d and δ_{dd} are given in the following example.

EXAMPLE PROBLEM 12.7 The statically determinate primary structure is as shown in Fig. 12.21(a).

The method of virtual work will be used to determine the deflections. The values of M and m are calculated from the forces on Fig. 12.21 (a and b).

	M	m	limits
$d \to c$	0	x	$0 \to 12$
$c \to e$	$8x$	12	$0 \to 9$
$e \to b$	72	12	$9 \to 18$
$a \to b$	$6x$	x	$0 \to 12$

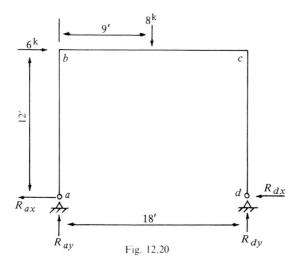

Fig. 12.20

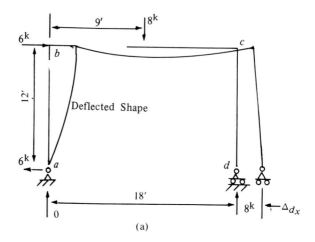

(a)

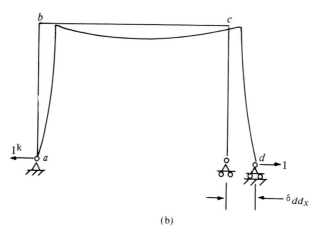

(b)

Fig. 12.21

$$EI\Delta_d = \int_0^9 (96x)\,dx + \int_9^{18} (864)\,dx + \int_0^{12} (6x^2)\,dx$$

$$= [48x^3]_0^9 + [864x]_9^{18} + [2x^3]_0^{12}$$

$$= 3888 + 7776 + 3456 = 15{,}120 \text{ ft}^3\text{-k}$$

$$EI\delta_{dd} = \int m^2\,dx = \int_0^{12} x^2\,dx + \int_0^9 144\,dx + \int_9^{18} 144\,dx + \int_0^{12} x^2\,dx$$

$$= 576 + 1296 + 1296 + 576 = 3744 \text{ ft}^3\text{-k}$$

$$\Delta_d + \delta_{dd}(R_{dx}) = 0$$

$$10{,}469 + 3744\,R_{dx} = 0$$

$$R_{dx} = -\frac{15{,}120}{3744} = -4.04\,\text{k}$$

The minus sign indicates that the reaction R_{dx} must be in the opposite direction of the unit load producing δ_{dd}.

It is of interest to note the effect on the value of the horizontal reactions at the supports if there were a yielding of the supports. Take the case of a yielding in the horizontal direction of one-fourth inch. Using a value of $E = 30 \times 10^3$ ksi, and $I = 2000$ in^4 the equation of consistant deflections is

$$\Delta_d + \delta_{dd}(R_{dx}) = 0.25 \text{ in.}$$

$$\frac{15{,}120(12)^3}{(30 \times 10^3)(2000)} + \frac{3{,}744(12)^3}{(30 \times 10^3)(2000)}R_{dx} = 0.25$$

$$15{,}120 + 3{,}744\,R_{dx} = 0.25\,(34{,}722) = 8680$$

$$R_{dx} = \frac{-6440}{3744} = -1.72\,\text{k}$$

12.4 Analysis of Trusses

Trusses can be either statically indeterminate externally or internally (See Section 5.2). In most actual cases, they are statically indeterminate externally. That is, they are continuous over several supports. Many long span bridges are of this nature. Once the reactions are determined, the bar forces can be determined from the equations of equilibrium.

The redundant reactions can be calculated from the equations of consistant deflections just as was done for the continuous beam. The deflections of the determinate primary structure can be calculated using the method of virtual work (Section 10.6).

As an example, consider the cantilever truss of Figure 12.22. This truss has a non-yielding roller support at point i. The objective is to determine the magnitude of the reaction at this support. The consistant deflection equation, using R_i as the redundant, is:

$$\Delta_i + \delta_{ii}R_i = 0$$

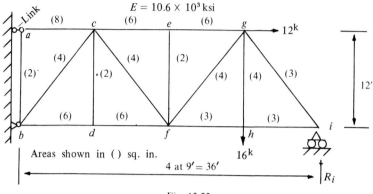

Fig. 12.22

The value of Δ_i is determined from the expression

$$\Delta_i = \sum \frac{SuL}{AE}.$$

This value is 0.807 in., as calculated in Table 10.3 The value of δ_{ii} will be determined from the equation

$$\delta_{ii} = \sum \frac{u^2 L}{AE}.$$

The deflection at i due to a unit load at i can be found from this expression where u is the bar force due to a unit load at point i. Since the values of the u bar forces have already been calculated, it is just a matter of summing up the values of $\frac{u^2 L}{A}$. The necessary calculations are shown in Table 12.1 (bars having $u = 0$ are omitted).

$$\Delta_i + \delta_{ii} R_i = 0$$

Since the net deflection is zero, the values do not have to be mutliplied by 12 or divided by E.

$$713.25 + 60.84 R_i = 0 \qquad R_i = -11.72\,\text{k (up)}$$

The bar forces in the statically indeterminate structure can now be easily calculated by superimposing on the S bar forces the values of u bar forces multiplied by the reaction R_i.

The structure could also have been reduced to a determinate truss by

Table 12.1

Member	u	L ft.	A in.^{2}	$\dfrac{u^2 L}{A}$
ac	+3.00	9	8	10.13
cg	+1.50	18	6	6.75
bf	−2.25	18	6	15.19
fi	−0.75	18	3	3.38
bc	−1.25	18	4	5.86
cf	+1.25	15	4	5.86
fg	−1.25	15	4	5.86
gi	+1.25	15	3	7.81

$$\Sigma = 60.84 \, \text{k-ft/in}^2$$

cutting bar *ac*. The concept of consistant deflection would then have been to calculate the separation of the ends of the bar *ac* (Δ_{ac}) due to the real loads and then to calculate the separation due to a unit load on bar *ac* (δ_{ac}). The equation could then be formulated as previously only as follows:

$$\Delta_{ac} + \delta_{ac}(F_{ac}) = 0$$

with F_{ac} equal to the force in bar *ac* due to the real loads on the statically indeterminate structure.

The redundants of continuous trusses of several spans can be either reactions or chord members, usually those over the supports. It is obvious that the numerical work in calculating deflections in a continuous truss of several spans would be considerable. Considerable work can sometimes be saved if the judicious choice of redundant would result in a repetition of displacement values. Consider the truss of Figure 12.23. If the redundants chosen are R_b and R_c then the deflections at *b* and *c* for the determinate structure would have to be calculated. In the case where the two end spans are similar $\Delta_b = \Delta_c$, and $\delta_{bb} = \delta_{cc}$. However, if this were not the case, then bar forces would have to be calculated as follows:

$$S = \text{bar forces for real loads}$$

$$u_b = \text{bar forces for unit load at } b$$

$$u_c = \text{bar forces for unit load at } c$$

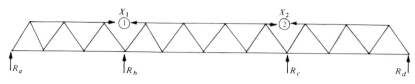

Fig. 12.23

Then all the coefficients of the redundant matrix R and the deflection matrix Δ can be calculated.

$$\begin{bmatrix} \delta_{bb} & \delta_{bc} \\ \\ \delta_{cb} & \delta_{cc} \end{bmatrix} \begin{bmatrix} R_b \\ \\ R_c \end{bmatrix} = \begin{bmatrix} \Delta_b \\ \\ \Delta_c \end{bmatrix}$$

The solution of R_b and R_c is then performed. It is obvious that much numerical work would be required. This is especially so if the structure is a bridge and influence line ordinates for each bar are needed.

If the redundants are taken as the bar forces X_1 and X_2 then the determinate structure is three independent simple span trusses. In determining the separation of bar 1 (Δ_1) the bar forces in only the first two trusses will have to be calculated. There will be no bar forces in the third truss. Likewise, the displacement of the cut ends of bar 1 for a unit force in bar 1 will only result in u bar forces in the first two trusses and zero values of force in the bars of the third truss. It is therefore apparent that there will be a considerable saving of time and effort if bar forces X_1 and X_2 are selected as the redundants instead of R_b and R_c. The saving of work by using bar forces as redundants would be further increased if the structure were not symmetrical about a center line.

By now it is apparent that a computer program could be a great time saver when designing statically indeterminate trusses. When this tool is used, a matrix method of analysis is usually preferable. This will be discussed further in Chapter 14.

12.5 Analysis of Two-hinged Arches

The arch has been used in structures for centuries. The Romans used the half-circular stone arch in their buildings, bridges, and aqueducts. The arch is still a popular form today for buildings and bridges.

The three hinged arch was presented in Chapter 6. The two hinged and fixed-end arch are statically indeterminate forms. There are two common arch shapes, the circular (full half-circle or segment of a circle) and the parabolic. The circular shape is more apt to be used for buildings because of increased head room. The parabolic shape is used more for bridges because of better economy and appearance.

The two-hinged circular arch will be analyzed first. Fig. 12.24 shows such an arch. This structure is statically indeterminate to the first degree. The primary structure can be selected as having the right support on a roller so that horizontal translation can take place. This selection makes H_R the redundant. The basic flexibility equation is then

$$D_1 + f_{11}(H_R) = \text{horizontal translation of right support}$$

The displacement of the right support of the primary structure is D_1 and is determined from the virtual work equation as

$$D_1 = \int \frac{Mmds}{EI}$$

The flexibility coefficient f_{11} is

$$f_{11} = \int \frac{m^2 ds}{EI}$$

The above two expressions are based upon displacements due to bending strains only. Neglecting the effect of shear and axial force (thrust) deformations, results in very minor differences in the value of the redundant.

For a circular curve, the values of M, m, and ds can be written in terms of the radius of the arch r and the variable angle α.

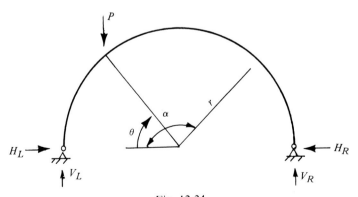

Fig. 12.24

The value of the redundant H_R will be determined for a concentrated vertical load P at any location θ on the arch axis.

The primary structure and its reactions are shown in Fig. 12.25. Also shown is the primary structure with the unit load in the x direction at the right support.

$$V_L = \frac{P(r + r\cos\theta)}{2r} = \frac{P}{2}(1 + \cos\theta)$$

$$V_R = P - \frac{P}{2}(1 + \cos\theta) = \frac{P}{2}(1 - \cos\theta)$$

Since the values of M are not continuous over the full half-circle, M will be evaluated from a to b and then from c to b.

$$a \text{ to } b: M = V_L(x) = V_L r(1 - \cos\alpha) = \frac{Pr}{2}(1 + \cos\theta)(1 - \cos\alpha)$$

$$c \text{ to } b: M = V_R(x) = V_R r(1 - \cos\alpha) = \frac{Pr}{2}(1 - \cos\theta)(1 - \cos\alpha)$$

The value of m is continuous over the entire arch

$$m = 1(r\sin\alpha)$$

The values of D_1 and f_{11} will now be determined.

$$ds = rd\alpha$$

$$D_1 = \int_0^s \frac{Mmds}{EI} = \int_0^\theta \frac{Pr^3}{2EI}(1 + \cos\theta)(1 - \cos\alpha)(\sin\alpha)d\alpha$$

$$+ \int_0^{(\pi-\theta)} \frac{Pr^3}{2EI}(1 - \cos\theta)(1 - \cos\alpha)(\sin\alpha)d\alpha$$

$$f_{11} = \int_0^s \frac{m^2 ds}{EI} = \int_0^\pi \frac{r^3}{EI}\sin^2\alpha d\alpha$$

The values of D_1 and f_{11} will be evaluated for a specific value of $\theta = (45°)$ and a constant EI over the entire arch.

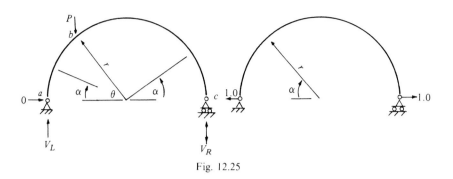

Fig. 12.25

$$EID_1 = 0.8535\,Pr^3 \int_0^{\pi/4} (\sin\alpha - \sin\alpha\cos\alpha)d\alpha$$

$$+\, 0.1465\,Pr^3 \int_0^{3\pi/4} (\sin\alpha - \sin\alpha\cos\alpha)d\alpha$$

$$= 0.8535\,Pr^3 \left[-\cos\alpha + \frac{\cos 2\alpha}{4} \right]_0^{\pi/4}$$

$$+\, 0.1465\,Pr^3 \left[-\cos\alpha + \frac{\cos 2\alpha}{4} \right]_0^{3\pi/4}$$

$$\cos(0) = 1.0; \quad \cos\left(\frac{\pi}{4}\right) = 0.7071; \quad \cos\left(\frac{\pi}{2}\right) = 0; \quad \cos\left(\frac{3\pi}{4}\right) = -0.7071$$

$$EID_1 = 0.8535\,Pr^3 [-0.7071 + 1.0 - 0.25]$$

$$+\, 0.1465\,Pr^3 [0.7071 + 1.0 - 0.25]$$

$$= Pr^3(0.0366 + 0.2135) = 0.25\,Pr^3$$

$$EIf_{11} = r^3 \int_0^\pi (\sin^2\alpha)d\alpha = r^3 \left[\frac{\alpha}{2} - \frac{\sin 2\alpha}{4} \right]_0^\pi$$

$$= r^3\left(\frac{\pi}{2}\right) = 1.57\,r^3$$

The value of H_R for no support translation is

$$H_R = -\frac{0.25}{1.571}(P) = -0.159P$$

The (−) sign indicates that the value of H_R is to the left, opposite to the direction of the unit force.

The two-hinged parabolic arch can be analyzed in a manner similar to the circular arch. However, in this case the independent variable is the distance along the x axis. Figure 12.26 shows a parabolic arch with rise of h and a concentrated load P. The equation of the arch axis is

$$y = 4h\left(\frac{x}{L} + \frac{x^2}{L^2}\right)$$

The equation for determining the horizontal reaction at one support neglecting shear and axial strains is:

$$H = -\frac{\displaystyle\int \frac{Mmds}{EI}}{\displaystyle\int \frac{m^2 ds}{EI}}$$

The values of M and m can be written in terms of x and y respectively. The values of y can be converted to x by the equation of the arch. However, it is necessary to convert ds to dx in order to perform the integration. This can be done by the trignometric relationship

$$ds = dx(\sec \alpha)$$

This, however, leaves a $\sec \alpha$ term in the integrals. The usual manner of getting around this difficulty is to introduce the relationship of the variation of the moment of inertia over the arch axis. This relationship is

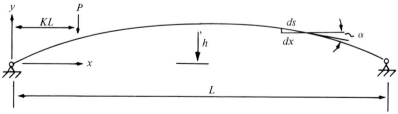

Fig. 12.26 Two-hinged parabolic arch.

$$I = I_c \sec\alpha.$$

Where I is the moment of inertia at any location and I_c is the moment of inertia at the crown. Designs of longer span parabolic arches have shown that the expression $I = I_c \sec\alpha$ is nearly correct where the cross-section of the arch is varied to meet the applied bending moments and thrusts. The equation can now be written

$$H = -\frac{\int_0^L My\,dx}{\int_0^L y^2\,dx}$$

The denominator in the above equation can be evaluated for any two-hinged parabolic arch with non-yielding supports located at the same elevation.

$$\int_0^L y^2\,dx = \frac{16h^2}{L^2}\int_0^L \left(x^2 - \frac{2x^3}{L} + \frac{x^4}{L^2}\right)dx$$

$$= \frac{16h^2}{L^2}\left[\frac{x^3}{3} - \frac{x^4}{2L} + \frac{x^5}{5L^2}\right]_0^L = 0.533\,h^2 L$$

If we now evaluate $\int_0^L My\,dx$ for an arch with a concentrated load at a distance KL from the left support

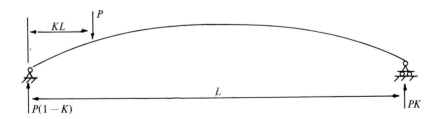

from a to b: $M = P(1 - K)(x)$

from c to b: $M = PK(x)$

$$\int Mydx = \frac{4hP}{L^2}(1-K)\int_0^{KL} x\left(x - \frac{x^2}{L}\right)dx$$

$$+ \frac{4hPK}{L^2}\int_0^{L(1-K)} x\left(x - \frac{x^2}{L}\right)dx$$

$$= \frac{PhL^2K(1-K)}{3}[K^3 + (1-K)^3 + 4K(1-K)]$$

$$H = -\frac{\int Mydx}{\int y^2 dx} = -\frac{5PLK(1-K)(1+K-K^2)}{8h}$$

It is desired to determine the reactions for the arch shown below. The value of H at the support is

$$H = -\frac{5PLK(1-K)(1+K-K^2)}{8h}$$

$$K = 0.3; \qquad 1 - K = 0.7; \qquad h/L = 0.12$$

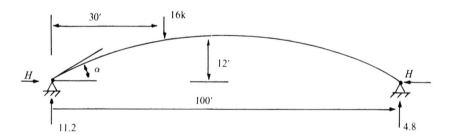

$$H = -\frac{5(16)(0.3)(0.7)(1 + 0.3 - 0.09)}{8(0.12)} = -21.2^k$$

To determine the thrust at the springing:

$$T = \sqrt{(11.2)^2 + (21.2)^2} = 24^k$$

To determine the shear at the springing the angle of the arch rib will have to be determined at this location.

$$\tan \alpha = \frac{dy}{dx} = \frac{4h}{L}\left(1 - \frac{2x}{L}\right)$$

at

$$x = 0 \qquad \frac{dy}{dx} = 4(0.12) = 0.48 = \tan \alpha$$

$$\alpha = 25.6°$$

$$\text{shear} = 11.2 \cos \alpha - 21.2 \sin \alpha = 0.93 \, k$$

If there is not sufficient strength at the supports to resist the thrust then an arch can be made by tying the ends together. This is called a *tied arch*. The equation is then

$$D_1 + f_{11}(H) = \frac{HL}{AE}$$

The 100 ft steel arch previously treated (Fig. 12.27) is to be changed to a tied arch with a steel tie rod one inch in diameter.

$$A = 0.785 \text{ in}^2 = 0.00545 \text{ ft}^2$$

$$E = 30 \times 10^3 \text{ ksi}$$

$$\frac{1}{EI_c}\int Mydx = \frac{16(12)(100)^2(0.3)(0.7)}{3EI_c}((0.3)^3 + (0.7)^3 + 4(0.3)(0.7))$$

$$= \frac{162{,}624}{EI_c}$$

$$\frac{1}{EI_c}\int y^2 dx = \frac{0.533(12)^2(100)}{EI_c} = \frac{7675}{EI_c}$$

$$\frac{162{,}624}{EI_c} + H\left(\frac{7675}{EI_c}\right) + \frac{H(100)}{0.00545E} = 0$$

If the I_c of the arch rib is equal to 1200 in^4 = 0.0579 ft^4

$$\frac{H(7675)}{0.0579} + \frac{H(100)}{0.00545} = -\frac{162.624}{0.0579}$$

$$H(132{,}560 + 18{,}350) = -2{,}809(10)^6$$

$$H = 18.6^k$$

Because of the elongation of the tie the horizontal reaction decreases 2.6 kips.

References

1. Carpenter, S. T., *Structural Mechanics,* John Wiley, 1960.
2. Firmage, D. A., and R. H. Chiu, *Influence Lines for Continuous Beams on Elastic Supports,* Brigham Young University Press, Provo, Utah, 1961.

Problems

1. Calculate the reactions for the beam, Fig. 12.27.

2. Calculate the reactions for the beam of Fig. 12.27. $EI = 3 \times 10^6 \text{k-in}^2$ Support *b* settles one-half inch.

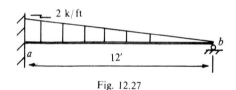

Fig. 12.27

3. Plot the bending moment diagram for the beam of Fig. 12.28. Use flexibility method to determine reactions.

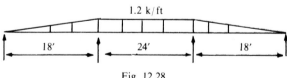

Fig. 12.28

4. Determine the reactions for the continuous beam of Fig. 12.29 using the three-moment equation.

Ans. $R_A = 6.7$ kips, $R_B = 24.9$ kips, $R_C = 6.4$ kips.

5. Draw the shear and moment diagrams for the beam of Fig. 12.30 (*E* is constant, no support settlement). Use three-moment equation.

6. Draw the shear and moment diagrams for the continuous beam of Fig. 12.31 (*E* is constant, no support settlement). Use three-moment equation.

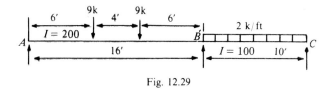

Fig. 12.29

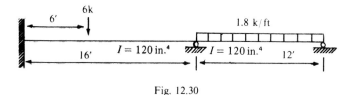

Fig. 12.30

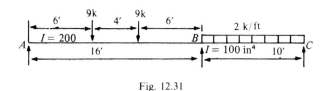

Fig. 12.31

7. If the beam in Fig. 12.31 has a constant E of 30×10^6 psi, what is the change in maximum shear and maximum moment if support B settles 0.24 in.?

8. Draw the shear and moment diagrams for the beam of Fig. 12.31 if both supports B and C settle 0.24 in. E and I are the same as for Problem 7.

9. Determine the reactions for a three-span continuous beam, with ends simple supported and all spans equal to 10 ft, for a 6-kip load at midspan of each span. The moment of inertia of the beam is constant at 800 in^4 and $E = 30 \times 10^6$ psi. The supports are elastic and have a coefficient of 0.25 in. per kip. $Ans.$ $R_a = 3.32$ kips, $R_b = 5.687$ kips.

10. A floating bridge is to be built with continuous girder supported by pontoons as shown in Fig. 12.32.

(a) How would a change in area (length × width) of the pontoons effect the bending moment in the girder due to live loads?

(b) What effect would increasing the moment of inertia of the girder have on the bending moment in the girder due to a truck loading?

(c) Would changing the girder from steel to aluminum (same I) require pontoons of larger or smaller area to maintain the same freeboard?

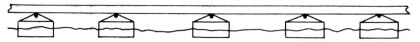

Fig. 12.32. Floating bridge.

11. Calculate the reactins for the structure shown in Fig. 12.33. Use flexibility method. Constant I.

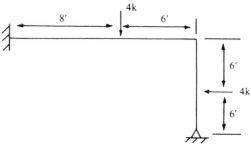

Fig. 12.33

12. Determine the reactions for the structure Fig. 12.34 using the flexibility method and plot the bending moment diagram on the tension side.

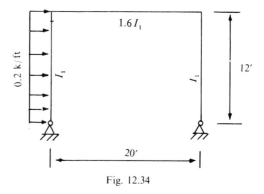

Fig. 12.34

13. For the frame in Fig. 12.35 calculate the values of the reactions. I constant.

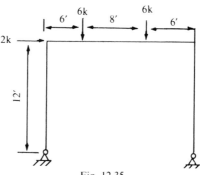

Fig. 12.35

14. Plot the bending moment diagram on the tension side for the frame of Fig. 12.36. *I* is constant.

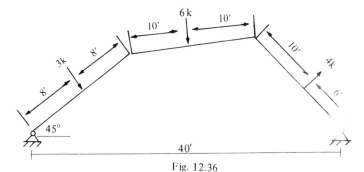

Fig. 12.36

15. Determine the force in the tie rod of the structure of Fig.

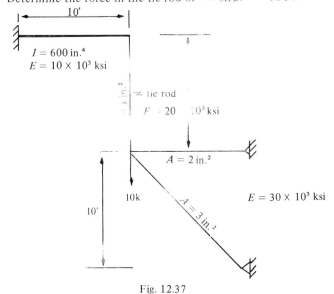

Fig. 12.37

16. Determine the reactions for the symmetrical truss of Fig. 12.38. Areas are in (in²). *E* constant.

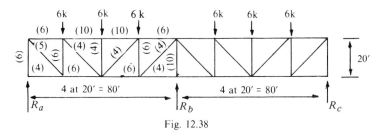

Fig. 12.38

17. Determine the thrust, shear and bending moment at the springings and crown for the circular arch of Fig. 12.39.

18. Determine the reactions for the arch of Fig. 12.39 for a 10 kip force in the radial direction instead of horizontal.

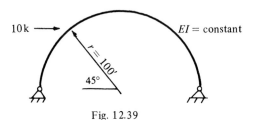

10k EI = constant

$r = 100'$

45°

Fig. 12.39

19. Determine the reactions for the segmental circular arch of Fig. 12.40.

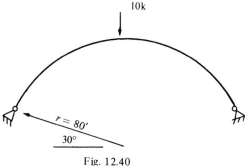

10k

$r = 80'$

30°

Fig. 12.40

20. Determine the thrust, shear, and bending moment at the crown for the parabolic arch of Fig. 12.41. $I = I_c \sec \alpha$

21. Work problem 20 if the supports were connected together by a steel cable of one inch diameter (see Appendix B). The value of $I_c = 1800 \text{ in}^4$. E of arch rib is 29.6×10^3 ksi. Footings can slide.

22. Determine the thrust, shear, and bending moment at the springing, one-quarter of the span and the crown for the arch of Fig. 12.41 with a uniform load of 0.6 k/ft of span.

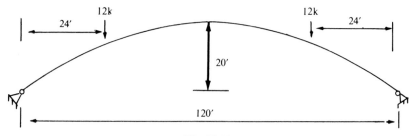

12k   12k

24'   24'

20'

120'

Fig. 12.41

23. Determine the reactions for the parabolic arch of Fig. 12.42. The force is perpendicular to the arch rib. $I = I_c \sec \alpha$

24. Find the reactions for the arch of Fig. 12.42 if the supports separated one-half inch. $E = 30 \times 10^3$ ksi $I_c = 1600 \text{ in}^4$

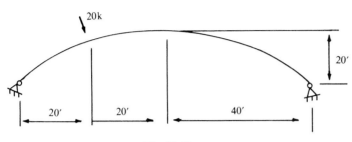

Fig. 12.42

SLOPE DEFLECTION AND MOMENT DISTRIBUTION

13.1 Introduction—Slope Deflection

As land became more valuable in the cities, economic necessity required buildings of increased story height. Multi-story buildings introduced special design problems. Load bearing walls were not useable for these taller structures. Lateral loads could be resisted by extensive lateral bracing in the form of diagonal ties or a braced frame where the girders were rigidly connected to the columns so that frame action could take place. Cross-diagonals provided architechural restrictions while rigid frames required sophisticated analysis procedures. In 1915 Professor G. A. Maney published a method of analysis that could be applied to rigid frame structures. This method was called the slope-deflection method and was derived from the basic concepts of the moments in a structural member that is due to loads, end rotations, and end displacements. The development of this method resulted in a set of simultaneous equations that related to the statical equilibrium of the structure. In this method the number of equations of statical equilibrium are equal to the number of degrees of freedom of the structure. For a structural frame of several stories and several lines of columns, there would be many simultaneous equations. Prior to about 1950, the slope-deflection method had limited practical use because of the difficulty of solving a large number of simultaneous equations. Nevertheless, this method was widely taught in structural analysis courses. The understanding of this method develops a good comprehension of the physical relationship of displacements to stresses in a building frame.

The slope-deflection method was the background on which Professor Hardy Cross developed his method of moment distribution. The method of moment distribution is just a numerical iteration of the solution of the slope deflection simultaneous equations. The later development of the digital high speed computer made it possible to solve large numbers of simultaneous equations with relative ease. This gave a practicality to the slope deflection method it never had before. So, now the tables are reversed, and the slope-deflection method is more practical than the moment distribution for the analysis of building frames with many degrees of freedom.

There is another approach to the solution of structures with many degrees of indeterminacy that is preferred by engineers who have the availability of a computer. This method will be given in the next chapter. However, the understanding of the slope deflection method will be of value to the student in understanding moment-distribution and in gaining a "feel" for the action of structures.

13.2   Development of Slope-Deflection Equations

The development of the basic slope deflection equation is relatively simple. Consider a structural element of constant section, having a length of L between fixed ends and acted upon by any load system (Fig. 13.1 (a)). Bending moments are induced in the member due to the applied loads. The moments at the support a and b are given the designation $M_{F_{ab}}$ and $M_{F_{ba}}$, respectively. These are called the fixed end moments (F.E.M.). The sign convention is that if the moments applied to the support (by the beam) is clockwise, the sign is positive, while the opposite is, of course, negative. If the moment applied to the support by the beam is clockwise (+) then the moment applied to the beam by the support is counterclockwise. In Figure 13.1 (a) the fixed end moment on the left end is positive and on the right end, negative.

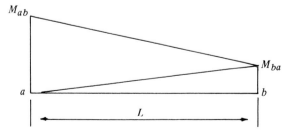

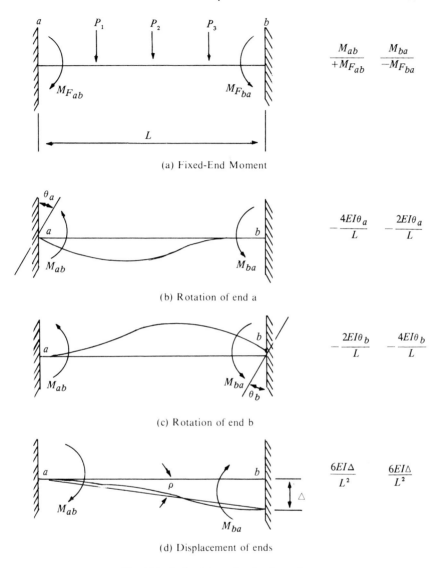

M_{ab}	M_{ba}
$+M_{F_{ab}}$	$-M_{F_{ba}}$

(a) Fixed-End Moment

$-\dfrac{4EI\theta_a}{L}$	$-\dfrac{2EI\theta_a}{L}$

(b) Rotation of end a

$-\dfrac{2EI\theta_b}{L}$	$-\dfrac{4EI\theta_b}{L}$

(c) Rotation of end b

$\dfrac{6EI\Delta}{L^2}$	$\dfrac{6EI\Delta}{L^2}$

(d) Displacement of ends

Fig. 13.1. Actions on a structural member.

The next step is to consider the same beam subjected only to a clockwise rotation of the left end an amount θ_a (Fig. 13.1 (b)). This rotation of end a will produce counterclockwise moments M_{ab} at end a and M_{ba} at end b. The magnitude of these moments are directly proportional to the magnitude of θ_a (elastic condition). For a prismatic beam, the values of moment can be easily determined by using the moment-area principles of rotation and deflection. The moment diagram for the condition of end rotation is to left.

The change is slope between a and b is equal to θ_a and is also equal to the area of the M/EI diagram between a and b. Therefore

$$EI\theta_a = \frac{M_{ab}(L)}{2} + \frac{M_{ba}(L)}{2}$$

Since this equation has two unknown moments, we will have to develop a second equation, from a physical relationship, that contains the unknown end moments. For this second equation, the second moment area principle is used. When there is no translation of ends a and b, the deflection of point a from the tangent to the elastic line at point b is equal to zero. In accordance with the second moment area principle, the moment of the M/EI diagram about point a is zero. This principle can be stated in equation form as:

$$\frac{M_{ab}(L)}{2} \times \frac{L}{3} + \frac{M_{ba}(L)}{2} \times \frac{2L}{3} = 0$$

Therefore, from the above equation

$$M_{ab} = -2M_{ba}$$

substituting in the first equation

$$EI\theta_a = -M_{ba}(L) + \frac{M_{ba}(L)}{2} = -\frac{M_{ba}(L)}{2}$$

or

$$M_{ba} = -\frac{2EI\theta_a}{L}$$

and

$$M_{ab} = \frac{4EI\theta_a}{L} \tag{13.1}$$

The above solution gives value of M_{ab} and M_{ba} in terms of $\theta_a, E, I,$ and L. It is also seen that the moment at the end where the rotation takes place is twice the moment at the opposite end. The signs also give the direction of the moments. The moment-area method gives signs in accordance with the beam sign convention; positive moment produces tension in the bottom of the beam and negative moment produces compression. Therefore, if θ_a is taken as positive, then M_{ba} is negative and causes compression in the bottom of the beam and is thus counterclockwise as shown in Fig. 13.1 (b). Observation of the shape of the elastic line confirms this direction of M_{ba}. The direction of M_{ab} can also be determined from the sign given in the moment-area solution. It is also counterclockwise for a clockwise rotation (+) of end a.

The next step in the slope-deflection development is to rotate the end b of the beam an amount θ_b in the clockwise (+) direction. From the developments of end moments for a rotation θ_a, it is obvious that the relationship of end moments to rotation θ_b are

$$M_{ab} = -\frac{2EI\theta_b}{L} \qquad M_{ba} = -\frac{4EI\theta_b}{L}$$

(signs relate to beam sign convention)

The last of the possible actions that may take place is a relative translation of one end of the member with respect to the other. This displacement (Δ) is measured perpendicular to the axis of the member. This displacement is shown in Fig. 13.1 (d). The sign convention for this translation is that if the line connecting the ends of the beam rotates through a clockwise angle ($\rho = \Delta/L$) the sign of Δ is positive.

The values of the end moments caused by the displacement Δ will now be developed, again using the moment-area principles. The change in slope of the tangents to the elastic line from a to b is equal to the area of the M/EI diagrams between a and b. Since the ends of the beam do not rotate, but only displace, then the change in slope and the area of the M/EI diagram between a and b is equal to zero. Writing this in equation form from the moment diagram

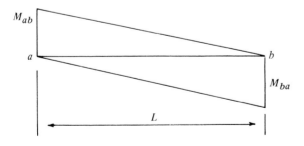

$$\frac{M_{ab}(L)}{2EI} + \frac{M_{ba}(L)}{2EI} = 0$$

$$M_{ab} = -M_{ba}$$

This relationship between M_{ab} and M_{ba} could have been deduced from symmetry.

Using the second moment-area principle, the value of Δ is equal to the moment about b of the area of the M/EI diagram between a and b. In equation form:

$$\frac{M_{ab}(L)}{2EI}\left(\frac{2L}{3}\right) + \frac{M_{ba}(L)}{2EI}\left(\frac{L}{3}\right) = \Delta$$

Substituting $-M_{ab}$ for M_{ba} in the above equation and solving for M_{ab}:

$$M_{ab} = \frac{6EI\Delta}{L} \qquad (13.2)$$

The positive value of Δ (clockwise rotation of the chord connecting the ends) will produce a clockwise moment at a and b.

Referring back to Figure 13.1, one can see the values, and directions, of the end moments due to the applied loads, plus the end rotations, plus the normal displacement of the ends of the members. The total end moments will be the sum of these four end moments. Making this summation, the equations for end moments are:

$$M_{ab} = M_{F_{ab}} - \frac{2EI}{L}\left(2\theta_a + \theta_b - \frac{3\Delta}{L}\right)$$

$$(13.3)$$

$$M_{ba} = -M_{F_{ba}} - \frac{2EI}{L}\left(\theta_a + 2\theta_b - \frac{3\Delta}{L}\right)$$

(the value of ρ can be used in place of $\frac{\Delta}{L}$).

In review the sign convention is given as follows:

> Clockwise moments applied to the joint are $+$
>
> Clockwise rotation of end tangents are $+$
>
> Clockwise rotation of axis of beam is $+$

13.3 Application of Slope-Deflection Method

The slope-deflection equations (13.3) can be used with the laws of equilibrium to determine the end-moments for all members of a statically indeterminate structure.

In order to use the slope-deflection equation, it is necessary to first evaluate the fixed-end moments (M_F). Some method of analysis of statically indeterminate structures has to be used to perform this operation. For a single span beam with fixed ends, the method of consistant deflections as presented in Chapter 12 can be used. The deflections can be easily determined by the moment-area or conjugate beam method. The following table gives the fixed-end moments for various loading conditions that are commonly encountered in structural analysis. Table 13.1 only shows F.E.M.'s for beams fixed at both ends. If a beam has one end on a simple support or pinned, the F.E.M. for the fixed continuous end can be determined by multiplying the value in Table 13.1 by 1.5. For instance, a uniformly loaded beam fixed at one end and simply supported at the other will have a F.E.M. of

$$\left(\frac{wL^2}{12}\right) 1.5 = \frac{wL^2}{8}.$$

Table 13.1

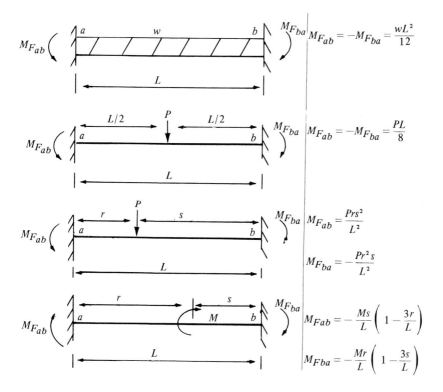

If a beam had a concentrated load at the midspan,

$$\text{F.E.M.} = \frac{PL}{8}(1.5) = \frac{3PL}{16}$$

at the fixed end.

An example of the application of the slope-deflection method is as follows:

EXAMPLE PROBLEM 13.1

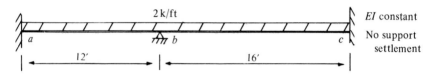

It is best to list the known values of displacement first and then the unknown displacements.

$$\text{Known:} \quad \theta_a = \theta_c = 0$$
$$\Delta_{ab} = \Delta_{bc} = 0$$
$$\text{Unknown:} \quad \theta_b$$

Because of only one unknown, only one equation of equilibrium is required. This will be the equation

$$M_{ba} + M_{bc} = 0.$$

The slope deflection equations for M_{ba} and M_{bc} will be written after the values of $M_{F_{ba}}$ are calculated.

$$M_{F_{ba}} = -\frac{wL^2}{12} = -\frac{2(12)^2}{12} = -24 \text{ ft-k}$$

$$M_{F_{bc}} = \frac{wL^2}{12} = \frac{2(16)^2}{12} = +42.7$$

$$M_{ba} = -24 - \frac{2EI}{12}(2\theta_b)$$

$$(\theta_a = \theta_c = \Delta\text{'s} = 0)$$

$$M_{bc} = +42.7 - \frac{2EI}{16}(2\theta_b)$$

$$M_{ba} + M_{bc} = 0$$

$$-24 - \frac{EI\theta_b}{3} + 42.7 - \frac{EI\theta_b}{4} = 0$$

$$\frac{7EI\theta_b}{12} = 18.7$$

$$EI\theta_b = 32.1$$

The moments at the ends of the members can now be calculated by substituting the value of $EI\theta_b$ into the slope deflection equations (13.3).

$$M_{ab} = +24 - \frac{2(32.1)}{12} = +18.7 \text{ ft-k}$$

$$M_{ba} = -24 - \frac{4(32.1)}{12} = -34.7 \text{ ft-k}$$

$$M_{bc} = +42.7 - \frac{4(32.1)}{16} = +34.6 \text{ ft-k}$$

$$M_{cb} = -42.7 - \frac{2(32.1)}{16} = -46.7 \text{ ft-k}$$

When the values of E and I are the same for all members, it is best to solve for $EI\theta$ and not use the specific values of E and I in the equilibrium equations.

EXAMPLE PROBLEM 13.2   The beam of Example Problem 13.1 has a settlement of support b of one inch. Member end moments will be calculated.

$$\text{Known: } \theta_a = \theta_c = 0$$
$$\Delta_{ba} = -\Delta_{bc} = +0.0833 \text{ ft}$$

$$\text{Unknown: } \theta_b$$

The same equilibrium equation can be written as previously.

$$M_{ba} + M_{bc} = 0$$

$$M_{ba} = -24 - \frac{EI}{6}\left(2\theta_b - \frac{3\Delta_{ba}}{12}\right)$$

$$M_{bc} = +42.7 - \frac{EI}{8}\left(2\theta_b - \frac{3\Delta_{bc}}{16}\right)$$

substituting $\Delta_{ba} = 0.0833 = -\Delta_{bc}$ and adding the two expressions for end moments

$$18.7 - EI(0.5833\theta_b - 0.001518) = 0$$

for $I = 400 \text{ in}^4$ and $E = 3 \times 10^3 \text{ ksi}; EI = 8333 \text{ k-ft}^2$

$$0.5833(EI\theta_b) - 12.65 = 18.7$$

$$EI\theta_b = 53.75$$

substituting in the slope deflection equation

$$M_{ab} = +24 - \frac{53.75}{6} + (0.00347)(8333) = 44 \text{ ft-k}$$

$$M_{ba} = -24 - \frac{53.75}{3} + (0.00347)(8333) = -13 \text{ ft-k}$$

$$M_{bc} = +42.7 - \frac{53.75}{4} - (0.001952)(8333) = +13 \text{ ft-k}$$

$$M_{cb} = -42.7 - \frac{53.75}{8} - (0.001952)(8333) = -65.7 \text{ ft-k}$$

Settlement of support b increases the magnitude of bending moments at a and c and reduces the bending moment at b.

EXAMPLE PROBLEM 13.3   The slope deflection equations can be used for the analysis of frames. In the bent Fig. 13.3 there are three displacement unknowns; the rotation of joints b and c and the lateral translation of the tops of the columns with respect to the base (same for both columns).

$$\text{Known:} \quad \theta_a = \theta_d = 0$$

$$\text{Unknown:} \quad \theta_b, \theta_c, \rho = \frac{\Delta}{26}$$

Fixed-end moments:

$$M_{Fbc} = +\frac{8(12)(6)^2}{(18)^2} = +10.67 \text{ ft-k}$$

$$M_{Fcb} = -\frac{8(6)(12)^2}{(18)^2} = -21.33 \text{ ft-k}$$

Equilibrium equations:

(1) $M_{ba} + M_{bc} = 0$

(2) $M_{cb} + M_{cd} = 0$

(3) $\dfrac{M_{ba} + M_{ab}}{L_{ab}} + \dfrac{M_{cd} + M_{dc}}{L_{dc}} = 0$

Writing the slope-deflection equations for end moments and writing $\frac{EI}{L} = K$

$$M_{ab} = -2K_{ab}(\theta_b - 3\rho)$$

$$M_{ba} = -2K_{ab}(2\theta_b - 3\rho)$$

$$M_{bc} = +10.67 - 2K_{bc}(2\theta_b + \theta_c)$$

$$M_{cb} = -21.33 - 2K_{bc}(\theta_b + 2\theta_c)$$

$$M_{cd} = -2K_{cd}(2\theta_c - 3\rho)$$

$$M_{dc} = -2K_{cd}(\theta_c - 3\rho)$$

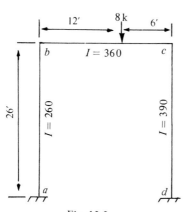

Fig. 13.3

Substituting the above values for end moments into the equilibrium equations and evaluating the values of K,

$$\frac{K_{ab}}{E} = \frac{260}{26(5)} = 2 \qquad \frac{K_{bc}}{E} = \frac{360}{18(5)} = 4 \qquad \frac{K_{cd}}{E} = \frac{390}{26(5)} = 3$$

(the value of 5 in the denominator is to reduce the K values to the lowest relative integer for convenience)

(1) $M_{ba} + M_{cb} = 0$

$$-80_b + 12\rho + \frac{10.67}{E} - 160_b - 80_c = 0$$

(2) $M_{cb} + M_{cd} = 0$

$$-\frac{21.33}{E} - 80_b - 160_c - 120_c + 18\rho = 0$$

(3) $M_{ab} + M_{ba} + M_{cd} + M_{dc} = 0$

$$-40_b + 12\rho - 80_b + 12\rho - 120_c + 18\rho - 60_c + 18\rho = 0$$

combining terms and setting in matrix form

$$E \begin{bmatrix} 24 + & 8 - 12 \\ 8 + & 28 - 18 \\ 12 + & 18 - 60 \end{bmatrix} \begin{bmatrix} \theta_b \\ \theta_c \\ \rho \end{bmatrix} = \begin{bmatrix} 10.67 \\ -21.33 \\ 0 \end{bmatrix}$$

Solving:

$$E\theta_b = 0.715$$

$$E\theta_c = -1.083$$

$$E\rho = -0.182$$

Substituting these values back into slope deflection equations

$$M_{ab} = -4[0.715 - 3(-0.182)] = -5.0 \text{ ft-k}$$

$$M_{ba} = -4[2(0.175) - 3(-0.182)] = -7.9$$

$$M_{bc} = 10.67 - 8[2(0.715) - 1.083] = +7.9$$

$$M_{cb} = -21.33 - 8[0.715 + 2(-1.083)] = -9.8$$

$$M_{cd} = -6[2(-1.083) - 3(-0.182)] = +9.8$$

$$M_{dc} = -6[-1.083 - 3(-0.182)] = +3.2$$

The moment diagram is

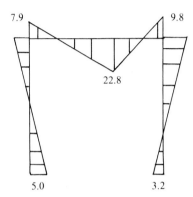

Moment diagram drawn on tension side of member

The slope deflection method can be used for any statically indeterminate structure. However, the equations 13.3 only apply to structures whose members have a constant value of I between the ends of each member. New equations could be developed for non-prismatic members by evaluating the end moments, for such members, due to loads, end rotations, and end translations. The slope deflection method is usually not used for structures with non-prismatic members.

13.4   Introduction to Moment Distribution

In both Chapter 12 and the slope-deflection method of this chapter, it was necessary, in multiple redundant structures, to solve sets of simultaneous equations. By hand calculations, this equation solving can be very time-consuming and tedious. The moment distribution method does not require the solving of simultaneous equations. It is a very practical procedure of analysis in that it produces the desired result by a method of successive approximations. This progression converges toward the correct solution and the process can be stopped anywhere depending upon the degree of accuracy desired.

Several engineers had presented the basic concept of solving simultaneous equations by successive approximations before Professor Hardy Cross introduced his method. His paper (1) in ASCE Transactions in 1936 was a great leap forward in the analysis of statically indeterminate structures. The paper generated a great amount of interest and certainly caused the rewriting of text books covering indeterminate structural analysis. It also gave great impetus to the use of continuous frames in both reinforced concrete and steel.

Referring back to the development of the slope-deflection equation (Figure 13.1), it is seen that the final end moments of a beam that is part of a continuous structure can be calculated if the end rotations and translations can be determined. In the mathematical procedure, it is possible to visualize a step-by-step, physical procedure. It is first considered that all members of a continuous beam or frame have their ends "locked" so that there is no rotation or translation. The member is next subjected to the real loads. Since the ends are "locked" from rotation and translation, the bending moments at the ends are equal in magnitude to the fixed-end moments (see Figure 13.1 (a)). Where two or more members meet, this will result in end moments at the matching ends that could have different values, Figure 13.4 shows two spans of a multiple span beam that have different span lengths and loadings. If the ends are considered "locked" against rotation and translation, the fixed end moments are then

$$M_{F_{bc}} = -M_{F_{cb}} = \frac{wL^2}{12} = \frac{0.8(20)}{12} = 26.7 \text{ ft-k}$$

$$M_{F_{cd}} = M_{F_{dc}} = \frac{PL}{8} = \frac{6(16)}{8} = 12.0 \text{ ft-k}$$

(clockwise moments applied by the beam to the joint are positive). It is noted that at support c there is a moment of -26.7 in beam cb and $+12.0$ in beam cd. These two moments do not satisfy equilibrium. There will likewise be an unbalance of end moments at all the other supports (values not shown). If the

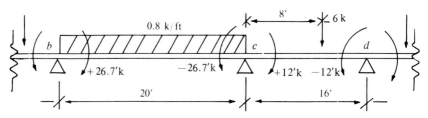

Fig. 13.4. Continuous Beam.

theoretical "lock" at support c is removed, then a rotation will take place at c until equilibrium is reached. If the magnitude of rotation were known or could be caluclated, then it would be possible to calculate the end-moments in the beams joining at point c. In the development of slope deflection, it was shown that the relationship of bending moment to rotation for a prismatic beam was (see Figure 13.1 (b)).

$$M = \frac{4EI\theta}{L}$$

Since, by continuity, the rotation of the two matching ends of the beams would be the same value, the moment introduced at the two ends of the beams would be proportional to the value of I/L of the two beam spans. This value of I/L is given the name *relative stiffness*.

It should be noted that since $M_{F_{cb}}$ is greater in value than $M_{F_{cd}}$ the tangent to the elastic line at support c will rotate counter-clockwise. This rotation will introduce a bending moment in end c of span cb that will be clockwise, thus reducing the numerical value of bending moment of -26.7. It will increase the value of $+12.0$ at end c of span cd. If the spans all had the same value of I, then the relative stiffness of span cb could be taken as $1/20 = 0.05$, and for span cd as $1/16 = 0.0625$. The unbalance of end-moments of $(-26.7 + 12.0)$ is -14.7. A bending moment of $+14.7$ ft-k must then be distributed to the end of each member meeting at c in proportion to the relative stiffness of the respective members meeting at that point. The distributed moments will then be

$$M_{cb} = 14.7\left(\frac{0.0625}{0.1125}\right) = +8.2 \text{ ft-k}$$

$$M_{cd} = 14.7\left(\frac{0.050}{0.1125}\right) = +6.5 \text{ ft-k}$$

The graphical portrayal of this operation is then

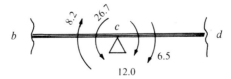

The end-moments after rotation will then be

$$M_{cb} = -26.7 + 8.2 = -18.5 \text{ ft-k}$$

$$M_{cd} = +12.0 + 6.5 = +18.5 \text{ ft-k}$$

Joint (*c*) is then temporarily "balanced." After this balance of moments takes place, the joint is again "locked."

From the slope-deflection development, it was seen that if the near end of a prismatic beam is rotated through an angle θ, there is a bending moment at the far end that is equal to

$$M = \frac{2EI\theta}{L}$$

Therefore the moment produced at the far end is just one-half that produced at the near end and from Figure 13.1 (c) it is seen that these two end-moments are in the same direction. That is, if the near end is rotated through a clockwise angle (+), the end moments produced will be counterclockwise (−).

In the example under consideration, the balancing of joint *c* produced balancing moments of +8.2 at *cb* and +6.5 at *cd*. The moments produced at *b* and *d* by this rotation at *c* would then be +4.1 and +3.25 respectively. The status of bending moments at this stage of the procedure are

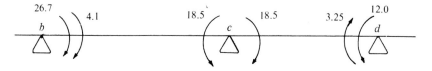

The end moments at *b* of span *bc* is then +30.8 and at *d* of span *dc* is −8.75. The next step in the procedure would be to move to the next joint (*b* or *d*) and "unlock" this joint and repeat the procedure that has just taken place at *c*. First, the unbalanced end-moments are then balanced and then 1/2 the bending moment assigned to the near end in the balancing process is "carried over" to the far end. It must be remembered that the 1/2 value for the *carry-over* factor is only correct for prismatic members. The determination of *carry-over factor* for non-prismatic members will be made later.

If there were a settlement of a support, then an additional operation would have to be performed. The magnitude of this settlement would have to be determined. Fixed-end moments for prismatic members could then be calculated from the equation:

$$M = \frac{6EI\Delta}{L^2}$$

These fixed-end moments would be superimposed on the fixed-end moments due to loads and the procedure would then be exactly as before.

When a member has one end on a simple support or pinned, the fixed-end moments should be determined for a member with one end fixed and the other pinned. The stiffness of this type of member is also different from a member with the far end fixed. As stiffness of one end of a member is defined as the moment required to rotate the end of the member through an angle of one radian, it is, for a prismatic member, equal to

$$\frac{4EI}{L}.$$

When the far end is pinned, the member is less stiff since the moment necessary to rotate the end through an angle of one radian is

$$\frac{3EI}{L}.$$

The student should prove this by applying the moment-area, or conjugate beam, principles. Therefore the *stiffness* of a member with the far end pinned is 3/4 of that for a member with the far end fixed or continuous.

13.5 Application of Moment-Distribution to Continuous Beams

The use of the moment-distribution method for several continuous beams will be shown in the following examples. The student should observe the order of the computations. It is more systematic to release each joint one at a time and perform the balancing at all the joints first. Then all the carry-overs are the next operation. The student should prove to himself that the final result will be the same as when the carry-over is made at a joint before the balancing is done at adjacent joints.

EXAMPLE PROBLEM 13.4

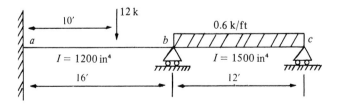

Relative stiffness of $ab = \dfrac{1200}{16} = 75$

Relative stiffness of end b of $bc = \dfrac{1500}{12}\left(\dfrac{3}{4}\right) = 93.75$

Distribution factor of end a of $ab = 0$ (support is infinitely rigid)

Distribution factor of end b of $ab = \dfrac{75}{168.75} = 0.44$

Distribution factor of end b of $bc = \dfrac{93.75}{168.75} = 0.56$

The fixed end moments are next calculated

$$M_{F_{ab}} = \dfrac{12(6)^2(10)}{(16)^2} = +16.9 \text{ ft-k}$$

$$M_{F_{ba}} = \dfrac{12(6)(10)^2}{(16)^2} = -28.1$$

$$M_{F_{bc}} = \dfrac{0.6(12)^2}{8} = +10.8$$

The carry-over in span ab will be $1/2$ and there is no carry-over in span bc since there is a simple support at c.

The usual solution format is as follows:

	a		b	c
Dist. factor	0	0.44	0.56	0
F.E.M.	+16.9	-28.1	+10.8	0
1st Dist.	0	+ 7.6	+ 9.7	0
1st carry-over	+ 3.8	0	0	0
2nd Dist.	0	0	0	0
Σ of moments	+20.7	-20.5	+20.5	0

The final end moments are determined by making an algebraic summation of the columns.

In the above example, there was a complete balance of all joints after two cycles of operation. Most problems do not converge this fast.

With the end-moments known, the reactions can be determined and the shear and bending moment diagrams drawn.

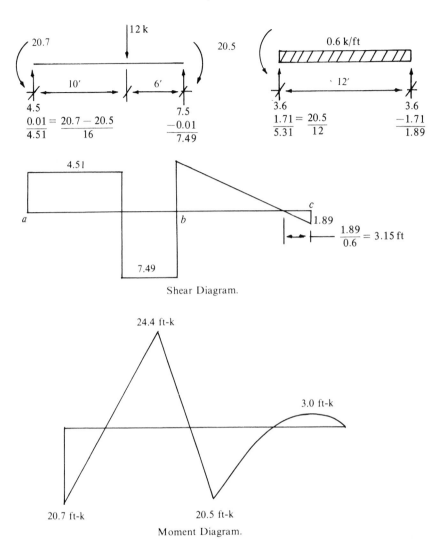

$$0.01 = \frac{20.7 - 20.5}{16}$$

$$1.71 = \frac{20.5}{12}$$

$$\frac{1.89}{0.6} = 3.15 \text{ ft}$$

Shear Diagram.

Moment Diagram.

EXAMPLE PROBLEM 13.5 In this problem there is a four span continuous beam of constant section throughout the length.

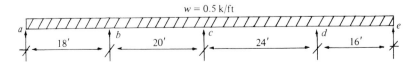

Relative Stiffnesses:

$$ba = \frac{1}{18}\left(\frac{3}{4}\right) = 0.042 \qquad bc = \frac{1}{20} = 0.050$$

$$cd = \frac{1}{24} = 0.0417 \qquad de = \frac{1}{16}\left(\frac{3}{4}\right) = 0.047$$

Fixed-end moments:

$$M_{F_{ba}} = -\frac{0.5(18)^2}{8} = -20.3 \text{ ft-k} \qquad M_{F_{bc}} = \frac{0.5(20)^2}{12} = 16.7 \text{ ft-k}$$

$$M_{F_{cd}} = \frac{0.5(24)^2}{12} = 24 \text{ ft-k} \qquad M_{F_{de}} = \frac{0.5(16)^2}{8} = 16 \text{ ft-k}$$

The distribution factors are next determined and the balancing and carry-over operation performed.

	a		b		c		d		e
D.F.	1.0	0.456	0.544	0.545	0.455	0.47	0.53	0	
F.E.M.	0	-20.3	+16.7	-16.7	+24	-24	+16.0	0	
1st Dist.	0	+16	+2.0	-4.0	-3.3	+3.8	+4.2	0	
C.O.	0	0	-2.0	+1.0	+1.9	-1.6	0	0	
2nd Dist.	0	+0.9	+1.1	-1.6	-1.3	+0.8	+0.8	0	
C.O.	0	0	-0.8	+0.6	+0.4	-0.6	0	0	
3rd Dist.	0	+0.4	+0.4	-0.5	-0.5	+0.3	+0.3	0	
C.O.	0	0	-0.2	+0.2	+0.2	-0.2	0	0	
4th Dist.	0	+0.1	+0.1	-0.2	-0.2	+0.1	+0.1	0	
Final M.	0	-17.3	+17.3	-21.2	+21.2	-21.4	+21.4	0	

It can be noted that the error would have been small (about 3%) if the process had been stopped after two distributions and practically no error if stopped after three cycles.

The student should contrast the amount of numerical work with that that would be required by any of the methods previously covered. It is very unlikely the engineer will use any other method than moment distribution for the analysis of continuous beams when only a hand calculator is available. An exception might be where there is only a single redundant.

EXAMPLE PROBLEM 13.6 In this problem, it will be considered that support b of the previous example settles 0.20 inches and support d settles 0.12

inches. It is now necessary to use actual values of E and I in determining the fixed-end moments due to joint translation. For distribution factors relative values of I are sufficient. Use $E = 30 \times 10^3$ ksi and $I = 600$ in^4.

The supplemental F.E.M. due to support settlements are:

$$M_{F_{ba}} = \frac{3EI\Delta}{L^2} = \frac{3(30 \times 10^3)(600)(0.2)}{(18)^2(12)^3} = 19.3 \text{ ft-k}$$

$$M_{F_{bc}} = -19.3\left(\frac{18}{20}\right)^2\left(\frac{6}{3}\right) = -31.3 \text{ ft-k} = M_{F_{cb}}$$

$$M_{F_{cd}} = +19.3\left(\frac{18}{24}\right)^2\left(\frac{6}{3}\right)\left(\frac{0.12}{0.20}\right) = +13.0 \text{ ft-k} = M_{F_{cd}}$$

$$M_{F_{de}} = -19.3\left(\frac{18}{16}\right)^2\left(\frac{0.12}{0.20}\right) = -14.7 \text{ ft-k}$$

	a	b		c		d		e
D.F.	1.0	0.456	0.544	0.545	0.455	0.47	0.53	0
	0	+19.3	-31.3	-31.3	+13.0	+13.0	-14.7	0
	0	+5.5	+6.5	+10.0	+8.3	+0.8	+0.9	0
	0	0	+5.0	+3.2	+0.4	+4.1	0	0
	0	-2.3	-2.7	-2.0	-1.6	-1.9	-2.2	0
	0	0	-1.0	-1.3	-1.0	-0.8	0	0
	0	+0.5	+0.5	+1.3	+1.0	+0.4	+0.4	0
	0	0	+0.7	+0.2	+0.2	+0.5	0	0
	0	-0.3	-0.4	-0.2	-0.2	-0.2	-0.3	0
Σ	0	+22.7	-22.7	-20.1	+20.1	+15.9	-15.9	0

Adding the bending moments due to the applied loads–Prob. 13.5

	0	-17.3	+17.3	-21.2	+21.2	-21.4	+21.4	. 0
Σ	0	+5.4	-5.4	-41.3	+41.3	-5.5	+5.5	0

It is observed that even small settlements of the supports can make appreciable changes in the bending moments. The stiffer the beam, the greater the effect of support settlements.

13.6   Application of Moment Distribution to Frames

A.   *Without Joint Translation*

The use of the moment-distribution method in the analysis of frames is very efficient where the frame has no translation of the joints. The procedure is similar to that of continuous beams. The relative stiffnesses are determined and the fixed end moments are calculated for all members that are loaded. The "locking" of the joints and the progressive unlocking and distributing of moments proceeds from one joint to the other in the same manner as the continuous beam. The following example problem gives the details.

EXAMPLE PROBLEM 13.7   Plot the moment diagram on the compression side for the frame shown below. All members have same $I = 400$ in^4.

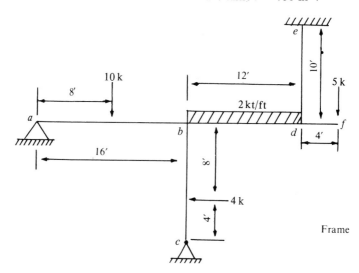

Frame

Stiffness and distribution factors:

Joint b

member ba: $K = \dfrac{400}{16}\left(\dfrac{3}{4}\right) = 18.75$ $\text{D.F.} = \dfrac{18.75}{70.83} = 0.265$

member bc: $K = \dfrac{400}{16}\left(\dfrac{3}{4}\right) = 18.75$ $\text{D.F.} = 0.265$

member bd: $K = \dfrac{400}{12} \quad = \underline{33.33}$ $\text{D.F.} = \underline{0.47}$

$\qquad\qquad\qquad\qquad\Sigma = 70.83 \qquad\qquad\qquad \Sigma = 1.00$

Joint d

member db: $K = \dfrac{400}{12} = 33.33$ D.F. = 0.45

member de: $K = \dfrac{400}{10} = \underline{40.00}$ D.F. = $\underline{0.55}$

$\Sigma = 73.33$ $\Sigma = 1.00$

Member df has no stiffness since the end f has no support.

Fixed-end moments

$$M_{F_{ba}} = -\frac{3(10)(16)}{16} = -30 \text{ ft-k}$$

$$M_{F_{bc}} = +\frac{Prs^2}{L^2} + \frac{Pr^2 s}{2L^2} = +\frac{4 \times 8(4)^2}{12^2} + \frac{4(8)^2(4)}{2(12)^2} = +7.11 \text{ (Table 13.1)}$$

$$M_{F_{bd}} = \frac{2(12)^2}{12} = +24 = -M_{F_{db}}$$

$$M_{F_{df}} = 5 \times 4 = +20$$

The mathematical operation can be performed on a line diagram of the frame or in a table as shown below. The table is used in this problem as it is simpler to designate the step by step operation. When using the tabular form, the student must take care to see that he places the carry-over values in the correct column. It is a little easier to avoid this error when the work is done on a line diagram of the frame.

	Joint b			Joint d			Joint e
	ba	bc	bd	db	de	df	ed
D.F.	0.265	0.265	0.47	0.45	0.55	0	0
F.E.M.	-30.0	+7.1	+24.0	-24.0	0	+20.0	0
1st Dist.	-0.3	-0.3	-0.5	+1.8	+2.2	0	0
C.O.	0	0	+0.9	-0.2	0	0	+1.1
2nd Dist.	-0.2	-0.2	-0.5	+0.1	+0.1	0	0
C.O.	0	0	0	+0.3	0	0	0
3rd Dist.	0	0	0	-0.1	-0.2	0	0
$\Sigma =$	-30.5	+6.6	+23.9	-22.1	+2.1	+20.0	+1.1

The moment diagram can be drawn by superimposing the end moments on the simple beam moments.

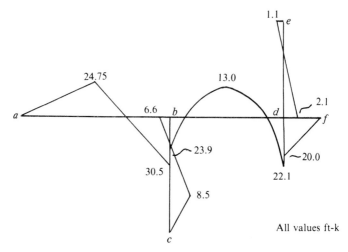

Final Moment Diagram

The previous example problem presented the solution for a frame where the joints were restrained from any translation. It is seen that the procedure was no different from that used in a continuous beam without support settlement.

B.   With Joint Translation

Most frames, especially those used in buildings, will have joint translation. A bent as shown in Fig. 13.3 will have translation of joints b and c because the bent is subjected to a lateral load as well as an unsymmetrical vertical load. This translation of the top of the bent is called sidesway.

Since the translation of the joints will cause bending moments in the frame, in addition to the bending moments that are produced by joint rotation, the effect of sidesway will have to be taken into account. The procedure is first to consider that the frame has an artificial restraint at the top that restricts the frame from any sidesway. The moment-distribution method is then applied in the same manner as in the previous example. A calculation of end shears in the columns from the end moments in the columns will give the value of the horizontal reactions at the base of the columns. If translation is possible, these horizontal reactions at the base of the columns will not produce static equilibrium of the entire frame. This unbalance of horizontal forces is in itself an indication of sidesway in the frame. Cor-

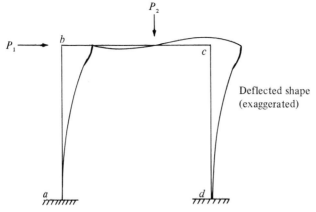

Fig. 13.3. Bent with Sidesway.

rective moments, due to sidesway, will have to be superimposed on the moments calculated for the case where the frame was restrained from sidesway.

If the numerical value of the sidesway were known, then it would be a relatively easy matter to determine the moments due to the amount of sidesway. Since this is not possible, an arbitrary amount of lateral translation is given to the top of the frame (keeping all joints "locked"). The fixed-end moments due to the displacement of one end of the member relative to the other end are calculated. This operation is performed for each member that is subjected to a relative displacement. This displacement is taken as that which is normal to the axis of the member

The next step is to "unlock" and balance each joint successively as previously explained. Joints are then "locked" and the carry-over moments written. Upon convergence, the numbers are totaled to obtain the end moments for the selected value of sidesway. From the end moments in the columns, the shears at the base of the columns are calculated. If these shears are algebraically added to the shears previously calculated for the case of the laterally restrained frame, the combined shears will not be equal to the sum of the applied lateral loads (that is, unless the arbitrarily assumed value of sidesway was a lucky guess). However, a fraction of these sidesway column shears could be added to the previous value of shear from the laterally restrained frame to achieve static equilibrium of the frames. This fraction (or factor) can be calculated. The final moment will then be the superposition of moments calculated from the restrained frame with the moments from the sidesway case multiplied by the factors.

The next example problem will be a demonstration of this procedure as applied to a single-story, single bay frame. The procedure would be the same for any number of bays. However, for multiple stories, the method requires additional steps. The multi-story case will be considered later in this chapter.

EXAMPLE PROBLEM 13.8 Determine the end moments in the frame of Fig. 13.4 for the loading shown.

Stiffness and distribution factors:

Joint b:

member ba: $K = \dfrac{200}{12} = 16.67$ D.F. $= \dfrac{16.67}{40} = 0.33$

member bc: $K = \dfrac{400}{12} = \underline{33.33}$ D.F. $= 0.67$

$\Sigma = 40.0$

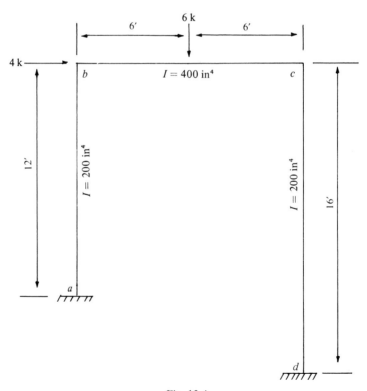

Fig. 13.4

Joint c:

member cb: $K = \dfrac{400}{12} = 33.3$ D.F. $= 0.73$

member cd: $K = \dfrac{200}{16} = 12.5$ D.F. $= 0.27$

$$\Sigma = 45.8$$

Fixed-end moments:

Since the 4^k lateral load is applied at the end of member ab it does not produce any fixed-end moments.

$$M_{F_{bc}} = \frac{6(12)}{8} = +9 \text{ ft-k} = -M_{F_{cb}}$$

The numerical work will be on a line diagram of the frame.

The end shears of the columns will next be calculated.

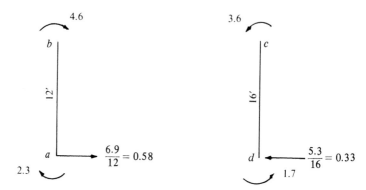

Since sidesway was neglected, $0.58 - 0.33 \neq 4.0$. An arbitrary amount of sidesway will be introduced at the top of the frame making the fixed end moments in the column ab equal to $+100$ ft-kips. Since pt. c on the frame will move laterally the same amount as pt. b, the moments in column dc will be equal to the moments in column ab multiplied by the ratio of the values of I and the inverse ratio of lengths squared. This is so because

$$M_F = \frac{6EI\Delta}{L^2}$$

therefore

$$M_{F_{ab}} = M_{F_{ba}} = +100 \text{ ft-k}$$

$$M_{F_{dc}} = M_{F_{cd}} = 100 \left(\frac{200}{200}\right)\left(\frac{12}{16}\right) = +56.2 \text{ ft-k}$$

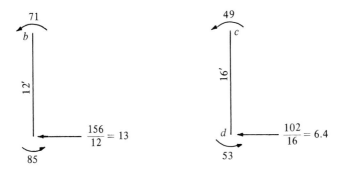

The equilibrium equation can be written

$$F_X = 0 = X(19.4) - (0.58 - 0.33) - 4.0 \qquad X = \frac{4.25}{19.4} = 0.22$$

The end moments are then

$$M_{ab} = -2.3 + 85(0.22) = 16.4 \text{ ft-k}$$

$$M_{ba} = -M_{bc} = -4.6 + 71(0.22) = +11.0$$

$$M_{cb} = -M_{cd} = -3.6 - 49(0.22) = -14.4$$

$$M_{dc} = +1.7 + 53(0.22) = +13.4$$

The final total shear in the columns is equal to

$$\frac{16.4 + 11.0}{12} + \frac{14.4 + 13.4}{16} = 4.0^k$$

This is a check on the accuracy. The student should verify the sign usage. It should be noted that in this example the greater part of the bending moments in the frame are due to sidesway.

C. Frames with Several Degrees of Freedom

The single story frame had only one degree of freedom. It had one independent joint translation. A frame with several stories will have joint translation equal to the number of stories. This is true as long as there is no diagonal bracing in any bay.

The moment distribution solution for multiple story frames consists of one distribution for the laterally restrained frame acted upon by the given loads applied between joints. Additional distributions for lateral displacements at the top of each story will be required. A displacement, of arbitrary amount, will be made at the top of the first story with the frame restrained at all other levels. The arbitrary displacement will in reality be arbitrary fixed end moments in the columns. After the cycles of distributing unbalanced moments is completed, the column shears at the base of each story are computed. This process is repeated for the translation of each story. When all the independent solutions have been performed, simultaneous equations that consist of unknown factors multiplied by the column shears for each story can be equated to the shear at each story due to the applied loads. The following example shows this procedure.

EXAMPLE PROBLEM 13.9 Plot the bending moment diagram on the compression side for the two story frame shown in Fig. 13.5. Take I of the girders as twice I of the columns.

Relative stiffness:

D.F.

Columns ab, fe, ih: $K = \dfrac{I}{20} = 0.050$ relative $K = 4$ 0.21

Columns bc, ed, hg: $K = \dfrac{I}{16} = 0.0625$ relative $K = 5$ 0.26

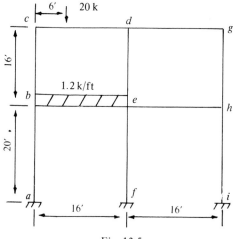

Fig. 13.5

Girders: $K = \dfrac{2I}{16} = 0.125$ relative $K = 10$ 0.53

Fixed-end moments:

$$M_{F_{cd}} = \dfrac{20(6)(10)^2}{(16)^2} = +46.9 \text{ ft-k}$$

$$M_{F_{dc}} = \dfrac{20(6)^2(10)}{(16)^2} = -28.1 \text{ ft-k}$$

$$M_{F_{bc}} = \dfrac{1.2(16)^2}{12} = +25.6 \text{ ft-k} = -M_{F_{eb}}$$

The operation of the moment distribution method is in the table on pages 394 and 395. The column shears are next calculated.

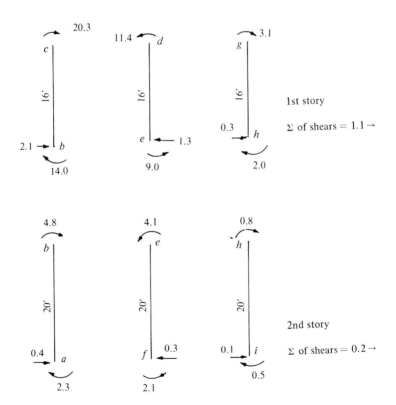

Joint	a	b			c		d		
Mem.	ab	ba	be	bc	cb	cd	dc	de	dg
K	4	4	10	5	5	10	10	5	10
D.F.	—	.21	.53	.26	.33	.67	.40	.20	.40
F.E.M.	0·	0	+25.6	0	0	+46.9	-28.1	0	0
	0	-5.4	-13.5	-6.7	-15.6	-31.3	+11.2	+5.7	+11.2
C.O.	-2.7	0	+4.4	-7.8	-3.4	+5.6	-15.6	+2.2	0
	0	+.7	+1.8	+.9	- 7	-1.5	+5.4	+2.7	+5.4
C.O.	+.4	0	+.7	-.3	+.4	+2.7	-0.8	+0.3	-1.9
	0	-0.1	-0.2	-0.1	-1.0	-2.1	+0.9	+0.5	+0.9
	-2.3	-4.8	+18.8	-14.0	-20.3	+20.3	-27.0	+11.4	+15.6

e				f	g		h			i
ed	eb	eh	ef	fe	gd	gh	hg	he	hi	ih
5	10	10	4	4	10	5	5	10	4	4
.17	.35	35	.13	—	.67	.33	.26	.53	.21	—
0	-25.6	0	0	0	0	0	0	0	0	0
+4.4	+8.8	+8.8	+3.6	0	0	0	0	0	0	0
+2.8	-6.7	0	0	+1.8	+5.6	0	0	+4.4	0	0
+.7	+1.3	+1.3	+.6	0	-3.8	-1.8	-1.2	-2.3	-.9	0
+1.3	+0.9	-1.1	0	+.3	+2.7	-.6	-.9	+.6	0	-.5
-0.2	-0.4	-0.4	-0.1	0	-1.4	-0.7	+0.1	+0.1	+0.1	0
+9.0	-21.7	+8.6	+4.1	+2.1	+3.1	-3.1	-2.0	+2.8	-0.8	-0.5

The fixed end moments due to sidesway in the first story are calculated. Since the values of I and L are the same for each column, a value of fixed end moment of $+100$ will be selected.

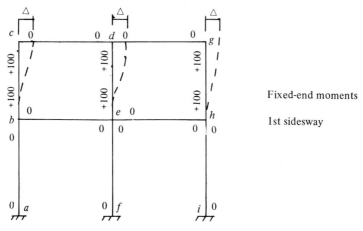

Fixed-end moments

1st sidesway

Sidesway in top story and fixed end moments

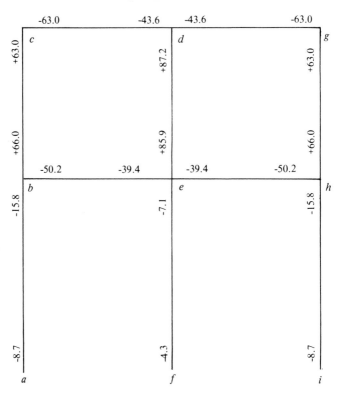

The detail moment distribution solution is not shown since the procedure would be similar to that already done for the applied loads. The resulting end moments are given at the bottom of page 396 and the column shears are calculated.

The student should perform the complete moment-distribution solution to check the above values. The column shears are next calculated.

Condition *X* column shears

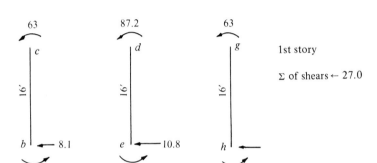

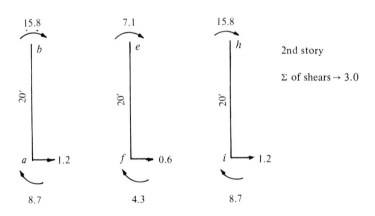

The second degree of freedom will be a translation of the second level (*beh*), but with the first story moving the same amount. The value of fixed-end moments in the columns of the second story are given an arbitrary amount of fixed-end bending moment of +100.

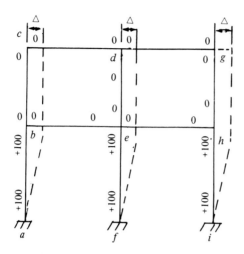

Fixed-end moments

2nd Sidesway

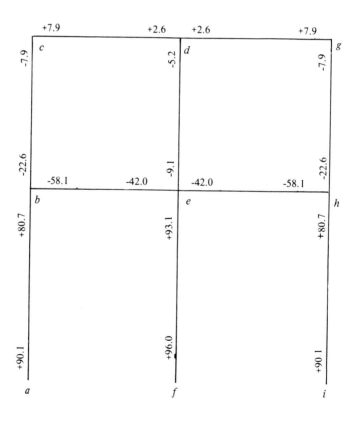

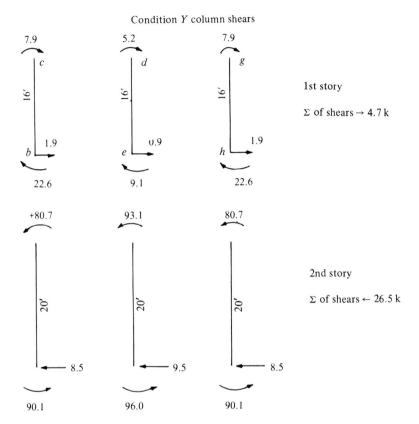

Condition Y column shears

1st story

Σ of shears $\rightarrow$ 4.7 k

2nd story

Σ of shears $\leftarrow$ 26.5 k

The final bending moment condition of the frame will consist of the moments due to the applied load plus moments due to top story sidesway plus moments due to the bottom story sidesway. The amount of sidesway in the top story will be designated X and the amount of the bottom story sidesway will be designated Y. Equilibrium equations representing the column shears for the simultaneous action can be written. Since there are no applied lateral loads, the column shears for the three simultaneous actions must equal zero.

1st story: $\Sigma F_X = 0$
$$1.1 - 27.0X + 4.7Y = 0$$

2nd story: $\Sigma F_X = 0$
$$0.20 + 3.0X - 26.5Y = 0$$

Solving these two simultaneous equations

$$X = 0.043$$

$$Y = 0.012$$

The end moments will then be the summation of the end moments due to the applied loads plus the end moments due to the first sidesway multiplied by 0.043 plus the end moments due to the second sidesway multiplied by 0.012. The student should be careful to give the correct signs to the bending moments.

Calculating the final end moments:

$$M_{ab} = -2.3 - 8.7(0.043) + 90.1(0.012) = -1.6 \text{ ft-k}$$

$$M_{ba} = -4.8 - 15.8(0.043) + 80.7(0.012) = -4.5$$

$$M_{be} = +18.8 - 50.2(0.043) - 58.1(0.012) = +15.9$$

$$M_{bc} = -14.0 + 66.0(0.043) - 22.6(0.012) = -11.4$$

$$M_{cb} = -20.3 + 63.0(0.043) - 7.9(0.012) = -17.7$$

$$M_{cd} = +20.3 - 63.0(0.043) + 7.9(0.012) = +17.7$$

$$M_{dc} = -27.0 - 43.6(0.043) + 2.6(0.012) = -28.9$$

$$M_{de} = +11.4 + 87.2(0.043) - 5.2(0.012) = +15.2$$

$$M_{dg} = +15.6 - 43.6(0.043) + 2.6(0.012) = +13.7$$

$$M_{ed} = +9.0 + 85.9(0.043) - 9.1(0.012) = +12.6$$

$$M_{eb} = -21.7 - 39.4(0.043) - 42.0(0.012) = -23.9$$

$$M_{eh} = +8.6 - 39.4(0.043) - 42.0(0.012) = +6.4$$

$$M_{ef} = +4.1 - 7.1(0.043) + 93.1(0.012) = +4.9$$

$$M_{fe} = +2.1 - 4.3(0.043) + 96.0(0.012) = +3.1$$

$$M_{gd} = +3.1 - 63.0(0.043) + 7.9(0.012) = +0.5$$

$$M_{gh} = -3.1 + 63.0(0.043) - 7.9(0.012) = -0.5$$

$$M_{hg} = -2.0 + 66.0(0.043) - 22.6(0.012) = +0.6$$

$$M_{he} = +2.8 - 50.2(0.043) - 58.1(0.012) = -0.1$$

$$M_{hi} = -0.8 - 15.8(0.043) + 80.7(0.012) = -0.5$$

$$M_{ih} = -0.5 - 8.7(0.043) + 90.1(0.012) = +0.2$$

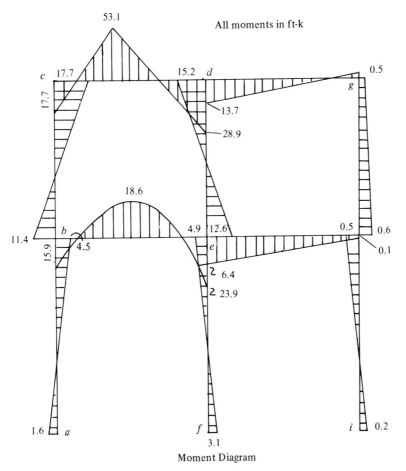

Moment Diagram

In this example problem the shears and moments were all rounded off to the nearest 1/10 value. In most problems, this will produce sufficient accuracy for engineering practice. The student should be careful, when the solution involves simultaneous equations, that sufficient significant numbers are used to avoid large errors.

The previous example problem showed the method of analysis for a frame with two degrees of freedom subjected to a nonsymmetrical gravity load. When a frame is acted upon by lateral loads, usually due to wind or earthquake, the same procedure can be used. In this case, the bending moment will be due entirely to sidesway since the lateral loads will be applied at the floor levels. The column shears will then be equal to the applied lateral loads above the level where columns shears are calculated. The following example problem will demonstrate this procedure.

EXAMPLE PROBLEM 13.10 The frame of Example Problem 13.9 will be analyzed for lateral loads as shown in Fig. 13.6.

It is possible to use the two sidesway distributions from Example Problem 13.9. The resulting equilibrium equations are then

$$\text{1st Story:}\quad \Sigma F_X = 0$$
$$10.0 - 27.0X + 4.7Y = 0$$

$$\text{2nd Story:}\quad \Sigma F_X = 0$$
$$26.0 + 3.0X - 26.5Y = 0$$

Solving these two simultaneous equations gives

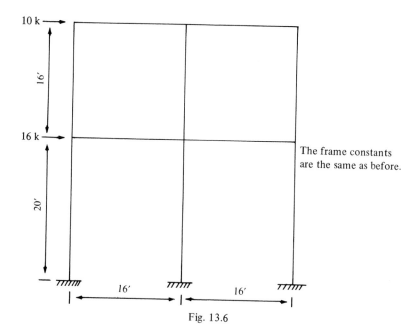

The frame constants are the same as before.

Fig. 13.6

$$X = 0.552$$

$$Y = 1.044$$

The final end moments are then determined by the summation of the moments in the first sidesway multiplied by 0.552 and the moments of the second sidesway multiplied by 1.044. A few of these final moments are given below. The student can calculate the other end moments

$$M_{ab} = -8.7(0.552) + 90.1(1.044) = +89.3 \text{ ft-k}$$

$$M_{ba} = -15.8(0.552) + 80.7(1.044) = +75.5$$

$$M_{be} = -50.2(0.552) - 58.1(1.044) = -88.3$$

$$M_{bc} = +66.0(0.552) - 22.6(1.044) = +12.8$$

A check can be made of the accuracy of the analysis by taking the sum of the final end moments at each joint. They must equal zero in order to satisfy equilibrium.

It is seen that the analysis of frames with many degrees of freedom can result in a large amount of numerical work, even with the moment-distribution method. It should be observed in Example Problem 13.9 that the correction for sidesway was relatively small. The maximum error neglecting sidesway would be about 15 percent. For frames with more bays, this would have been even smaller. For preliminary design of multiple bay frames, the effects of sidesway for vertical loads could be neglected. Another procedure for reducing the volume of computation would be to end the solution at two cycles of distribution. Even with the moment distribution method a lot of numbers have to be generated when the structure has several degrees of freedom. It is easily seen why structural engineers were quick to adopt the computer for analysis of highly redundant structures.

D. Gabled Frames

The gable frame as shown in Figure 13.7 has two degrees of freedom. Each corner can move independent of the other. If the frame has an axis of symmetry, then the displacement at one corner (b) will produce the mirror image of bending moments caused by the same magnitude of displacement at corner (d). If the frame meets this condition, then only one sidesway distribution is necessary in order to produce the two symmetrical simultaneous equations of equilibrium. From the structural action of the frame, any displace-

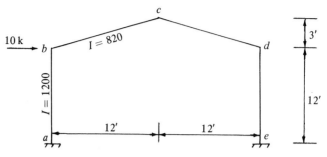

Fig. 13.7. Gabled frame.

ment of corner b will not only produce fixed-end moments in column ab with corner d restrained, but fixed end moments will be produced on girders bc and cd. If corner b is moved laterally to the left, the point c will also move to the left and vertically downward. The fixed end moments in girders bc and cd will be a function of the moments of inertia of these two members, their lengths, and the relative displacements of the ends, perpendicular to the axis of the girders. It then becomes necessary to determine the displacements of joints b and c with a given horizontal displacement of joint b. These displacements can be calculated by considering the geometry of the structure.

Consider the frame of Figure 13.8 with a given lateral displacement of corner b of an arbitrary amount Δ. The corner d is restrained from displace-

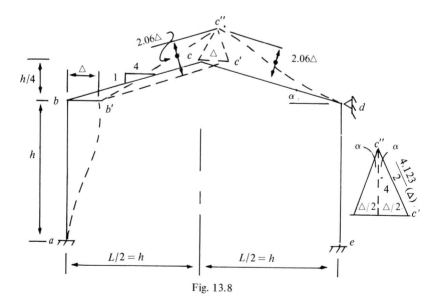

Fig. 13.8

ment by an artificial restraint. The deflected shape of the structure is shown in Figure 13.8 by the dashed line.

Since all joint displacements are so small with respect to the dimensions of the frame, all displacements can be taken as being perpendicular to the respective member. Joint b is given a displacement of amount Δ so that the new location of b is at b'. Neglecting any change in length of members, the horizontal movement of joint c would also be the amount Δ. However, end c of member cd must move perpendicular to member cd as well as perpendicular to member bc. The new location of c is c''. This point (c'') is located by drawing a displacement triangle whose base (horizontal dimension) is Δ. From the ends of this base (cc'), perpendiculars to members bc and cd are drawn. The intersection of these perpendiculars (c'') is the displaced position of joint c. From the displacement triangle, the relative displacements, perpendicular to their axes, of all the members of the frame can be determined. Giving member ab a unit displacement the displacements of bc and cd will be 2.06. If the frame was not symmetrical then the displacements corresponding to each roof girder would be different.

With the displacements known, the fixed end moments in all the members due to unit displacement of joint b can be calculated. The fixed end moments due to displacements will be directly proportional to displacements and moments of inertia and inversely proportional to the square of the lengths. Making the F.E.M of member ab equal to $+100$:

$$M_{F_{ab}} = 100(+)$$

$$M_{F_{bc}} = 100(2.06)\left(\frac{12}{12.37}\right)^2\left(\frac{820}{1000}\right) = 159(-)$$

$$M_{F_{cd}} = 159(+)$$

$$M_{F_{de}} = 0$$

The signs of the F.E.M. are determined from the direction of the moments as applied to the joints.

The moment distribution procedure can take place next. The shears at the base of each column can be calculated from the column end moments.

The resulting end moments after the moment distribution solution has been performed is given in Fig. 13.9.

The shears at the base of the columns are:

$$H_a = \frac{112 + 121}{12} = 19.42^k \text{ (to the left)}$$

$$H_e = \frac{34 + 68}{12} = 8.50^k \text{ (to the right)}$$

With the member end-moments as determined from a sidesway of joint b, an artificial horizontal force at b and d can be calculated from the equilibrium of free-bodies of the frame.

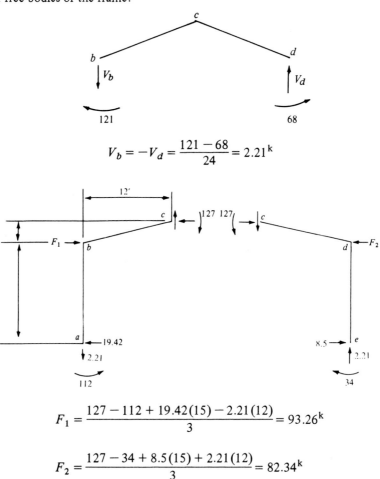

$$V_b = -V_d = \frac{121 - 68}{24} = 2.21^k$$

$$F_1 = \frac{127 - 112 + 19.42(15) - 2.21(12)}{3} = 93.26^k$$

$$F_2 = \frac{127 - 34 + 8.5(15) + 2.21(12)}{3} = 82.34^k$$

If joint d was displaced laterally the same amount as b was displaced the resulting member end-moments would be the mirror image of the moments obtained when b was displaced. The values of F_1 and F_2 would then be just reversed.

The simultaneous equations for lateral forces at b and d with the two degrees of freedom displacements are then obtained by taking a summation of lateral forces at b and d. If the real load on the frame was 10 kips to the right at b, the equations of summation of lateral forces at b and d are:

$$\Sigma F_x @b: \quad 93.26X + 82.34Y = 10$$

$$\Sigma F_x @d: \quad -82.34X - 93.26Y = 0$$

Solving these two equations gives

$$X = +0.486$$

$$Y = -0.429$$

The member end-moments for a 10 kip horizontal force at b are:

$$M_{ab} = +112(0.486) + 34(-0.429) = +40.2 \text{ ft-k}$$

$$M_{ba} = +121(0.486) + 68(-0.429) = +29.6$$

$$M_{bc} = -121(0.486) - 68(-0.429) = -29.6$$

$$M_{cb} = -127(0.486) - 127(-0.429) = -7.2$$

$$M_{cd} = +127(0.486) + 127(-0.429) = +7.2$$

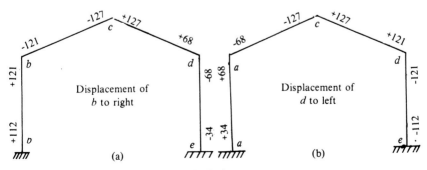

Fig. 13.9 End moments.

$$M_{dc} = +68(0.486) + 121(-0.429) = -18.9$$

$$M_{de} = -68(0.486) - 121(-0.429) = +18.9$$

$$M_{ed} = -34(0.486) - 112(-0.429) = +31.5$$

If a solution were required for vertical forces on the girder the procedure would be as follows:

(1) Determine fixed-end moments for the vertical girder loads and with a restraint of sidesway complete the moment distribution analysis.

(2) In a similar manner as previously done, determine, from equilibrium, the artificial horizontal forces at b and d.

(3) Combine these artificial restraint forces due to the girder loads with the forces due to sidesway. Then write the equations for summation of horizontal forces at b and d. These equations will now contain the horizontal restraint forces at b and d due to girder loads instead of the 10k and 0 values as for the previous lateral force solution.

(4) Solve for X and Y.

(5) Determine the total member end-moments by the summation of the moments due to girder loads with the moments due to the two sidesway moments.

E. Non-Prismatic Members

It is common in structures of all types to encounter members that have changes in the moment of inertia between the ends. This is definitely the case in continuous bridges. The mathematical expressions for stiffness, carry-over factor, and fixed end moments that have been used for prismatic members are no longer valid. In this book, those values were developed from the moment-area principles. Values of stiffness, carry-over factor, and fixed end moments for non-prismatic members can be calculated by the same method. Only, wherein for prismatic members the moment diagram was used, the M/I diagram must be used for non-prismatic members. The end result is more numerical work in finding these constants but after that operation the moment-distribution method of procedure is the same. Several publications (1) (2) have given the frame constants in tabular or graphic form for a wide range of structural shapes. The following example problem shows the procedure for determining the carry-over and stiffness for one type of non-prismatic member (Fig. 13.10).

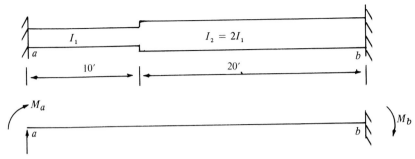

Fig. 13.10

EXAMPLE PROBLEM 13.11

(a) Determine the carry-over factor from a to b and also b to a for the variable section member shown below.

(b) Determine the relative stiffness of both ends of member ab.

With a moment applied at $a(M_a)$, there will be an induced moment at $b(M_b)$. The carry-over will be equal to the ratio M_b/M_a.

M_b is determined by the relationship that the deflection at a from the tangent to the elastic line at b is equal to zero, and is also equal to the moment about point a of the area of the M/EI diagram between a and b. An arbitrary value of 100 is assigned to M_a.

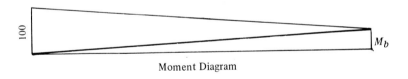

Moment Diagram

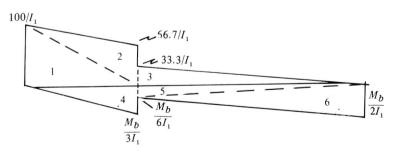

M/I Diagram

Triangle	Area $\times I_1$	Moment Arm	Moment of (Area of I_1)
1	$100(5) = 500$	3.33 ft	1665
2	$66.7(5) = 333$	6.67 ft	2222
3	$33.3(10) = 333$	16.67 ft	5556
4	$\dfrac{M_b}{3}(5) = 1.67M_b$	6.67 ft	$11.11M_b$
5	$\dfrac{M_b}{6}(10) = 1.67M_b$	16.67	$27.78M_b$
6	$\dfrac{M_b}{2}(10) = 5.0M_b$	23.33	$\underline{116.67M_b}$

$$\Sigma = 9443 + 155.6M_b$$

$$\Delta_a = 0 = 9443 + 155.6M_b$$

$$M_b = -\frac{9443}{155.6} = -60.7 \qquad \begin{array}{l}\text{$-$sign by beam sign convention}\\ \text{shows tension on top}\end{array}$$

$$\text{carry-over factor } a \text{ to } b = \frac{60.7}{100} = 0.607 \qquad \begin{array}{l}\text{$+$ by moment-distribution}\\ \text{convention}\end{array}$$

If a moment of 100 is induced at b and a determination made for M_a the M/I diagram will be the same as previously but with M_b multiplied by 100 and the M_a values divided by 100. Therefore, we can use the same diagrams.

The area of the M_a/I diagram is 1166.

The moment about a of the M_a/I diagram is 9443,

$$\text{c.g. from } a = \frac{9443}{1166} = 8.1 \text{ ft}$$

$$\text{c.g. from } b = 30 - 8.1 = 21.9 \text{ ft.}$$

$$\text{Moment of } \frac{M_a}{100I} \text{ diagram about } b = \frac{1166M_a}{100} \times 21.9 = 225.4M_a.$$

The area of the $\dfrac{M_b}{I}$ diagram $= 8.34M_b$,

$$\text{c.g. from } a = \frac{155.6}{8.34} = 18.6 \text{ ft,}$$

$$\text{c.g. from } b = 30 - 18.6 = 11.4 \text{ ft.}$$

Moment of $\left(\dfrac{M_b}{I} \times 100\right)$ diagram about $b = 834 \times 11.4 = 9507$,

$$M_a = \frac{9507}{255.4} = 37.2,$$

carry over factor b to $a = 0.372$.

(b) The stiffness of each end of the member can be determined using the first moment-area principle. The stiffness will be the moment required to rotate the end of the beam through a one radian angle. This angle is equal to the area of the M/EI diagram between a and b. Therefore, the same M/I diagrams can be used with a 100 ft-k moment applied at a. $M_b = -60.7$ ft-k and the area of the M/I diagram $= 1166 + 8.34(-60.7) = 660$ ft^2-k.

$$\theta_a = \frac{660}{EI_1}.$$

The stiffness is then

$$= \frac{100}{660/EI_1} = \frac{EI_1}{6.60}.$$

If the value of I_1 is in in^4 and E in k/in^2 then

$$\text{stiffness (end } a) = \frac{EI_1}{6.60(12)^2}$$

with

$$M_b = 100 \text{ ft-k} \qquad M_a = -37.2 \text{ ft-k}$$

The area of the M/I diagram is $= 834 + 11.66(-37.2) = 400$ ft^2-k. The

$$\text{stiffness (end } b) = \frac{EI_1}{4(12)^2}.$$

It is seen that

$$\text{end } b \text{ is } \frac{6.6}{4} = 1.65$$

times stiffer than end a.

References

1. Cross, H., "Analysis of Continuous Frames by Distributing Fixed-End Moments, "Trans., ASCE, Paper No. 1793, Vol. 96, 1936.
2. Gere, J M., *Moment Distribution*, Van Nostrand, New York, 1963.

Problems

1. Using the slope deflection method, determine the support reactions for the beam in Fig. 13.11. EI is constant.

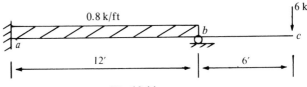

Fig. 13.11

2. Determine the bending moments over the supports for the continuous beam of Fig. 13.12 by the slope deflection method. Plot the bending moment diagram. $I_{ab} = 300 \text{ in}^4$; $I_{bc} = 360$; $I_{cd} = 240$.

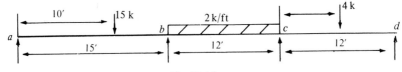

Fig. 13.12

3. Using the slope deflection method, determine the bending moments at the ends of all the members of the frame of Fig. 13.13. EI is constant.
4. If the frame of Fig. 13.13 had an I of bc equal to 1.6 times the I of the columns, what are the bending moments at the ends of the members? What are the percentage changes from the constant section frame?
5. Change the frame of Fig. 13.13 to a hinge support at a and d. What rotation would take place at a and d if $E = 30 \times 10^3$ ksi; I of columns $= 150 \text{ in}^4$, and I of girder $= 180 \text{ in}^4$.

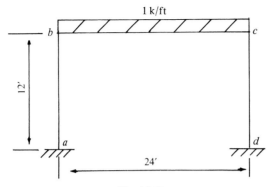

Fig. 13.13

6. By slope deflection method, determine the end moments and plot the bending moment diagram for the frame shown in Fig. 13.14. *EI* is constant.

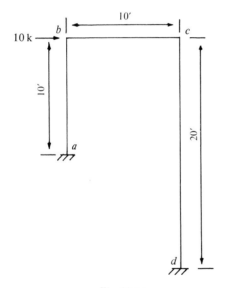

Fig. 13.14

7. Rework Problem No. 1 but with a vertical settlement of support *b* of 1/4 inch. $EI = 6 \times 10^6$ k-in^2.

8. Solve Problem No. 1 by the moment distribution method.

9. Determine the reactions of the beam in Fig. 13.15 using moment distribution.

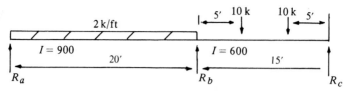

Fig. 13.15

10. Solve Problem No. 3 using moment distribution.
11. Determine member end moments for the frame of Fig. 13.16.

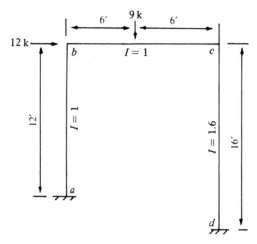

Fig. 13.16

12. Determine member end moments for the frame of Fig. 13.17.

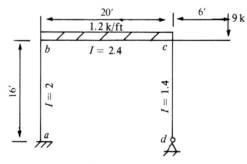

Fig. 13.17

13. Determine the member end moments for the frame of Fig. 13.18. I of girders equals 1.6 times I of columns.

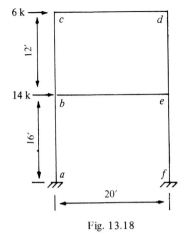

Fig. 13.18

14. Determine the member end moments for the gabled frame of Fig. 13.19.

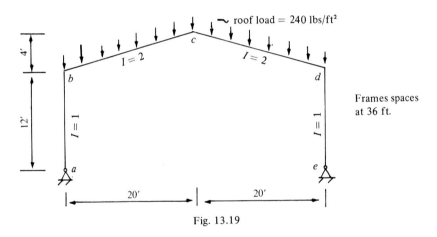

Fig. 13.19

15. Determine the member end moments for the frame of Fig. 13.19 with a horizontal load to the right of 8 kips at *b* only.

16. Determine the carry-over factor, absolute stiffness, and fixed-end moments for the structural element of Fig. 13.20. $E = 30 \times 10^3$ ksi.

17. Determine the carry-over factor from *a* to *b* for the beam section shown in Fig. 13.21.

18. Determine the stiffness of the end *a* of beam section *ab* of Fig. 13.21.

19. If the beam in Fig. 13.20 has one end displaced 1/2 inch normal to the beam axis what are the values of the bending moments at the ends of the beam.

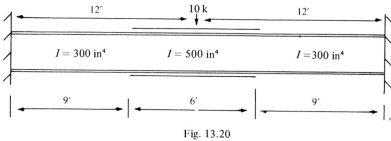

Fig. 13.20

Fig. 13.21

MATRIX ANALYSIS

14.1 Introduction

The availability of the electronic computer has brought about many changes in structural engineering. Notably so, the analysis of statically indeterminate structures. Methods that involved long and tedious computations for solution have become practical methods when programmed and solved by the computer. The general method presented in Chapter 12 shows that when this method is used for analysis of structures it is necessary to solve the same number of simultaneous equations as the number of redundants. For structures with more than about four redundants, the numerical work becomes time-consuming and tedious if processed by hand solution. However, the electronic computer can handle the solutions of many simultaneous equations with great ease and rapidity.

Two systematic methods have been formulated that simplify the numerical work required for solution by the computer. Mention has been made of one method in Chapter 12. This is the *flexibility method*, as generally spoken of, although *force method* has been used for the name of the same procedure. In this method the structure is evaluated as having *n* number of redundants. Equations are written for the total displacement at the location of each redundant. The total displacements are the displacements due to the given loading system acting on the structure with the redundants removed (primary structure), plus the displacements due to the redundants acting on the primary structure. This was the method in Chapter 12. As can be seen, this method requires the calculation of displacements in the primary structure.

A procedure that is preferred by engineers for complex structures is a different approach than that used in the flexibility method. This second procedure is called the *stiffness method*. It has also been referred to as the *displacement method*, but most engineers refer to it as the *stiffness method*. In this method, all joints are fixed against displacements. Each joint is then successively displaced a unit amount corresponding to the unknown displacements. Equations of joint equilibrium are written, producing a set of simultaneous equations. The solution of these equations gives the joint displacements. In this method, the actions (forces) required to produce unit displacements have to be computed. The magnitude of these actions are the *stiffness coefficients.*

It will not be possible in this book to develop the two matrix methods to great depth. The methods are developed to the level of hand computation. Through this process the student can then understand how and why the mathematical models work. Blindly using computer programs can be a dangerous road to follow. At the present time, there are a multitude of programs available to the structural engineer. The more sophisticated ones place the analysis at the fully automatic level. At this level, the program generates all matrices and performs all mathematical routines required for a solution, The computer input data are the geometry of the structure, the cross-sectional dimensions of all the members, the identification of the loads, and the boundary conditions. In using fully automated programs, the engineer is restrained from using programs that exceed the capacity of the computer. The published literature pertaining to the development and use of matrix methods has been preponderous in the last decade. Many books (1) (2) (3) are available that present these methods in considerable detail. The student of structural engineering can find detailed information covering a wide range of problems in these books just referenced or in others with similar titles.

Structural engineers should be familiar with many approaches to the solutions of statically determinate and indeterminate structures. Hand calculation methods and computer solved methods should be a part of the engineer's background knowledge. The student should beware of developing a procedural rote in the solution of engineering problems. Such traps are easy to fall into. Every solution is based on a mathematical model of a real structure. If the mathematical model does not fit the real structure, then the answer, no matter how smart the solution, may be of little worth. Practical solutions to all engineering problems will require assumptions to a lesser or greater degree. These assumptions usually relate to a physical parameter but may also be based on the structural response of the structure. The validity of these assumptions will require engineering judgment and skill. Such can never be replaced by the computer or the blind use of any method.

14.2 Flexibility Method—General.

As stated in the previous section, the first step in applying the method is to select the redundant forces or moments. These redundants are then removed from the structure to give a statically determinate stable structure in which displacements at the point of application of the redundants, and in their direction, can be determined. This structure is called the primary structure. The judicious selection of the redundants can expedite the solution.

The displacements at the positions of the redundants and in their corresponding directions are computed for the primary structure acted upon by the given real loads. Matrix analysis has developed a set of nomenclature separate from that usually used in the literature in the presentation of the classical methods. A switch to this accepted nomenclature will be made in this chapter. Displacements of the primary structure due to the given real loads are referenced by the letter D. A subscript number will accompany the letter D to indicate the location of this displacement. Displacements can be either a translation (deflection) or rotation.

The second step is to successively place unit forces corresponding to the redundants on the primary structure and calculate the displacements, resulting from these unit forces, at the points of application of these redundants and in their respective direction. These displacements are termed the *flexibility coefficients* and are represented by the letter f. Two subscript numbers will accompany the letter f. For example, f_{12} is the displacement at point 1 on the structure due to a unit force at point 2 in the direction of the redundant at point 2.

The next step in the procedure is to write the compatibility equation corresponding to the true action of the structure. If the redundants are support actions of the structure and there is no movement of the structure at the points of support of the real structure, then the compatibility equations will equal zero. If a movement is known, then the equation is equal to this value.

The equations are set in matrix form and the solution performed to determine the numerical value of the selected redundants.

14.3 Flexibility Method—Beams.

An example of the procedure just explained will be performed for the beam structure shown in Figure 14.1.

Equations can now be written expressing the compatibility of deflections at points 1 and 2 on the beam. It is assumed that there are no support displacements.

$$D_1 + f_{11}R_1 + f_{12}R_2 = 0$$

$$D_2 + f_{21}R_1 + f_{22}R_2 = 0$$

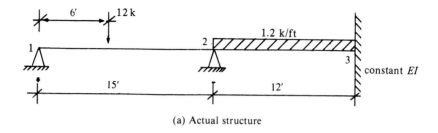

(a) Actual structure

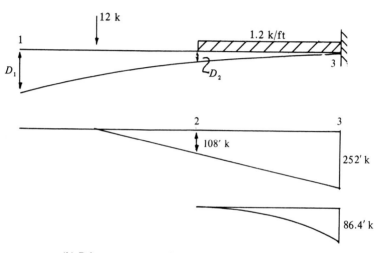

(b) Primary structure with bending moment diagrams

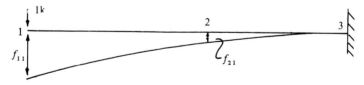

(c) Unit load at 1, corresponding flexibility coefficients

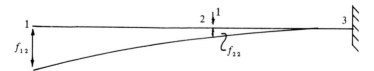

(d) Unit load at 2, corresponding flexibility coefficients

Fig. 14.1. Continuous beam.

These equations can now be written in matrix form. There is the square matrix containing the flexibility coefficients and the two column

$$\begin{bmatrix} f_{11} & f_{12} \\ f_{21} & f_{22} \end{bmatrix} \begin{bmatrix} R_1 \\ R_2 \end{bmatrix} = - \begin{bmatrix} D_1 \\ D_2 \end{bmatrix} \tag{14.1}$$

matrices. It is now necessary to determine the flexibility coefficients and the values of D. They will be determined by the moment area method. Since EI is constant for the length of the beam and since this term would appear in the denominator of both the f and D terms, it can be cancelled out.

To determine EID_1, the moment of the area of the moment diagram for the real loads is taken about point 1.

$$EID_1 = \frac{252}{2}(12 + 9)(6 + 14) + 86.4 \left(\frac{12}{3}\right) \left(15 + \frac{3 \times 12}{4}\right) = 61,214 \text{ ft}^3 \cdot \text{k}$$

The value of EID_2 is determined by taking moments about point 2 of the bending moment diagram that lies between points 2 and 3.

$$EID_2 = (108)(12)(6) + 144(6)(8) + 86.4(4)(9) = 17,798 \text{ ft}^3 \cdot \text{k}$$

The flexibility coefficients are next determined. These values will be determined in general form so that they can be easily evaluated for any similar problem although for this specific problem it would be quicker to evaluate them in terms of the given dimensions. The required moment diagrams are shown in Fig. 14.2.

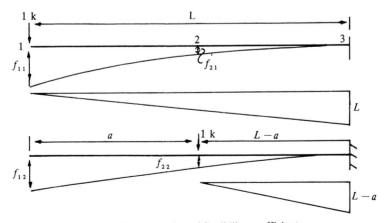

Fig. 14.2. Determination of flexibility coefficients.

The value of Elf_{11} is equal to the moment of the area of the moment diagram, with unit load at 1, taken about 1.

$$Elf_{11} = L\left(\frac{L}{2}\right)\left(\frac{2L}{3}\right) = \frac{L^3}{3}$$

The value of Elf_{21} is equal to the moment of the area of the moment diagram between 2 and 3 taken about 2.

$$Elf_{21} = \frac{(L-a)^2}{2}\left(a + \frac{2}{3}(L-a)\right)$$

From Maxwell's theorem of reciprocal deflections

$$f_{21} = f_{12}$$

The value of Elf_{22} is the moment about 2 of the bending moment diagram between 2 and 3 with the unit load at 2.

$$Elf_{22} = \frac{(L-a)^3}{3}$$

The evaluations of the flexibility coefficients for the beam in Fig. 14.1 are

$$Elf_{11} = \frac{(27)^3}{3} = 6561 \text{ ft}^3\text{-k}$$

$$Elf_{12} = Elf_{21} = 1656$$

$$Elf_{22} = \frac{1728}{3} = 576$$

Equation 14.1 can be written

$$[F][R] = -[D]$$

or

$$[R] = -[F]^{-1}[D] \qquad (14.2)$$

Where $[F]^{-1}$ is the inverse of the flexibility matrix. The values of R can be found from solving this last equation, 14.2, or the preceeding equation.
Writing the matrix, Eq. 14.1

$$\frac{1}{EI}\begin{bmatrix} 6561 & 1656 \\ 1656 & 576 \end{bmatrix}\begin{bmatrix} R_1 \\ R_2 \end{bmatrix} = \frac{1}{EI}\begin{bmatrix} 61,214 \\ 17,798 \end{bmatrix}$$

The determinantal equations of the matrix 14.1 are

$$R_1 = \frac{f_{22}(-D_1) - f_{12}(-D_2)}{f_{11}(f_{22}) - f_{12}(f_{21})} = \frac{-576(61,214) + 1656(17,798)}{6561(576) - 1656(1656)}$$

$$= \frac{-5,785,776}{1,036,800} = -5.58\,k$$

$$R_2 = \frac{f_{11}(-D_2) - f_{21}(-D_1)}{f_{11}(f_{22}) - f_{12}(f_{21})}$$

$$= \frac{-6561(17,798) + 1656(61,214)}{1,036,800} = -14.86\,k$$

The minus sign indicates R values are upward, opposite to the direction of the unit load.

The final force diagram for the structure is shown in Fig. 14.3.

The matrix 14.1 can easily be expanded to fit any number of equations corresponding to the number of redundants. This general matrix is

$$\begin{bmatrix} f_{11}f_{12}f_{13} & \cdots & f_{1n} \\ f_{21}f_{22}f_{23} & \cdots & f_{2n} \\ f_{31}f_{32}f_{33} & \cdots & f_{3n} \\ \cdots & \cdots & \cdots \\ f_{n1}f_{n2}f_{n3} & \cdots & f_{nn} \end{bmatrix}\begin{bmatrix} R_1 \\ R_2 \\ R_3 \\ \cdots \\ R_n \end{bmatrix} = -\begin{bmatrix} D_1 \\ D_2 \\ D_3 \\ \cdots \\ D_n \end{bmatrix}$$

The flexibility matrix is always a square matrix and is always symmetric about its main diagonal. The flexibility coefficients are functions of the

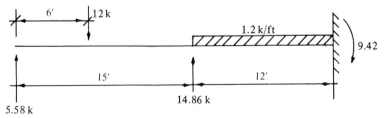

Fig. 14.3. Reactions.

geometry and the elastic properties of the structure. They remain the same for any variation of loading condition. The values in the D matrix are dependent upon the loading condition.

The solution to large matrices requires a computer. There are special techniques for the most efficient solution of sets of linear equations. Accuracy is best achieved by having large coefficient terms on the main diagonal and smaller values for the off-diagonal terms. This possibly could be achieved by the proper selection of the primary structure. In the structure just analyzed there are several other possible primary structures. The primary structure selected gave easy solutions to the flexibility coefficients and the D displacements.

14.4 Flexibility Method—Frames

The matrix method can be readily applied to frame structures. Again, the proper selection of the primary structure can expedite the solution. The frame in Fig. 14.4 has three redundants. The frame will be released at the right support. The redundants are then the moment, horizontal reaction, and the vertical reaction at that support.

The displacements at the right support as well as the flexibility coefficients will be determined by the method of virtual work.

Evaluating the values for the D matrix

Segment	$x = 0$ at	m_1	m_2	m_3	M
$e \rightarrow d$	e	$-x$	0	-1	0
$d \rightarrow c$	d	-12	x	-1	0
$c \rightarrow b$	c	-12	$9 + x$	-1	$-10x$
$b \rightarrow a$	b	$-12 + x$	18	-1	$-90 - 4x$

$$EI_1 D_1 = \int m_1 M dx = \frac{1}{1.6} \int_0^9 120x dx$$

$$+ \int_0^{12} (1080 - 42x - 4x^2) dx = 10,670 \text{ ft}^3\text{-k}$$

$$EI_1 D_2 = \int m_2 M dx = \frac{1}{1.6} \int_0^9 (-90x - 10x^2) dx$$

$$+ \int_0^{12} (-1620 - 72x) dx = -28,421 \text{ ft}^3\text{-k}$$

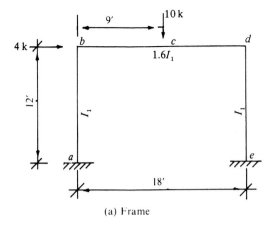

(a) Frame

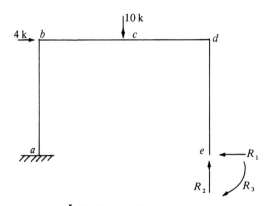

(b) Primary Structure

Fig. 14.4. Frame analysis.

$$EI_1D_3 = \int m_3 M dx = \frac{1}{1.6} \int_0^9 10x dx + \int_0^{12} (90 + 4x)dx = 1{,}621 \text{ ft}^3\text{-k}$$

The displacement D_3 is an angular rotation.
The flexibility matrix is as follows

$$\begin{bmatrix} f_{11} & f_{12} & f_{13} \\ f_{21} & f_{22} & f_{23} \\ f_{31} & f_{32} & f_{33} \end{bmatrix}$$

It should be remembered that the first subscript denotes the direction of the displacement; i.e., 1 is in the horizontal direction, 2 in the vertical direction, 3 in the angular direction. The second subscript denotes the type of action, corresponding to the action of the redundants. The unit load will be in the direction of the redundants and the same sign convention for bending moments that was used for evaluating the D displacements will be used; i.e., tension on the inside of the frame is positive. (See figure 14.5.)

Segment	Action 1 m_1	Action 2 m_2	Action 3 m_3
$e \to d$	$-x$	0	-1
$d \to b$	-12	x	-1
$b \to a$	$-12 + x$	18	-1

$$EI_1 f_{11} = \int (m_1)^2 dx = \int_0^{12} x^2 dx + \frac{1}{1.6} \int_0^{18} 144 dx$$

$$+ \int_0^{12} (144 - 24x + x^2)dx = 2772 \text{ ft}^3\text{·k}$$

$$EI_1 f_{12} = \int (m_1)(m_2) dx = \frac{1}{1.6} \int_0^{18} - 12x dx$$

$$+ \int_0^{12} (-216 + 18x)dx = -2511$$

$$EI_1 f_{13} = \int (m_1)(m_3) dx = \int_0^{12} x dx + \frac{1}{1.6} \int_0^{18} + 12 dx$$

$$+ \int_0^{12} (12 - x)dx = 279$$

$$EI_1 f_{22} = \int (m_2)^2 dx = \frac{1}{1.6} \int_0^{18} x^2 dx + \int_0^{12} 324 dx = 5103$$

$$EI_1 f_{23} = \int (m_2)(m_3) dx = \frac{1}{1.6} \int_0^{18} - x dx + \int_0^{12} - 18 dx = -317.3$$

$$EI_1 f_{33} = \int (m_3)^2 dx = \int_0^{12} dx + \frac{1}{1.6} \int_0^{18} dx + \int_0^{12} dx = 35.25$$

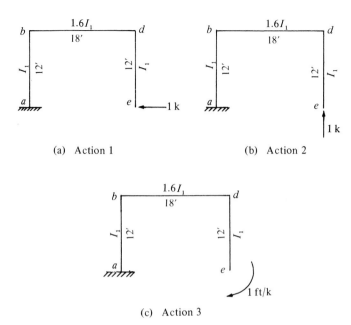

(a) Action 1

(b) Action 2

(c) Action 3

Fig. 14.5. Primary structure with action for determining flexibility coefficients.

The above values can then be inserted into the matrix.

$$EI_1 \begin{bmatrix} +2772 & -2511 & +279 \\ -2511 & +5103 & -317.3 \\ +279 & -317.3 & +35.25 \end{bmatrix} \begin{bmatrix} R_1 \\ R_2 \\ R_3 \end{bmatrix} = -\frac{1}{EI_1} \begin{bmatrix} 10{,}670 \\ -28{,}421 \\ 1{,}621 \end{bmatrix}$$

The solution of this matrix is

$$R_1 = 3.83 \text{ kips}$$

$$R_2 = 6.15 \text{ kips}$$

$$R_3 = -20.88 \text{ ft-kips}$$

The "$-$" sign for R_3 indicates that the moment is counterclockwise, op-posite to that assumed. This could have been recognized at the beginning of the problem and R_3 taken as counterclockwise, but the procedure indicates that it is not necessary to know the correct direction of the redundants at the start of the solution.

This problem also indicates that the flexibility coefficients can be rotational as well as translational.

14.5 Flexibility Method—Trusses

The flexibility method can be used for truss structures. If the structure is statically indeterminate externally, then the reactions can be selected as the redundants as was done for the beam. The D displacements are calculated for the primary structure by any selected method. The virtual work method is probably most applicable. The flexibility coefficients can also be determined by calculating the displacements for the unit forces in the direction of the redundants. As was explained in Section 12.5 the redundants can be selected as internal bars. This can sometimes reduce the amount of numerical work.

For trusses that are statically indeterminate internally, then the selected redundants are bars of the truss. The basic approach to the determination of the forces in the bars of the statically indeterminate force system is to first cut the redundant bars. The system with the cut bars is the primary

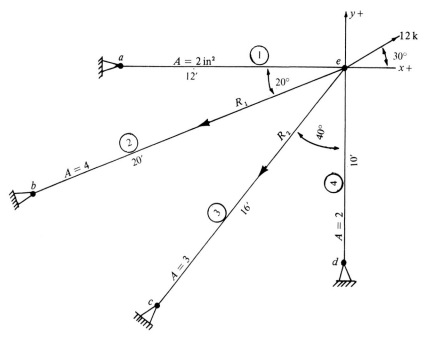

Fig. 14.6. Redundant bar system.

structure. The loads are then applied to the primary structure and the displacements of the cut ends of the redundant bars are calculated. These displacements are called the D displacements. They will, of course, be in the direction of the redundant bars. These displacements are easily calculated by the method of virtual work. As an example the four bar system of Fig. 14.6 is subjected to a load of 12 kips at an angle of 30 degrees with the x axis. Figure 14.7 shows the primary structure and the flexibility coefficients for each action. The actions for determining the flexibility coefficients are a unit load at the cut ends of each redundant bar. These actions are also required to determine the D displacements.

The basic displacement equations for the bars of Fig. 14.7 are

$$D_1 + f_{11}R_1 + f_{12}R_2 = 0$$

$$D_2 + f_{21}R_1 + f_{22}R_2 = 0$$

or in matrix form

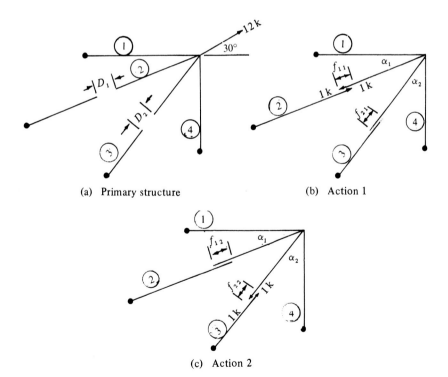

(a) Primary structure   (b) Action 1

(c) Action 2

Fig. 14.7. Actions to determine flexibility components.

$$\begin{bmatrix} f_{11} & f_{12} \\ f_{21} & f_{22} \end{bmatrix} \begin{bmatrix} R_1 \\ R_2 \end{bmatrix} = -\begin{bmatrix} D_1 \\ D_2 \end{bmatrix}$$

$$ED_1 = \Sigma u_1 \left(\frac{PL}{A} \right)$$

$$ED_2 = \Sigma u_2 \left(\frac{PL}{A} \right)$$

$$Ef_{11} = \Sigma u_1 \left(\frac{u_1 L}{A} \right)$$

$$Ef_{12} = Ef_{21} = \Sigma u_1 \left(\frac{u_2 L}{A} \right)$$

$$Ef_{22} = \Sigma u_2 \left(\frac{u_2 L}{A} \right)$$

The solution of the above values of displacement can best be determined by the formulation in Table 14.1. Tension values are given a plus sign.

Table 14.1

Bar	L(ft)	A(in²)	L/A(ft/in)	P(kips)
1	12	2	72	+10.4
2	20	4	60	0
3	16	3	64	0
4	10	2	60	+ 6.0

Bar	u_1 (kips)	u_2 (kips)	$\dfrac{u_1 PL}{A}$	$\dfrac{u_2 PL}{A}$	$\dfrac{(u_1)^2 L}{A}$	$\dfrac{u_1 u_2 L}{A}$	$\dfrac{(u_2)^2 L}{A}$
1	$-\cos\alpha_1$	$-\sin\alpha_2$	-703.6	-481.3	63.58	43.49	29.75
2	+1.0	0	0	0	60.0	0	0
3	0	+1.0	0	0	0	0	64.0
4	$-\sin\alpha_1$	$-\cos\alpha_2$	-123.1	-275.8	7.02	15.72	35.21
		$\Sigma =$	-826.7	-757.1	+130.60	+59.21	+128.96

$$ED_1 = -826.7$$

$$ED_2 = -757.1$$

$$Ef_{11} = 130.60$$

$$Ef_{12} = 59.21$$

$$Ef_{22} = 128.96$$

The compatability equations are

$$\frac{1}{E}\begin{bmatrix} 130.60 & 59.21 \\ 59.21 & 128.96 \end{bmatrix}\begin{bmatrix} R_1 \\ R_2 \end{bmatrix} = \frac{1}{E}\begin{bmatrix} 826.7 \\ 757.1 \end{bmatrix}$$

$R_1 = 4.63$ k (Force in bar 2) $R_2 = 3.74$ kips (Force in bar 3). The forces in bars 1 and 4 can then be determined from statics. They are 3.64 kips and 1.55 kips respectively.

This same procedure could be used for determining the redundant bar forces for any trussed structure.

14.6 Stiffness Method—General

It is seen that the flexibility or force method is simple of concept and the formulation of the equations based upon compatibility of displacements is very straightforward. The determination of D displacements requires considerable computational work when the structure is statically indeterminate to several degrees. In the basic matrix formulation the number of equations was equal to the number of redundants (degrees of indeterminacy).

The basic load deformation relationship in the flexibility method was equation 14.2.

$$D = f \cdot R \qquad\qquad (14.2) \text{ repeated}$$

This equation can be related to a spring. The displacement of the end of a spring will be equal to the flexibility f of the spring (displacement per unit load) multiplied by the force of R applied to the spring.

The stiffness method is the reciprocal of the flexibility method. Again relating to a spring

$$R = K \cdot D \qquad\qquad (14.3)$$

The force required to displace a spring is the displacement D multiplied by the stiffness K (force per unit displacement). Both of these above mathematical models are based upon the spring having a linear relationship between force and displacement.

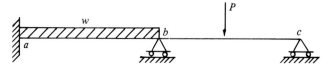

Fig. 14.8. Continuous beam.

In the flexibility method, it was necessary to determine the number of redundants in the structure. In the stiffness method, it is necessary to know the degree of kinematic indeterminacy. Kinematic indeterminacy is not exactly analogous with statical indeterminacy since there are no equations of kinematic equilibrium. Kinematic indeterminacy can be illustrated by referring to the two span beam of Fig. 14.8. This beam is restrained from translating or rotating at support a. If the change in the length of the beam is neglected, then the beam only rotates at supports b and c with no translation taking place. The beam can then be said to be kinematically indeterminate to the second degree.

The same reasoning can be applied to the frame of Fig. 14.9. The change in the lengths of the members is neglected. At support a there is no translation or rotation while there is rotation only at joint f. At all other joints there is rotation making five possible rotations. There is also lateral translation of the ends of the top girder cd (equal at both ends if change in length is neglected), and lateral translation of girder be. With the two translations and the five rotations, the kinematic indeterminacy of the frame is seven. The student may believe he has heard something like this before. He has, in the previous chapter, when he studied the slope deflection method. The number of unknowns in the set of slope deflection equations was the degree of kinematic

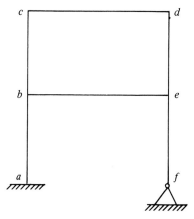

Fig. 14.9. Frame.

indeterminacy. The slope deflection method is a form of the stiffness method. The slope deflection equation

$$M = M_F - \frac{4EI}{L}\theta_a - \frac{2EI}{L}\theta_b + \frac{6EI\Delta}{L^2}$$

is of the same form as equation 14.3, only with the M_F (fixed end moment) term added. The displacements θ_a, θ_b, and Δ are the D values and

$$\frac{KEI}{L}$$

are the stiffness coefficients.

Trussed structures are readily analyzed by the stiffness method, In ideal trusses where the ends of all members are pin-connected and the loads are all applied at the joints, then the displacements are all translations and there are no unknown rotations. Considering the truss of Fig. 14.10, it has nine joints that can translate in the x and y direction. The other two joints at the supports have no movement. Therefore the truss is 18 times kinematically indeterminate.

At this stage of the discussion, the student may wonder the need of learning both the flexibility and the stiffness method. And, if both are necessary tools for the structural engineer, which one should be used when? The order of equations in the flexibility method is the degree of statical indeterminacy while in the stiffness method it is the degree of kinematic indeterminacy. It then appears that if hand calculations are to be used it would be best to use the method that would produce the fewest simultaneous equations. This

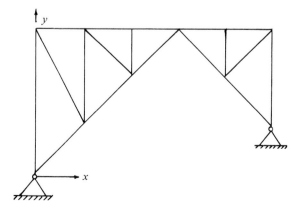

Fig. 14.10. Trussed arch.

would then result in comparing the degree of static indeterminacy against the degree of kinematic indeterminacy. The orders of the f matrix and the K matrix would correspond respectively. However, this evaluation would not necessarily always be the criterion for the selection of method. The amount of effort and time involved in determining the values of the f and K matrices would also be a factor in the selection of method.

Before proceeding further, it is best to once again be sure that a sign convention is firmly established. A coordinate system is shown in Fig. 14.11. The plus directions of x, y, and z are shown. A moment applied to the joint in a clockwise direction is positive. The internal moment on the end of a member is then positive if it is counterclockwise. Internal forces acting on the ends of members in the positive direction of coordinates are positive. Internal forces acting on the joints will be equal and opposite to the direction of the forces acting on the ends of the members.

Since it will be convenient (and also necessary) to equate external forces to internal forces for obtaining joint equilibrium, the positive direction of external forces are shown in Fig. 14.11(b). Translations of the ends of the members will be positive in the positive direction of the axes. Rotations of the ends of members will be positive if they are counterclockwise. Positive (counterclockwise) rotations will produce positive moments (counterclockwise) in the ends of the members.

In matrix analysis, the location on the structure where forces or displacements are equated are called *nodes*. In most cases, the *nodes* will be the joints and support locations of the structure. However, other locations can also be selected as nodes. The points where members change section (non-

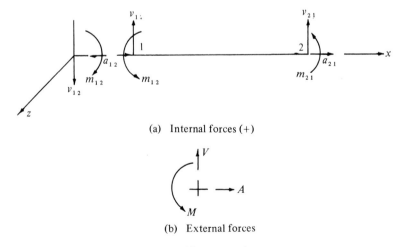

(a) Internal forces (+)

(b) External forces

Fig. 14.11. Sign conventions.

prismatic members) are usually selected as nodal points also. All joints and support locations are nodes, but not all nodes are joints or support points.

14.7 Stiffness Method—Beams

The stiffness method will now be applied to the beam of Fig. 14.12. This is also the same beam as analyzed in Example Problem 12.3. The student should compare methods.

This beam is kinematically indeterminate to the fourth degree. That is, it has rotation at each support, but no translation if the supports are unyielding. If the bending moments in the beam at the supports are known then the bending moments and shears at all locations along the beam can be determined by the equations of static equilibrium. The R values (Eq. 14.3) are the actions corresponding to the kinematic indeterminates. In this problem, they will be the external moments applied to the supports. These actions are shown in Fig. 14.13 (a).

The procedure is to first place an artificial restraint to each kinematic action (Fig. 14.13 (b)). In doing this, there will be moments applied to these artificial restraints that will be the fixed end moments of a single span beam. The M_F values are easily evaluated from previously developed equations

$$M_{F_{ab}} = -M_{F_{ba}} = \frac{wL^2}{12} = \frac{1.6(24)^2}{12} = 76.8 \text{ ft-k}$$

$$M_{F_{bc}} = \frac{Pab^2}{L^2} = \frac{12(12)(6)^2}{(18)^2} = 16 \text{ ft-k}$$

$$M_{F_{cb}} = -\frac{12(12)^2(6)}{(18)^2} = -32 \text{ ft-k}$$

$$M_{F_{cd}} = -M_{F_{dc}} = \frac{Pa}{L}(L-a) = \frac{6(6)(12)}{18} = 24 \text{ ft-k}$$

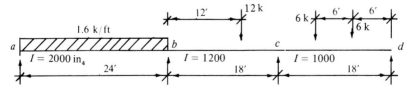

Fig. 14.12. Continuous beam.

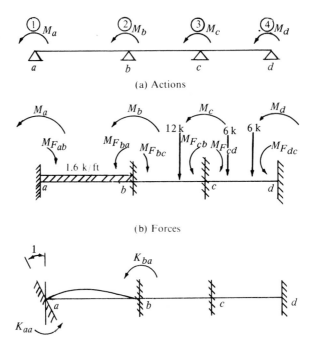

(a) Actions

(b) Forces

(c) Unit rotation at a and corresponding stiffness factors.

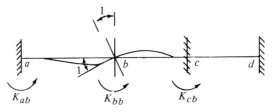

(d) Unit rotation at b and corresponding stiffness factors.

Fig. 14.13.   Stiffness factors.

Equation 14.3 can now be written in specific form as

$$M - M_F = K \cdot D \qquad (14.4)$$

If there are no external applied moments, as is the case in the example problem, then the values of M are zero and equation 14.4 can then be written

$$[K] \, [D] = -[M_F] \qquad (14.5)$$

The values of D can then be found from the equation

$$[D] = [K]^{-1}[-M_F]$$

The values of K_{ij} are next evaluated. The subscript i refers to the location where the moment is being considered and j is the location where the unit rotation is taking place. In Fig. 14.13 (c) it is seen that the a end of beam ab is rotated through a counterclockwise angle of one radian. The external moment required to produce this rotation is called the stiffness. From Chapter 13 it was seen that for a prismatic beam this moment was equal to

$$\frac{4EI}{L}$$

Therefore

$$K_{aa} = \frac{4EI_{ab}}{L_{ab}}$$

The internal moment produced in the end of member ab and the moment applied to the joint are shown in Fig. 14.14. All values are positive.

When a rotation of one radian is given to end a then a moment is produced at the restrained end b. If the member is prismatic then the moment at b is equal to one-half the moment at a. This was proven in developing the *carry-over factor* in moment distribution. The stiffness coefficient K_{ba} is then

$$K_{ba} = \frac{2EI_{ab}}{L_{ab}}$$

The remaining stiffness coefficients will now be evaluated. At node b a rotation of one radian is placed on the restrained support at this location (Fig. 14.13 (d)). The stiffness coefficient K_{bb} will be the moment required to rotate the b ends of both beams ba and bc through an angle of one radian. Therefore

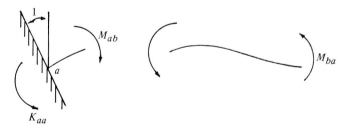

Fig. 14.14. Stiffness coefficient.

$$K_{bb} = \frac{4EI_{ab}}{L_{ab}} + \frac{4EI_{bc}}{L_{bc}}$$

and

$$K_{ab} = \frac{2EI_{ab}}{L_{ab}}$$

$$K_{bc} = \frac{2EI_{bc}}{L_{bc}}$$

By like manner, the other stiffness coefficients can be written

$$K_{cc} = \frac{4EI_{bc}}{L_{bc}} + \frac{4EI_{cd}}{L_{cd}}$$

$$K_{cb} = \frac{2EI_{bc}}{L_{bc}}$$

$$K_{dc} = \frac{2EI_{cd}}{L_{cd}}$$

$$K_{dd} = \frac{4EI_{cd}}{L_{cd}}$$

$$K_{cd} = \frac{2EI_{cd}}{L_{cd}}$$

From Maxwell's reciprocal theorem, it is noted that in all cases

$$K_{ij} = K_{ji}$$

It should also be noted that

$$K_{ca} = K_{da} = K_{db} = K_{ac} = K_{ad} = K_{bd} = 0$$

Because of the zero rotation at all restraints except the node being rotated, stiffness coefficients at other locations than the span being acted upon will be zero. This is one of the advantages of the stiffness method. Another advantage is the simplicity of evaluating the values of the stiffness coefficients. Every coefficient in the above case was based upon a beam (or flexural member in the case of a frame) fixed at one end and given a unit rotation at the other end. The evaluation of the stiffness coefficients was a lot easier than the flexibility coefficients would be (see Example Problem 12.3).

It is now seen that the general matrix (Eq. 14.5) for the solution of this problem can be written

$$
\begin{bmatrix}
K_{aa} & K_{ab} & K_{ac} & K_{ad} \\
K_{ba} & K_{bb} & K_{bc} & K_{bd} \\
K_{ca} & K_{cb} & K_{cc} & K_{cd} \\
K_{da} & K_{db} & K_{dc} & K_{dd}
\end{bmatrix}
\begin{bmatrix}
\theta_a \\
\theta_b \\
\theta_c \\
\theta_d
\end{bmatrix}
= -
\begin{bmatrix}
M_{Fa} \\
M_{Fb} \\
M_{Fc} \\
M_{Fd}
\end{bmatrix}
$$

The quantative values of

$$
K_{aa} = \frac{4EI_{ab}}{L_{ab}} = 4E(12)^{-4}\left(\frac{20}{24}\right)(10)^2 = N(3.333)
$$

The value of $N = E(12)^{-4}(10)^2$ will be common to all stiffness coefficients and is a numerical convenience.

$$
K_{ba} = \frac{2EI_{ab}}{L_{ab}} = N\left(\frac{2 \times 20}{24}\right) = N(1.667)
$$

$$
K_{bb} = \frac{4EI_{ab}}{L_{ab}} + \frac{4EI_{bc}}{L_{bc}} = N\left(\frac{4 \times 20}{24}\right) + N\left(\frac{4 \times 12}{18}\right) = N(6)
$$

$$
K_{cb} = \frac{2EI_{bc}}{L_{bc}} = N\left(\frac{2 \times 12}{18}\right) = N(1.333)
$$

$$
K_{cc} = \frac{4EI_{bc}}{L_{bc}} + \frac{4EI_{cd}}{L_{cd}} = N\left(\frac{4 \times 12}{18}\right) + N\left(\frac{4 \times 10}{18}\right) = N(4.889)
$$

$$
K_{dc} = \frac{2EI_{cd}}{L_{cd}} = N\left(\frac{2 \times 10}{18}\right) = N(1.111)
$$

$$
K_{dd} = \frac{4EI_{cd}}{L_{cd}} = N\left(\frac{4 \times 10}{18}\right) = N(2.222)
$$

$$MF_a = +76.8 \text{ ft-k}$$

$$MF_b = MF_{ba} + MF_{bc} = -76.8 + 16.0 = -60.8$$

$$MF_c = MF_{cb} + MF_{cd} = -32.0 + 24.0 = -8.0$$

$$MF_d = MF_{dc} = -24.0$$

The matrix can now be written with the specific values inserted.

$$N \begin{bmatrix} 3.333 & 1.667 & 0 & 0 \\ 1.667 & 6.000 & 1.333 & 0 \\ 0 & 1.333 & 4.889 & 1.111 \\ 0 & 0 & 1.111 & 2.222 \end{bmatrix} \begin{bmatrix} \theta_a \\ \theta_b \\ \theta_c \\ \theta_d \end{bmatrix} = - \begin{bmatrix} +76.8 \\ -60.8 \\ -8.0 \\ -24.0 \end{bmatrix}$$

It should be noted that the K matrix is a square matrix and the largest values are along the principle axis.

Solving the above matrix gives

$$N\theta_a = -33.60$$

$$N\theta_b = +21.12$$

$$N\theta_c = -7.42$$

$$N\theta_d = +14.51$$

A visual check of the signs of the angle rotations is always a good policy at this stage of the solution. The results show that the tangent at a rotates clockwise which is correct. The tangent to the slope at b rotates counterclockwise since $N\theta_b$ is +. This appears correct upon a comparison of loads and lengths of spans ab and bc. The direction of rotation at c is clockwise from the $-$ sign of the answer. The rotation at d would of course be counterclockwise which agrees with the + sign on the result.

Once the displacements at the ends of a member are known, the action in the terms of bending moments and shears can be readily calculated. These bending moments and shears can be superimposed on the bending moments and shears due to the imposed loading. These steps were shown in Fig. 13.1 when the slope-deflection method was presented. However, they will be repeated here. Fig. 14.15 (a) (b) (c) shows the superposition of the moments due to (a) the given loading on a fixed-end beam; (b) a + rotation of end a; (c) a + rotation of end b. The sign convention is as previously explained.

It can be noted that the direction of support rotations in Fig. 14.15 are taken as counterclockwise for + sign while in the development of the slope-deflection equations in Fig. 13.1, the opposite was used. This just reflects the switch in convention between the time period in the development of the two methods.

A general equation can be written to give the superposition of all three actions shown in Fig. 14.15.

$$M_a = M_{F_{ab}} + \frac{4EI}{L}(\theta_a) + \frac{2EI}{L}(\theta_b)$$

$$(14.6)$$

$$M_b = -M_{F_{ba}} + \frac{2EI}{L}(\theta_a) + \frac{4EI}{L}(\theta_b)$$

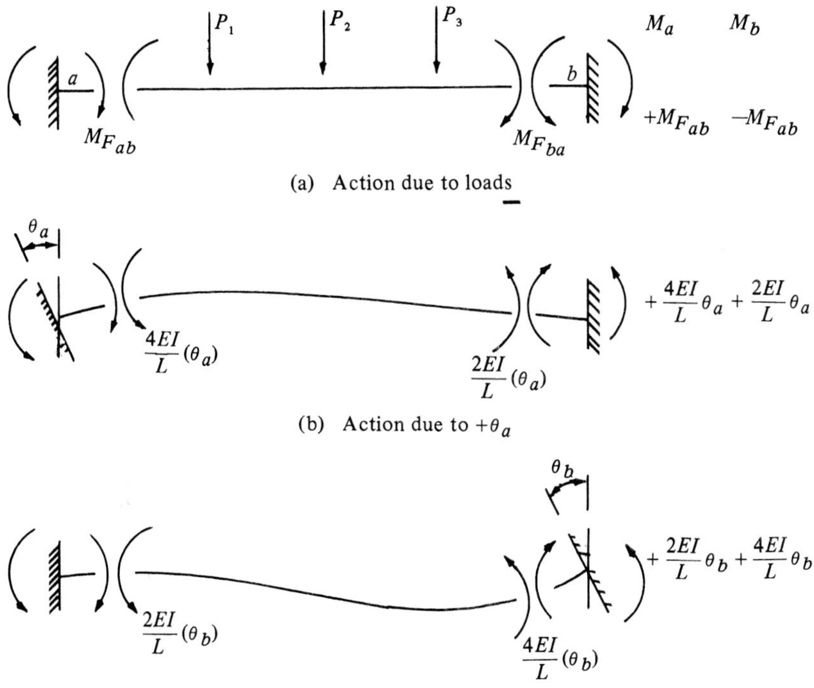

(a)   Action due to loads

(b)   Action due to $+\theta_a$

(c)   Action due to $+\theta_b$

Fig. 14.15.   Superposition of forces.

The above equations (14.16) hold true for any prismatic member whether a part of a continuous beam or a member of a frame and where the subscript a denotes the left end and b the right end of the member. In general terms, equations 14.6 can be written

$$M = M_F + S_{11}\theta_1 + S_{12}\theta_2 \qquad (14.7)$$

where the S terms are the stiffness influence coefficients.

The bending moments in the ends of the beam of Fig. 14.12 will now be found using equation 14.7. The stiffness influence coefficients will first be calculated.

$$S_{aa} = \frac{4EI_{ab}}{L_{ab}} = 3.333N$$

where $N = E(12)^{-4}(10)^2$.

$$S_{ba} = 1.667N = S_{ab}$$

$$S_{bb_1} = 3.333N$$

$$S_{bb_2} = \frac{4EI_{bc}}{L_{bc}} = 2.667N$$

$$S_{bc} = 1.333N = S_{cb}$$

$$S_{cc_1} = 2.667N$$

$$S_{cc_2} = 2.222N$$

$$S_{cd} = 1.111N = S_{dc}$$

$$S_{dd} = 2.222N$$

$$M_{ab} = +76.8 + S_{aa}(\theta_a) + S_{ab}(\theta_b)$$

$$= +76.8 + 3.333(-33.60) + 1.667(21.12) = 0$$

$$M_{ba} = -76.8 + 1.667(-33.60) + 3.333(21.12) = -62.4 \text{ ft-k}$$

$$M_{bc} = +16.0 + 2.667(21.12) + 1.333(-7.42) = +62.4$$

$$M_{cb} = -32.0 + 1.333(21.12) + 2.667(-7.42) = -23.6$$

$$M_{cd} = +24.0 + 2.222(-7.42) + 1.111(14.51) = +23.6$$

$$M_{dc} = -24.0 + 1.111(-7.42) + 2.222(14.51) = 0$$

The solution of this problem by the stiffness method should be compared to the flexibility method. The stiffness method required the solution of a 4 × 4 matrix while the flexibility method only required a 2 × 2 matrix solution. The stiffness matrix, even though 4 × 4, had several zero terms. The calculation of the stiffness coefficients K for prismatic members can be easily determined from the value of stiffness of a beam

$$\left(\frac{4EI}{L} \right)$$

or the carry-over moment

$$\left(\frac{2EI}{L} \right)$$

or the summation of the stiffness of all the members meeting at a node. These values can be easily calculated where the calculation of the flexibility coefficients can be considerably involved and time-consuming. It should be apparent that for a complex structure, the stiffness method is much quicker. It can be programmed for ready solution.

14.8 Stiffness Method—Frames

The stiffness method will now be demonstrated for frames. In a frame the kinematic unknowns are the rotations and the translations of the joints. These will be given the θ and Δ symbols as used in the classical methods. If the change in length of the members is neglected, then the Δ terms are only due to lateral displacements of the ends of the girders due to sidesway and any movement of the supports. The general equation is then

$$[K][D] + [M_F] = [M] \qquad\qquad \text{(14.4) repeated}$$

where K is the stiffness matrix, D is the displacement matrix, M_F is the member end moment matrix due to the given loads, and M will be the external nodal moment matrix.

As a comparison with the flexibility method, the solution to the frame of Fig. 14.4 will be developed. The frame is again shown in Fig. 14.16 (a)

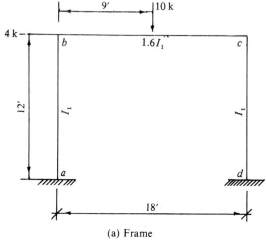

(a) Frame

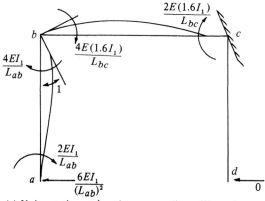

(b) Frame with artificial restraints at b and c.

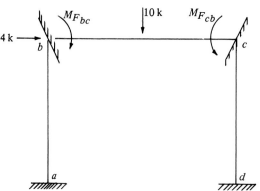

(c) Unit rotation at b and corresponding stiffness factors.

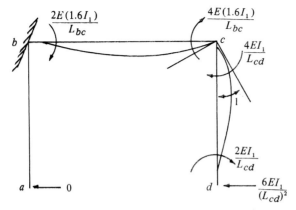

(d) Unit rotation at c and corresponding stiffness factors.

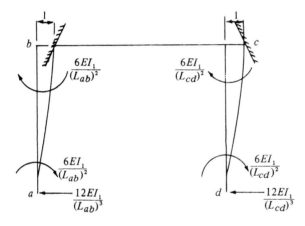

(e) Unit translation of bc and corresponding stiffness factors.

Fig. 14.16. Stiffness method.

and the artificial restraints are shown in (b). This frame is kinematically indeterminate to the third degree. That is, there is an unknown rotation at node b and c and also a horizontal displacement at b and c. This translation of b and c will be equal when the change in length of member bc is neglected. It is again assumed that there is no rotation or translation at either a or d. How to handle a problem, when support movement takes place, will be covered later in this chapter.

Part (c) of Fig. 14.16 shows the moments applied to the joints with a unit rotation of joint b. It also shows the shears applied to the bottom of the columns. Since joint c has an artificial restraint there is no bending in cd

and no shear at d. Part (d) of Fig. 14.16 shows the same values with a unit rotation of joint c. Part (e) of this same figure shows the bending moments and column shears for a unit translation of b and c. The mathematical expressions for moments and shears should now be familiar to the student.

The next step is to write the equations for equilibrium of moments at joints b and c and to also write the equation for equilibrium of total horizontal forces on the frame. These equations will be of the form of Eq. 14.4 with three variables. The value of M is the external moment applied to the joint for which equilibrium is being considered. In the frame of Fig. 14.16 there are no external moments so this term is zero. The M_F values are the moments applied to the restrained joints due to the given loading conditions. This is the now familiar "fixed end moment."

When equilibrium of horizontal forces are being considered, Equation 14.4 takes the form of

$$KD = P$$

where P is the sum of the external horizontal forces applied above the bottoms of the columns under consideration. There is no equivalent M_F term in this equilibrium equation.

It now becomes necessary to determine the values of the K matrix. This matrix is written by finding the values of K_{ij}. The K matrix for the frame will be

$$K = \begin{bmatrix} K_{bb} & K_{bc} & K_{b\Delta} \\ K_{cb} & K_{cc} & K_{c\Delta} \\ K_{\Delta b} & K_{\Delta c} & K_{\Delta\Delta} \end{bmatrix}$$

K_{bb} is the moment applied to the joint b when a unit rotation is given to this joint. This is easily evaluated by referring to Fig. 14.15 (c) and adding the moments on joint b.

$$K_{bb} = \frac{4EI_1}{L_{ab}} + \frac{4E(1.6I_1)}{L_{bc}}$$

making $EI_1 = N$

$$K_{bb} = \left(\frac{4}{12} + \frac{6.4}{18}\right)N = 0.6889N$$

The value of K_{bc} is the moment on joint b due to a unit rotation at joint c.

From Fig. 14.15 (d), this factor is determined

$$K_{bc} = \frac{2E(1.6I_1)}{L_{bc}} = 0.1788N$$

The stiffness factor $K_{b\Delta}$ is the moment produced at b due to the lateral displacement of member bc a unit distance. This value is seen on part (e) of Fig. 14.16.

$$K_{b\Delta} = \frac{6EI_1}{(L_{ab})^2} = \frac{6}{(12)^2}(N) = 0.0417N$$

From Maxwell's theorem

$$K_{cb} = K_{bc}$$

$$K_{\Delta b} = K_{b\Delta}$$

$$K_{\Delta c} = K_{c\Delta}$$

These equalities can easily be verified from the bending moments in Fig. 14.16. The stiffness factor K_{cc} is the sum of the moments at c due to a unit rotation at this point.

$$K_{cc} = \frac{4E(1.6I_1)}{L_{bc}} + \frac{4EI_1}{L_{cd}}$$

$$= \left(\frac{6.4}{18} + \frac{4}{12} \right) N = 0.6889N$$

The stiffness factor K is the sum of the shears at the bottoms of the columns due to a unit lateral displacement Δ. These values are shown in part (e) of Fig. 14.16.

$$K_{\Delta\Delta} = 2\left(\frac{12EI_1}{(L_{ab})^3} \right) = \frac{24}{1728}(N) = 0.0139N$$

The only values now needed are M_{Fb} and M_{Fc}. Since the 10 kip load is at mid-span, these two values are

$$M_{Fb} = -M_{Fc} = \frac{PL}{8} = \frac{10(18)}{8} = 22.5 \text{ ft-k}$$

The equilibrium equations can now be written.

(1) $K_{bb}\theta_b + K_{bc}\theta_c + K_{b\Delta}(\Delta) + 22.5 = 0$

(2) $K_{cb}\theta_b + K_{cc}\theta_c + K_{c\Delta}(\Delta) - 22.5 = 0$

(3) $K_{\Delta b}\theta_b + K_{\Delta c}\theta_c + K_{\Delta\Delta}(\Delta) - 4.0 = 0$

$$N \begin{bmatrix} 0.6889 & 0.1778 & 0.0417 \\ 0.1778 & 0.6889 & 0.0417 \\ 0.0417 & 0.0417 & 0.0139 \end{bmatrix} \begin{bmatrix} \theta_b \\ \theta_c \\ \Delta \end{bmatrix} = - \begin{bmatrix} 22.5 \\ -22.5 \\ -\ 4.0 \end{bmatrix}$$

The solution of this matrix gives the following values for the displacements:

$$N\theta_b = -63.5$$

$$N\theta_c = +24.6$$

$$N\Delta = +404.6$$

The values for bending moments applied to the joints at the end of each member are easily determined by a summation of the product of the stiffness influence coefficients and the displacements. These bending moments are:

$$M_{ab} = \frac{2EI_1}{L_{ab}}(\theta_b) + \frac{6EI_1}{(L_{ab})^2}(\Delta)$$

since $EI_1 = N$

$$M_{ab} = \frac{2}{12}(N\theta_b) + \frac{6}{144}(N\Delta)$$

$$= 0.167(-63.5) + 0.0417(404.6) = +6.3 \text{ ft-k}$$

$$M_{ba} = \frac{4EI_1}{L_{ab}}(\theta_b) + \frac{6EI_1}{(L_{ab})^2}(\Delta)$$

$$= 0.333(-63.5) + 0.0417(404.6) = -4.3 \text{ ft-k}$$

$$M_{bc} = +22.5 + \frac{4E(1.6I_1)}{L_{bc}}\theta_b + \frac{2E(1.6I_1)}{L_{bc}}\theta_c$$

$$= +22.5 + 0.356(-63.5) + 0.178(24.6) = +4.3 \text{ ft-k}$$

$$M_{cb} = -22.5 + \frac{2E(1.6I_1)}{L_{bc}}(\theta_b) + \frac{4E(1.6I_1)}{L_{bc}}(\theta_c)$$

$$= -22.5 + 0.178(-63.5) + 0.356(24.6) = -25.1 \text{ ft-k}$$

$$M_{cd} = \frac{4EI_1}{L_{cd}}(\theta_c) + \frac{6EI_1}{(L_{cd})^2}(\Delta)$$

$$= 0.333(24.6) + 0.0417(404.6) = 25.1 \text{ ft-k}$$

$$M_{dc} = \frac{2EI_1}{L_{cd}}(\theta_c) + \frac{6EI_1}{(L_{cd})^2}(\Delta)$$

$$= 0.167(24.6) + 0.0417(404.6) = 21.0 \text{ ft-k}$$

All of the above mathematical operations could be programmed for automatic computation. With such a program, then only the frame dimensions, load characteristics, and member sizes would have to be input to the computer.

From an understanding of the development of the stiffness matrix (see Fig. 14.16) the following general stiffness matrix can be written for any frame, symmetrical or unsymmetrical, where the change in lengths of the members are neglected and there is no support settlement.

$$K = E \begin{bmatrix} \dfrac{4I_{ab}}{L_{ab}} + \dfrac{4I_{bc}}{L_{bc}} & \dfrac{2I_{bc}}{L_{bc}} & \dfrac{6I_{ab}}{(L_{ab})^2} \\[3ex] \dfrac{2I_{bc}}{L_{bc}} & \dfrac{4I_{bc}}{L_{bc}} + \dfrac{4I_{cd}}{L_{dc}} & \dfrac{6I_{cd}}{(L_{cd})^2} \\[3ex] \dfrac{6I_{ab}}{(L_{ab})^2} & \dfrac{6I_{cd}}{(L_{cd})^2} & \dfrac{12I_{ab}}{(L_{ab})^3} + \dfrac{12I_{cd}}{(L_{cd})^3} \end{bmatrix} \begin{bmatrix} \theta_b \\[3ex] \theta_c \\[3ex] \Delta \end{bmatrix}$$

The first row relates to the moment equilibrium of joint b, the second row is the moment equilibrium of joint c, and the third row relates to the sum of the shears at the bottom of the columns.

If there were support translation or rotation, then the moments produced at each node would have to be determined for the specific value of support displacements. Fig. 14.17 shows a vertical displacement of Δ_{dy} at support d. Moments would then be produced in girder bc in the positive direction. The value of these bending moments would be

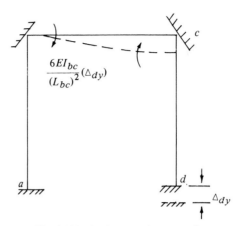

Fig. 14.17. Settlement of support d.

$$\frac{6EI_{bc}}{(L_{bc})^2}(\Delta_{dy}).$$

In order to calculate this bending moment, the numerical value of Δ_{dy} would have to be known. This bending moment will then be added as a moment in the equilibrium equations just as the fixed end moment. Because of the kinematic restraints at joints b and c there is no bending moment in the columns so there is likewise no shears in the columns. It is thus seen that the evaluation for support displacement is easily made with a minimum of extra computation.

The development of the stiffness matrix for a multiple bay and story frame is just an extension of the procedure used in the single bent. Equations for each degree of freedom are written. It is now apparent that such equations will contain only a limited number of stiffness coefficients. Because of the artificial joint restraints, nodal displacements further away than the far end of each structural element will not have any effect on the moments and shears at the node under consideration.

The effect of axial strain on the bending moments in a frame is minimal. This effect can be evaluated by the stiffness method. It will require the calculation of several more stiffness coefficients. Such a development is left to more complete texts on matrix analysis.

14.9 Stiffness Method—Trusses

The stiffness method can be applied to the analysis of statically determinate or indeterminate truss structures. Large span bridge trusses have many members. The analysis of such involves the generation of a large amount of numerical work. When such bridges have multiple spans, as they usually do when a truss is the selected structure, then the numerical work becomes very extensive. A computer is the "only way to go" under such circumstances. Brief mention was made of the flexibility method of redundant truss analysis in Section 14.5. An introduction into the stiffness method for trusses will be presented. As an introduction, and also a means of comparing the stiffness and flexibility methods, the bar assemblage of Fig. 14.6 will be analyzed by the stiffness method. This figure is repeated below in Fig. 14.18. The bars are numbered from 1 to 4, and the cross-sectional areas and lengths are given.

The presentation here is based on the structure being an ideal truss; i.e., all loads are applied at the joints and the ends are free to rotate. In this example, the effect of the weight of the bars has been neglected.

The first requirement is to develop the stiffness matrix of the bars. Then the force equilibrium equations are written. These will contain the unknown displacements in the x and y direction. The equilibrium equations are then solved for the unknown displacements. With the displacements known then the bar forces can be evaluated. The first step is to displace the joint e a unit displacement in first the $+x$ direction and the $+y$ direction. The steps are shown in Fig. 14.19. It is assumed that the unit displacement is so small that the original directions of the members can be used in the calculations.

It is now necessary to determine the force in each bar due to the unit displacements. General expressions will now be developed from Fig. 14.20.

If the end b of bar ab is given a unit displacement in the x direction then the bar will be elongated an amount equal to $1(\cos\alpha)$ or $\cos\alpha$. The axial force produced in the bar will then be

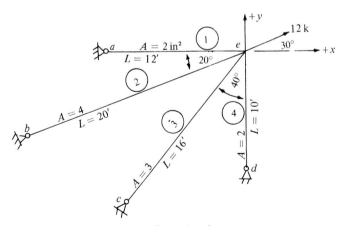

Fig. 14.18. Redundant bar system.

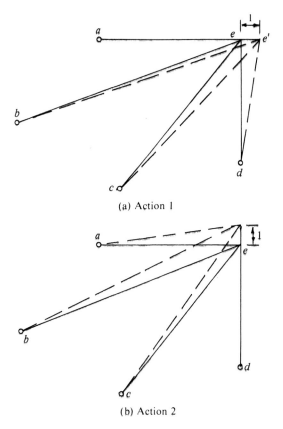

(a) Action I

(b) Action 2

Fig. 14.19. Actions.

$$F_1 = \frac{AE}{L}(\cos\alpha)$$

The x component of the force in the bar is

$$Fx_1 = \frac{AE}{L}(\cos\alpha)^2$$

and the y component is

$$Fy_1 = \frac{AE}{L}(\cos\alpha)(\sin\alpha)$$

Likewise for the unit displacement in the y direction, the bar elongation is equal to $\sin\alpha$ and

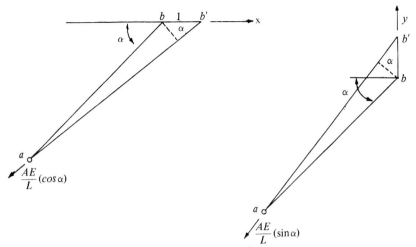

Fig. 14.20. Stiffness coefficients.

$$F_1 = \frac{AE}{L}\sin\alpha$$

$$Fx_2 = \frac{AE}{L}(\sin\alpha)(\cos\alpha) \qquad\qquad (14.6)$$

$$Fy_2 = \frac{AE}{L}(\sin\alpha)^2$$

The stiffness factor K_{11} is the summation of the forces in the x direction of all bars due to the action 1 displacement (unit displacement in x direction). The stiffness K_{21} is the sum of the component forces in the y direction due to action 1. The stiffness K_{22} is the sum of the component forces in the y direction due to action 2 and K_{12} is the sum of the component forces in the x direction due to action 2. Of course, $K_{21} = K_{12}$. These stiffness values are determined from Table 14.2. It is assumed that all bars are made from the same material so E is constant. From Table 14.2 it is seen that;

$$EK_{11} = 0.4208(12)^{-1}$$

$$EK_{21} = EK_{12} = 0.1567(12)^{-1}$$

$$EK_{22} = 0.3334(12)^{-1}$$

The applied force at e in the x and y directions are

Table 14.2

Bar	A in²	L ft	A/L in²/ft	α	cos α	sin α
1	2	12	0.1667	0	1.00	0
2	4	20	0.200	20°	0.9397	0.3420
3	3	16	0.1875	50°	0.6428	0.7660
4	2	10	0.200	90°	0	1.00

Bar	F_1	Fx_1	Fy_1	F_2	Fx_2	Fy_2
1	0.1667	0.1667	0	0	0	0
2	0.1879	0.1766	0.0643	0.0684	0.0643	0.0234
3	0.1205	0.0775	0.0923	0.1436	0.0923	0.1100
4	0	0	0	0.200	0	0.200
Σ		0.4208	0.1567		0.1567	0.3334

$$P_x = 12(0.8667) = 10.40 \text{ k}$$

$$P_y = 12(0.500) = 6.0 \text{ k}$$

The equilibrium equations are

$$\frac{E}{12}\begin{bmatrix} K_{11} & K_{12} \\ K_{21} & K_{22} \end{bmatrix}\begin{bmatrix} D_1 \\ D_2 \end{bmatrix} = \begin{bmatrix} P_x \\ P_y \end{bmatrix}$$

$$\frac{E}{12}\begin{bmatrix} 0.4208 & 0.1567 \\ 0.1567 & 0.3334 \end{bmatrix}\begin{bmatrix} D_1 \\ D_2 \end{bmatrix} = \begin{bmatrix} 10.4 \\ 6.0 \end{bmatrix}$$

The solution of these equations gives

$$\frac{ED_1}{12} = 21.84$$

$$\frac{ED_2}{12} = 7.74$$

The forces in the bars can now be determined by the addition of equation 14.5 and 14.6.

$$F = \frac{A}{L}(ED_1)\cos\alpha + \frac{A}{L}(ED_2)\sin\alpha \tag{14.7}$$

The forces in the four bars are

$$F_1 = 3.64 \text{ k} \qquad F_3 = 3.74 \text{ k}$$

$$F_2 = 4.63 \text{ k} \qquad F_4 = 1.55 \text{ k}$$

These answers compare with the solution by the flexibility method.
If the actual displacements of joint e are required, they are (for steel):

$$D_1 = \frac{21.84(12)}{30 \times 10^3} = 0.0087 \text{ in.}$$

$$D_2 = \frac{7.74(12)}{30 \times 10^3} = 0.0031 \text{ in.}$$

A three dimensional truss system could be solved in the same manner only with three displacements and three equations. The angles the bars make with the three orthogonal axes would have to be determined similar to the method in Chapter 8.

In the example just completed, hand calculation was easily performed. However, the analysis consisted of only one joint. In an actual truss, it would be necessary to calculate the displacement of all joints if the bar forces in all the members were to be calculated by the stiffness method. The final bar forces for any given loading could be determined from multiplying the relative displacements, in the direction of the bar axis of the two ends of the member, by the value of

$$\frac{AE}{L}.$$

It is apparent that this would involve the processing of a great many calculations. A computer would be necessary as well as a systematic program so that once the input data was inserted into the program the computer could perform the routine calculations and the solution of the bar forces. The input data would consist of the geometry of the structure, the

$$\frac{AE}{L}$$

of all members, and the magnitude of applied loads, as well as their location and direction. Such programs are available to the practicing engineer.

References

1. Vanderbilt, M. D., *Matrix Structural Analysis,* Quantum Publishers, Inc., 1974.
2. Martin, H. C., *Introduction to Matrix Methods of Structural Analysis,* McGraw-Hill Book Company, 1966.
3. Gere, J. M., and W. Weaver, Jr., *Analysis of Framed Structures,* D. Van Nostrand Company, Inc., 1965.

Problems

1. Determine the bending moment diagram for the beam shown in Fig. 14.21. *EI* is constant. Use flexibility method. $EI = 12 \times 10^6$ k-in^2.

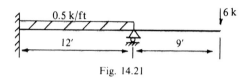

Fig. 14.21

2. Determine the bending moments at the supports for the beam in Fig. 14.22. Use flexibility method.

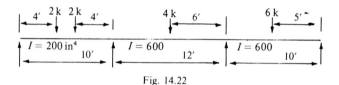

Fig. 14.22

3. Work Problem No. 1 by flexibility method but with a settlement of the roller support of 1/4 inch.
4. Work Problem No. 4 of Chapter 12 by the flexibility method.
5. Solve Problem No. 1 by stiffness method.
6. Determine the bending moments at the supports using stiffness method for the beam of Fig. 14.23.

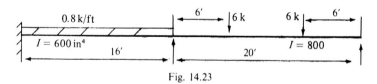

Fig. 14.23

7. What could be the change in support bending moments if the fixed support at the left end of the beam in Fig. 14.23 rotated 0.001 radians. The beam is steel with $E = 29 \times 10^3$ ksi.

8. Using flexibility method, determine reactions in the frame of Fig. 14.24. *EI* is constant.

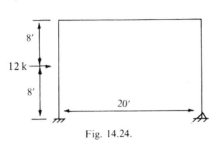

Fig. 14.24.

9. Work Problem No. 8, but change the right support to a fixed condition.
10. Work Problem No. 9 by the stiffness method.
11. Write the flexibility matrix for the structure of Fig. 14.25.

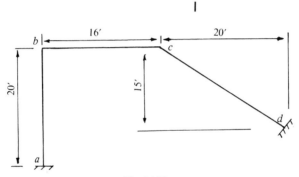

Fig. 14.25

12. Write the stiffness matrix for the structure of Fig. 14.25.
13. Determine the bending moments in the end of all members of the structure in Fig. 14.26 by the stiffness method. *EI* constant.
14. Determine the forces in all the bars of the structure of Fig. 14.27. Areas are in ( ). *E* constant.

   With point *e* as origin, the coordinates of the other ends of the members are: $a = (-6, -8, -10)$; $b = (-6, -8, +4)$; $c = (+6, -6, +2)$; $d = (+5, -6, -8)$.

   (a) Work by flexibility method.
   (b) Work by stiffness method.
15. Work Problem No. 14 if *ae*, and *ce* have $E = 29 \times 10^3$ ksi and the other two bars have an *E* of 10.6×10^3 ksi.

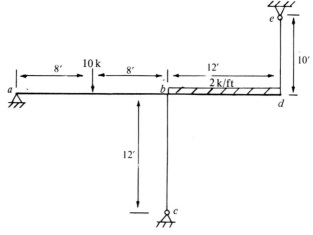

Fig. 14.26.

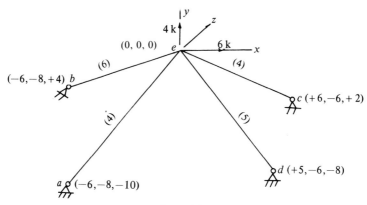

Fig. 14.27.

ANALYSIS BY PLASTIC
HINGE METHOD

15.1 Introduction

The methods of analysis in the previous chapters of this book have been based on the concept that the material of the structure is never stressed beyond the proportional limit. This is essentially true for structures that are properly designed and are not subjected to loads greater than the design loads. It can then be said that for structures properly designed and subjected to loads equal to or less than the design loads the methods used in the previous chapters will produce satisfactory results. However, the question can be asked, "What about the validity of the analysis when the loads are greater than the design loads?" A similar question is, "How can I predict the failure load of a structure?"

The limit of stress in most all design codes for steel structures ranges between 55% and 65% of the proportional limit of the steel. This then means that for applied loads up to about 1.6 times the design load, the methods previously presented are valid. However, for loads greater than this, the methods can no longer be used because the material is no longer obeying Hooke's law. The structures are not likely to fail at loads 1.6 times the design load. The collapse load may vary from about 1.75 to several times the design load. The value of collapse load depends upon several factors, including the shape of the cross-section, the manner in which the load is distributed to the structure, the location of the load or loads on the structure,

463

and whether the structure is statically determinate or indeterminate. This then means that if the failure load for various structures can vary from about 1.75 times the design load to several times the design load that there is no consistent factor of safety, based on failure, for the many types of structures. This truth has bothered structural engineers for many years.

In the early years of the twentieth century, Dr. G. Kaginczy of Hungary, initiated the concept of plastic design. About two decades later, Van den Brock in the U.S.A. and J. F. Baker in England promoted plastic design in their publications.

Shortly after the middle of the twentieth century, a start was made to develop a method of design of steel structures that would consider the entire range of the stress-strain diagram in predicting the collapse load of any structure. The main effort was directed toward the design of building frames. Through research and analytical studies sponsored by the American Institute of Steel Construction and other organizations, a logical and rational procedure has been developed that is called "the plastic design method." It only applies to structures that are fabricated from steel that has a large plastic deformation range.

There is not space in this text to treat the entire plastic design procedure. Only the basic concepts and procedures will be covered. References (1), (2), (3) as given at the end of this chapter have additional treatment of this subject.

As an introduction to the basic concept, the structure in Figure 15.1 will be considered. This structure is a steel beam, of constant section, that is supported by three steel rods that have their upper ends attached to a rigid structure. The rods are of equal length and equal cross-section ($A = 2$ in^2). The beam is subjected to two concentrated loads, symmetrically placed. Several different investigations will be made. First, the magnitude of load P at the time the stress in any one rod reaches the yield point ($\sigma_y = 36$ ksi) will be determined. For this case, it will be assumed that the beam is supported by three non-yielding supports and that the beam is of sufficient cross-section that the maximum stress in the beam is below the yield point at any value of P. Applying the moment distribution method to a prismatic beam.

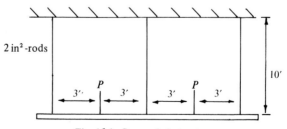

Fig. 15.1. Suspended structure.

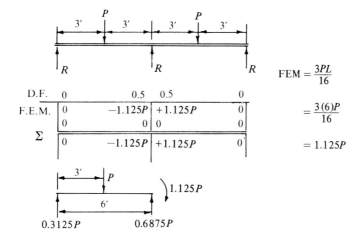

$$\text{FEM} = \frac{3PL}{16}$$

$$= \frac{3(6)P}{16}$$

$$= 1.125P$$

From this analysis the middle rod will be subjected to a load of $1.375P$. The outer rods will only carry a $0.3125P$ load. If the center rod is two square inches in area and has a yield point of 36 ksi, the value of P is

$$1.375P = 2(36)$$

$$P = 52.4\,\text{k}.$$

The question can be asked, "Is this value correct?" It is not so because the rods, being elastic, are not non-yielding supports. If the supports are elastic, it will be necessary to use some method (Section 12.1(d)) that considers such. It will also be necessary to know the value of EI for the beam. If the beam has an I of 300 in^4 and $E = 30 \times 10^3$ ksi the stiffness factor for a continuous beam on elastic supports is determined from Eq. 12.2.

$$W = \frac{6EIk}{L}$$

$$k = \frac{L}{AE}$$

$$W = \frac{6 \times 30 \times 10^6 \times 300 \times 0.002(10^{-3})}{(72)^3} = 0.29$$

from the equation for the middle reaction of a two span beam (pg. 339) when subjected to a single load at midspan,

$$R_b = \frac{4 - 6(r)^2 + 2(r)^3 + 2W}{4 + 6W}$$

$$= \frac{4 - 6(0.5)^2 + 2(0.5)^3 + 2(0.29)}{4 + 6(0.29)} = 0.58$$

for two P loads

$$R_b = 2(0.58P) = 1.16P$$

The limit of P when the stress in the middle rod reaches 36 ksi is

$$1.16P = 2(36)$$

$$P = 62 \, k$$

It is observed that the load can truly reach 62 kips before yielding of the middle rod takes place. There would be a fifteen percent error when the elastic action of the rods are neglected.

Would reaching the yield stress in the middle rod mean that additional loads above $P = 62 \, k$ result in large deformations (plastic yielding) of this center rod? It is seen that this would not take place because the end rods can carry additional loads since the force in these rods would be 26 kips only. The structure would be stable under an increase in load. From simple reasoning, it can be seen that if the limit of load is based on the point when excessive deflection of the structure takes place, the limit of P will be reached when the stress in all three rods have reached the yield point. As the load P is increased above 62 kips the center rod will resist no more load and the two end rods will have to carry the additional loads above $P = 62$ kips.

The analysis is very simple in this plastic condition.

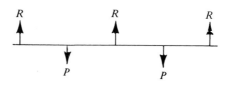

$$p = 1.5R$$

$$R = 2(36) = 72 \, k$$

$$P = 1.5(72) = 108 \, k$$

It is seen that the load P could be increased by $108 - 62 = 46$ from the load when the yield stress was first reached in the middle rod to the time when excessive yielding would take place. If it were desired that the stress due to the design load could not exceed 22 ksi, then the design load is

$$P = 62\left(\frac{22}{36}\right) = 37.9\,k$$

The factor of safety at this design load is

$$F.S. = \frac{108}{37.9} = 2.85$$

In this illustration, the plastic design concept was applied to tension members. It has been developed and used most extensively for members subject to bending only or to members subject to bending plus compression. The remainder of this chapter will be devoted to that type of member.

15.2 Plastic Hinge Moment

The commonly used structural steels produce a definite yield point and an elastic strain region that is only about one-tenth the plastic region. Figure 15.2 shows the stress-strain diagram for ASTM A-36 steel. It is seen that the strain at the yield point of 36 ksi is about 0.00122 in/in while the unit strain at the failure load is over 0.037 in/in.

This large plastic region means that when a beam is subject to a bending

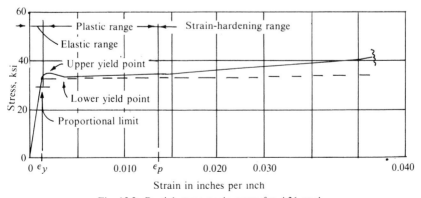

Fig. 15.2. Partial stress-starin curve for A36 steel.

moment slightly above that value which will just produce a yield point stress in the extreme fiber, the steel at the extreme fiber will be in the plastic region. Therefore, the cross-section of the beam at the point of maximum bending moment will have the outer fibers strained into the plastic region while those areas of steel nearer the centrodial axis will have magnitudes of strain still in the elastic range. Although the strain will still be increasing with increase in bending moment, the stress in the outer fibers will remain essentially at the yield point stress.

From a practical stand point, and simplicity of analysis, the stress-strain diagrams can be idealized as consisting of two straight lines. This idealized stress-strain diagram is represented by the dashed lines in Figure 15.2. Neglecting the increase of stress in the strain-hardening region will be conservative and will also help balance the negative effects of residual stresses built into the steel during the rolling and fabricating process. Therefore, in plastic design analysis, it is assumed that the stress in a beam in bending will increase in proportion to the bending moment applied to the cross-section until the yield point stress is reached and then the stress will remain constant for any further increase in bending moment. As the bending moment increases

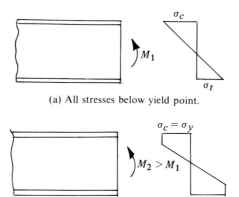

(a) All stresses below yield point.

(b) Stress in outer fibers have reached yield point.

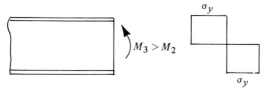

(c) All stresses have reached yield point.

Fig. 15.3. Development of Plastic hinge.

a condition will be reached where the stress on the entire cross-section will have reached the yield point stress. The cross-section cannot take any more bending moment. This limit of bending moment is called the *plastic hinge moment*.

The diagrams in Fig. 15.3 show the stress condition on a beam cross-section with increases in bending moment. In part (a) of this figure the bending moment is less than the value that would cause yielding at the extreme fiber. In part (b) the moment has been increased in magnitude so that yield point stress has been reached in the outer portions of the cross-section only. In part (c) the bending moment has increased to a value that just produces yield point stresses.in the entire cross-section. This value of bending moment is the upper limit of moment the section can resist. The section would then act as a hinge at this location on the member thus resisting a bending moment of M_p (*plastic hinge moment*) but no more even though the structural member was subjected to more load.

The concept of the formation of plastic hinge moments is basic to the *plastic design* procedure.

It is of interest to investigate the difference in bending moment between the bending moment that will just produce the yield point stress at the extreme fiber to the bending moment that will produce a plastic hinge. First consider a beam of rectangular cross-section of b width and d depth. The resisting moment when the extreme fiber has just reached the yield point is

$$M_y = \sigma_y \, \frac{bd^2}{6}$$

The plastic hinge moment is

$$M_p = C\left(\frac{d}{2}\right) = \sigma_y \, \frac{bd^2}{4}$$

$$\frac{M_p}{M_y} = \frac{\dfrac{\sigma_y bd^2}{4}}{\dfrac{\sigma_y bd^2}{6}} = 1.5$$

The value of M_p/M_y for any cross-section is defined as the *shape factor*.

For rectangular sections the shape factor is 1.5. The shape factor varies for different shapes of cross-section. It can also be defined as the plastic section modulus (Z) divided by the elastic section modulus.

The plastic section modulus can be calculated for any cross-section. It is equal to the sum of the first moments of the compression and tension areas as taken about the centroidal axis. The shape factor for the following "*I*" section is determined.

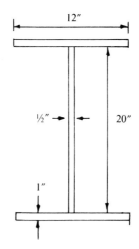

$$I = \frac{0.5(20)^2}{12} + 2(12)(10.5)^2 + \frac{2(12)(1)^3}{12} = 2981 \text{ in}^4$$

$$S = \frac{2081}{11} = 271 \text{ in}^3$$

$$Z = 2[12(10.5) + 5(5)] = 302 \text{ in}^3$$

$$\text{Shape Factor} = \frac{302}{271} = 1.11$$

The shape factor for "*I*" sections will be the lowest of any shape. The student can calculate the shape factor for other shapes. The plastic hinge moment for the above shape with a 36 ksi yield point is:

$$M_p = \sigma_y(Z) = 36(302) = 10{,}872 \text{ in-kips}$$

Values of plastic modulus for all rolled shapes as well as values of elastic modulus are given in the AISC Manual of Steel Construction.

15.3   Plastic Hinge Development in Beams

In the previous section it was shown how a plastic hinge develops at a cross-section of a steel beam. A full plastic hinge will not develop until the steel in the entire cross-section has reached the yield point stress. It is now necessary to see how and where plastic hinges develop along a beam. First to be studied will be a simply supported beam subjected to a single concentrated load (Fig. 15.4).

The maximum moment will be at the point of load. If the magnitude of the load is thought of as gradually increasing, the stress at the cross-section will increase until a plastic hinge forms. The beam can sustain this load, called the ultimate load (P_u), but any additional load will cause excessive deflection of the beam. The load-deflection diagram (Fig. 15.4 (c)) shows a linear relationship until the extreme fiber reaches the yield point stress. The load at this point is designated P_y. As the cross-section moves into the plastic range, the deflection is no longer linear and as the load increases, the value of P_u is soon reached. The ratio of P_u/P_y is the value of the shape factor. The maximum bending moment is

$$\frac{P_u(ab)}{L}$$

If this value is set equal to the maximum resisting moment of the cross-section, it is possible to determine P_u from the beam dimensions and the dimensions of the cross-section.

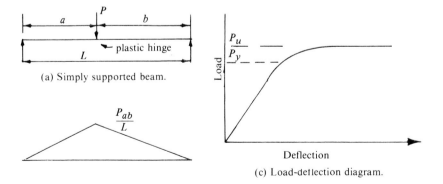

(a) Simply supported beam.

(c) Load-deflection diagram.

Fig. 15.4. Plastic hinge in beam.

$$M_p = \frac{P_u(ab)}{L}$$

$$P_u = \frac{M_p L}{ab}$$

$$M_p = \sigma_y Z$$

$$P_u = \frac{\sigma_y Z L}{ab}$$

For a concentrated load in the center:

$$P_u = \frac{4\sigma_y Z}{L}$$

Now consider the case of a fixed-end beam, first with a concentrated load in the center and then with a uniformly distributed load (Fig. 15.5).

The beam with the concentrated load has the same value of bending moment at the ends as at the mid-span. This means that plastic hinges will form at all three points simultaneously. The moment at these points will be equal to M_p and the free-body diagram of half the beam will be

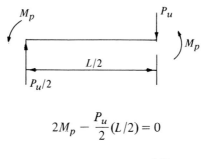

$$2M_p - \frac{P_u}{2}(L/2) = 0$$

$$P_u = \frac{8M_p}{L}$$

The fixed-end beam with the uniformly distributed load does not have all three hinges forming simultaneously. Since the bending moment at the supports is twice the bending moment at the mid-span, plastic hinges will form first at the ends of the beam. When the bending moment at the ends reach the value of M_p the beam then acts as a beam with a constant value of end moment of M_p at the ends for any further increase in the applied load.

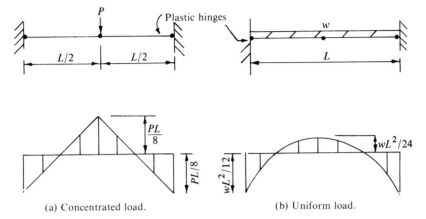

(a) Concentrated load.   (b) Uniform load.

Fig. 15.5. Fixed-end beams.

During this increase in load, the beam is stable and will carry additional load until a plastic hinge forms at mid-span. No additional load can be carried by the beam, since with three hinges the structure would be unstable.

The value of w_u, the ultimate distributed load, can be determined from statics of half of the beam as a free body.

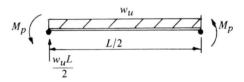

$$2M_p - \frac{w_u L}{2}(L/2) + \frac{w_u L}{2}(L/4) = 0$$

$$w_u = \frac{16M_p}{L^2}$$

It would be well to compare the reserve of strength in a steel beam, under various loadings, from the time the steel first reaches the allowable elastic design bending stress until all the plastic hinges form (ultimate load condition).

The following example problem will show this.

EXAMPLE PROBLEM 15.1   Determine the reserve of strength in the three beams shown below from the bending moment when the extreme fiber stress is at an allowable elastic design stress of 24 ksi to the plastic hinge moment.

(a) Simple Beam:

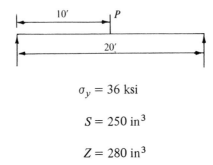

$$\sigma_y = 36 \text{ ksi}$$

$$S = 250 \text{ in}^3$$

$$Z = 280 \text{ in}^3$$

The value of $M_e = 24(250) = 6{,}000$ in-k

$$M = \frac{\dot{P}_e(20)(12)}{4} = 60P_e \text{ ft-k}$$

$$P_u = \frac{4\sigma_y Z}{L} = \frac{4(36)(280)}{20 \times 12} = 168 \text{ k}$$

$$P_e = \frac{6000}{60} = 100 \text{ k}$$

The reserve strength $= 168 - 100 = 68 \text{ k}$

$$\text{reserve} = \frac{68}{100}(100) = 68\%$$

(b) Fixed-end Beam—Concentrated Load:

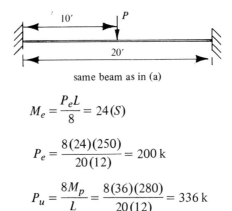

same beam as in (a)

$$M_e = \frac{P_e L}{8} = 24(S)$$

$$P_e = \frac{8(24)(250)}{20(12)} = 200 \text{ k}$$

$$P_u = \frac{8M_p}{L} = \frac{8(36)(280)}{20(12)} = 336 \text{ k}$$

The reserve strength is $336 - 200 = 136$ k

$$\text{reserve} = \frac{136}{200}(100) = 68\%$$

(c) Fixed-end Beam—Uniformly Distributed Load:

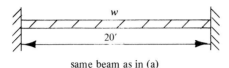

same beam as in (a)

$$M_e = \frac{w_e L^2}{12} = 24\,(S)$$

$$w_e = \frac{12(24)(250)}{(20)^2(12)} = 25 \text{ k/ft}$$

$$w_u = \frac{16 M_p}{L^2} = \frac{16(36)(280)}{(20)^2(12)} = 33.6 \text{ k/ft}$$

The reserve strength is $33.6 - 15 = 18.6$ k/ft

$$\text{reserve} = \frac{18.6}{15}(100) = 124\%$$

Example Problem 15.1 shows the variation in reserve strength between beams of different loading and support conditions. If the size of these beams were selected on a basis of elastic design the factor of safety in beams (a) and (b) would be 1.68 while in beam (c) it would be 2.24. Many engineers believe that the type of support and whether the load is concentrated or distributed should not be a basis for variation of factor of safety. In the plastic design approach, the allowable working load is determined by taking the ultimate load and dividing it by a safety factor—generally termed the load factor. It must be recognized that in plastic design analysis, the calculation of the ultimate load is the prime objective. Stresses at loads below the ultimate load are not determined. Also the magnitude of bending moment at those sections that remain elastic is of no interest.

In reviewing the procedure just presented for the plastic analysis the steps are as follows:

(1) The locations of the plastic hinges are first determined. These will occur at locations of peak bending moments. The number of plastic hinges will be such as to just render the structure unstable. At this

stage a mechanism has formed. It is possible in structures that are highly redundant to have several possible mechanisms. The mechanism that results in the lowest value of ultimate load will be the correct mechanism. This will be illustrated in the next section.

(2) The values of bending moment at the plastic hinges are equal to a known value of M_p which is dependent upon the cross-section size and shape, and the yield point stress. With the bending moment at the plastic hinges known, the reactions and the bending moment diagram can be drawn using the principles of static equilibrium only.

(3) The value of the ultimate load in terms of M_p and span dimensions can be solved from the equilibrium equations.

(4) A check should be made to see that the moment everywhere on the structure is equal to or less than M_p.

15.4   Plastic Analysis by Virtual Work

(a)   Beams

In the previous section, analyses were developed for beams and ultimate loads were determined in terms of the plastic hinge moment of a beam cross-section. This development was based on the equilibrium equations only. For more highly redundant structures as well as more complex loadings, the use of the equilibrium equations only is not the easiest approach. The use of the cirtual work concept is a simple and direct approach to the problem.

The first step is to select the location of the plastic hinges. A sufficient number, but not an excess, of plastic hinges are selected so that the structure would be unstable for any increase in load over that magnitude of load that was necessary to develop the last plastic hinge. That is, a mechanism is selected. As an example the fixed-end beam with concentrated load at the mid-span will be analyzed.

The beam in Fig. 15.6 will require plastic hinges at the ends and point of application of the load to form a mechanism. The limit of load required to form this mechanism is the ultimate load P_u. The beam is next given a virtual displacement such that the mechanism moves. This movement will require the rotation of the rigid parts of the mechanism about the plastic hinges. It should be remembered from the virtual work development in Chapter 10, that the virtual displacement is caused by an extraneous force that is independent of the force system that is applied to the structure.

In the virtual displacement the segments of the beam will rotate about the plastic hinges and also about the interior plastic hinge. The applied load

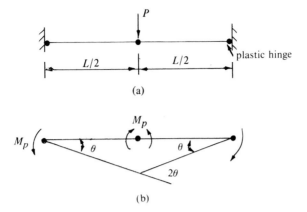

Fig. 15.6. Fixed-end beam with plastic hinges.

which now has reached the magnitude P_u will dispalce a vertical distance. The virtual displacement is shown in Fig. 15.6 (b). Applying the principle of virtual work, the internal virtual work done during the virtual displacement is equal to the external work done.

The internal virtual work will be equal to the plastic hinge moments multiplied by the angle they turn through. The external virtual work is equal to the product of the magnitude of the applied loads and the displacements they pass through in the virtual displacement. In the fixed-end beam, the solution is as follows:

$$\text{internal work} = M_p(\theta + 2\theta + \theta) = 4M_p\theta$$

$$\text{external work} = P_u\left(\frac{L}{2} \times \theta\right) = \frac{P_u L\theta}{2}$$

$$\frac{P_u L\theta}{2} = 4M_p\theta$$

$$P_u = \frac{8M_p}{L}$$

Since θ can be a very small value the arc is considered equal to the chord.

A comparison of the value of P_u just calculated with that value previously determined by equilibrium will show agreement.

If the same virtual work method is applied to the fixed-end beam with a uniformly distributed load, the procedure will be similar. The external virtual work will be equal to the intensity of load multiplied by the area of the triangle formed by the mechanism after the virtual displacement. In this case, there will be three plastic hinges also and the internal virtual work will be of the same magnitude.

$$\text{internal work} = 4M_p\theta$$

$$\text{external work} = w_u\left(\frac{L}{2} \times \frac{L\theta}{2}\right) = \frac{w_uL^2\theta}{4}$$

$$\frac{w_uL^2\theta}{4} = 4M_p\theta$$

$$w_u = \frac{16M_p}{L^2}$$

This, of course, is the same result as previously. This method will now be applied to continuous beam.

Continuous Beams:

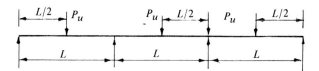

The three span continuous beam can have two possible independent mechanisms. A mechanism could form in the outside spans or the inside span. The mechanisms and virtual displacements can be as follows:

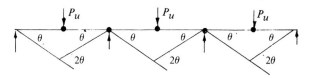

Plastic hinges would form over the interior supports and under the loads. Equating the virtual work in the end spans first

$$\frac{P_uL\theta}{2} = M_p(2\theta + \theta) = 3M_p\theta$$

$$P_u = \frac{6M_p}{L}$$

Interior span computations would be the same as determined for the fixed-end beam:

$$P_u = \frac{8M_p}{L}$$

The value of P_u would be controlled by the end span. It is apparent that if a continuous beam was to have a constant section, then it would be more economical to have the end spans shorter than the middle span.

The case of the continuous span with a uniformly distributed load will next be considered. The spans will be made equal.

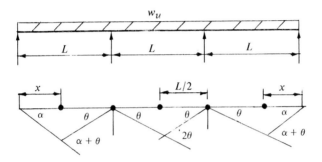

There will be five possible plastic hinges. Each mechanism will be considered separately. The center span mechanism will be considered first. The determination of w_u for this span will be the same as for the fixed-end beam:

$$w_u = \frac{16M_p}{L^2}$$

The end spans will have a plastic hinge at the interior support, a natural hinge at the end support and a plastic hinge at an interior point. This interior plastic hinge will be near mid-span but not right at mid-span. The end span does not have a symmetry of supports. The location of this plastic hinge can be considered at a distance of x from the outside support. If θ is the angle turned by the mechanism near the interior support the angle at the exterior support is

$$\alpha x = \theta (L - x)$$

$$\alpha = \frac{\theta (L - x)}{x}$$

The internal virtual work is then:

$$M_p\theta\left(1 + 1 + \frac{L-x}{x}\right) = M_p\theta\left(2 + \frac{L-x}{x}\right)$$

The external virtual work is

$$w_u(L/2)(\alpha x) = w_u(L/2)(L-x)(\theta)$$

equating

$$w_u = \frac{2M_p\left(2 + \frac{L-x}{x}\right)}{L(L-x)}$$

$$w_u = \frac{2M_p}{L}\left[\frac{L+x}{x(L-x)}\right]$$

The correct value of w_u is the minimum possible value. Therefore, the value of x that would give the minimum value of w_u is the correct location of the plastic hinge. Solving for correct value of x gives

$$x = 0.415L$$

substituting this value into the expression for w_u gives

$$w_u = \frac{11.66M_p}{L^2}$$

This value is less than w_u for the center span so plastic hinges would form in the end spans before the center span. Here again a better design would be to increase the length of the middle span and decrease the length of the end span until w_u for both spans are equal.

As in elastic design, it may be more economical to design a structural member of variable cross-section. There is no difficulty in the analysis of beams of variable plastic section modulus.

In a beam with fixed ends and concentrated load at mid-span only, plastic hinges form simultaneously at both ends and load point. The ultimate load could be increased by adding cover plates at the location of any or all plastic hinges. The necessary length of cover plates can be easily obtained by the use of the bending moment diagram at ultimate load.

As an example of this design situation, the following beam is analyzed.

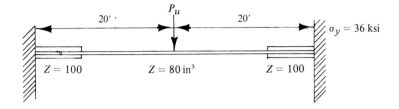

If there were no cover plates at the ends of the beam to increase the value of Z

$$P_u = \frac{8M_p}{L} = \frac{8(80)(36)}{40(12)} = 48\,k$$

With the cover plates at the ends, the plastic hinge moments will be 1.25 times the plastic hinge moment at the load point. P_u for this case will be

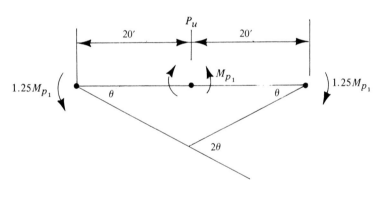

$$20P_u(\theta) = M_{p_1}(2 \times 1.25 + 2)(\theta) = 4.5M_{p_1}(\theta)$$

$$P_u = \frac{4.5M_{p_1}}{20} = \frac{4.5(80)(36)}{20(12)} = 54\,k$$

The value of P_u has been increased from 48 k to 54 k by increasing the section at the ends of the beam. It would be necessary to make the cover plates sufficiently long so that the plastic hinges would form at the ends of the beam and not at the ends of the cover plates. This required length of cover plate x can be found by drawing the bending moment diagram.

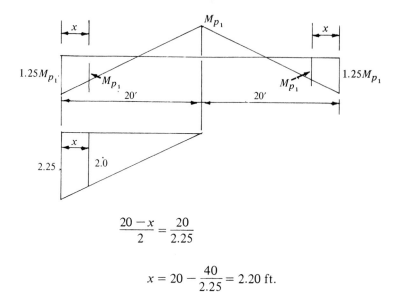

$$\frac{20-x}{2} = \frac{20}{2.25}$$

$$x = 20 - \frac{40}{2.25} = 2.20 \text{ ft.}$$

The same process could be used in the case of a fixed end beam with a uniform load and cover plates at the end as in the previous example.

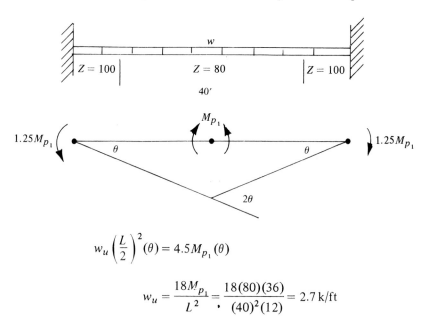

$$w_u \left(\frac{L}{2}\right)^2 (\theta) = 4.5 M_{p_1} (\theta)$$

$$w_u = \frac{18 M_{p_1}}{L^2} = \frac{18(80)(36)}{(40)^2(12)} = 2.7 \text{ k/ft}$$

To determine length of cover plate, draw bending moment diagram

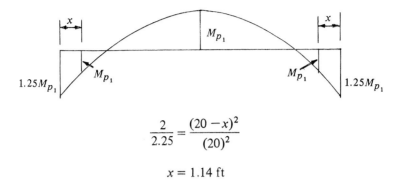

$$\frac{2}{2.25} = \frac{(20-x)^2}{(20)^2}$$

$$x = 1.14 \text{ ft}$$

(b) Frames

The virtual work method provides a quick solution to determining the ultimate loads in frames. Before the procedure for plastic analysis of frames is developed, a study of mechanisms in frames should be made. A frame may develop several different independent mechanisms. The mechanism that will develop in a frame will of course depend on the location of the plastic hinges. The location of the plastic hinges will depend upon the location and magnitudes of the loads as well as the lengths and cross sections of the members.

A single bent frame will be used as an example. Figure 15.7 shows such a frame with two applied concentrated loads. Both of the independent mechanisms are shown in (b) and (c) of Fig. 15.7. The first mechanism could form before the second mechanism would form if the load P_1 was relatively large, the columns were long, and/or the column section was smaller than the girder section. The plastic hinges necessary for this first mechanism would be at the tops and bases of the columns. This first mechanism is referred to as the sway mechanism.

The second mechanism could form if the load P_2 was large relative to P_1 and if the beam span was large. This mechanism would require the three plastic hinges as shown; at the ends of the beam and at the point of application of the load. This is the same mechanism as any single span beam fixed at the ends and loaded with a concentrated load. This mechanism is called the beam mechanism.

A simple procedure is available for determining the number of independent mechanisms possible in a structure. The number of independent mechanisms n is determined from the equation

$$n = N - X \tag{15.1}$$

where N is the number of possible plastic hinges and X is the degree of redundancy in a structure. In the frame of Fig. 15.7 $N = 5$ and $X = 3$.

$$n = 5 - 3 = 2$$

If the base of the columns were naturally pinned then $N = 3$ and $X = 1$ and

$$n = 3 - 1 = 2$$

It should be noted that at the corner of a frame the plastic hinge would form in either the column or the beam which ever had the smallest plastic section modulus. If the column and beam had materials of different yield points, then M_p would have to be found for both members in order to locate the plastic hinge and determine the magnitude of the plastic hinge at the corner.

The value of ultimate load can now be determined for both independent mechanisms. For the first mechanism (sway mechanism) the value of P_u is determined by equating the internal work to the external work.

In the following analysis it is assumed that the $M_{pcol} = M_{p_1} < M_{pbeam} = M_{p_2}$.

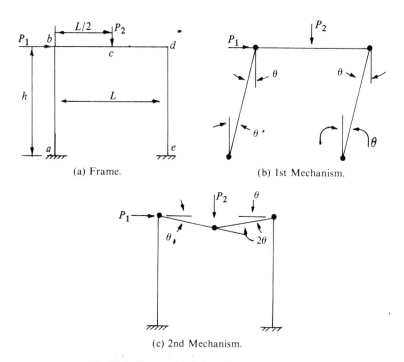

(a) Frame.   (b) 1st Mechanism.

(c) 2nd Mechanism.

Fig. 15.7. Frame with independent mechanisms.

$$\text{Internal work} = M_{p_1}(4\theta)$$

$$\text{External work} = P_{u_1}(h)(\theta)$$

$$P_{u_1} = \frac{4M_{p_1}}{h}$$

For the second mechanism

$$\text{Internal work} = M_{p_1}(2\theta) + M_{p_2}$$

$$\text{External work} = P_{u_2}\frac{L}{2}(\theta)$$

$$P_{u_2} = \frac{4(M_{p_1} + M_{p_2})}{L}$$

Now to determine the relative values of P_{u_1} and P_{u_2} choosing $M_{p_2} = 1.5M_{p_1}$ and $h = L$

$$P_{u_1} = \frac{4M_{p_1}}{L}$$

$$P_{u_2} = \frac{4(2.5M_{p_1})}{L} = \frac{10M_{p_1}}{L}$$

For the selected dimensions

$$P_{u_2} = 2.5P_{u_1}$$

It should now be recognized, that it may be possible to have some hinges from the first mechanism form at the same time some of the hinges from the second mechanism form. This would be a combined mechanism. Such a combined mechanism is shown in Fig. 15.8. With reasoning it is apparent that the loads P_1 and P_2 cause bending moments at corner b that are opposite to each other while the bending moments at corner d due to these two loads are in the same direction. Therefore, with plastic hinges at a, c, d, and e, a mechanism is formed. Using the same relationships, $M_{p_2} = 1.5M_{p_1}$, $h = L$, and $P_1 = P_2$

$$\text{Internal work} = M_{p_1}(\theta + 2\theta + \theta) + M_{p_2}(2\theta)$$

$$= M_{p_1}(7\theta)$$

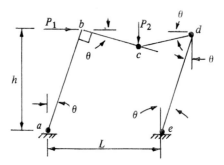

Fig. 15.8. Combined mechanism.

$$\text{External work} = P_{u_1} = (h\theta) + P_{u_2}\left(\frac{L\theta}{2}\right)$$

$$= P_{u_1}\left(\frac{3L\theta}{2}\right)$$

$$P_{u_1} = \frac{14M_{p_1}}{3L}$$

Comparing all three mechanisms for the dimensions selected and for the case where $P_1 = P_2$, then the lowest value of P_u is given by the beam mechanism where

$$P_u = \frac{4M_{p_1}}{L}$$

For different ratios of M_{p_2}/M_{p_1} and h/L one of the other two possible mechanisms may give the minimum value of P_u. The following example shows this.

EXAMPLE PROBLEM 15.2   Determine the value of P_u for the frame shown at the top of page 387 for each possible mechanism. $\sigma_y = 36$ ksi.

$$M_{p\text{col}} = 345 \text{ ft-k} = M_{p_1}$$

$$M_{p\text{beam}} = 432 \text{ ft-k} = M_{p_2}$$

$$\frac{M_{p_2}}{M_{p_1}} = 1.25$$

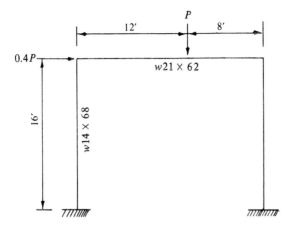

Sway mechanism

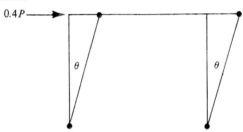

$$16(0.4P_u)(\theta) = 4M_{p_1}(\theta)$$

$$P_u = \frac{4M_{p_1}}{6.4} = \frac{4(345)}{6.4} = 215.6\,\text{k}$$

Beam mechanism

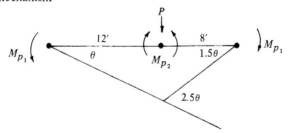

$$12P_u(\theta) = M_{p_1}(\theta + 1.5\theta) + M_{p_2}(2.5\theta)$$

$$P_u = \frac{2.5(M_{p_1} + M_{p_2})}{12} = \frac{2.5}{12}(345 + 432) = 162\,\text{k}$$

Combined mechanism

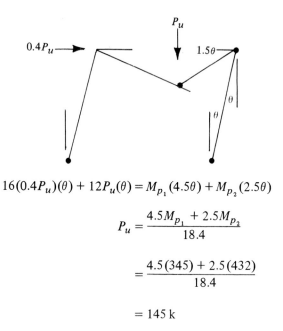

$$16(0.4P_u)(\theta) + 12P_u(\theta) = M_{p_1}(4.5\theta) + M_{p_2}(2.5\theta)$$

$$P_u = \frac{4.5M_{p_1} + 2.5M_{p_2}}{18.4}$$

$$= \frac{4.5(345) + 2.5(432)}{18.4}$$

$$= 145 \text{ k}$$

15.5 Gabled Frame Analysis

A gabled frame can best be analyzed by the use of "instant centers" in conjunction with the virtual displacement method. The use of an instant

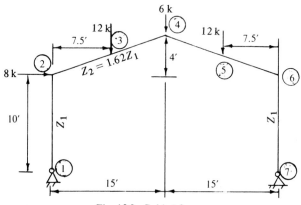

Fig. 15.9. Gabled frames.

center will aid in determining the rotations and displacements during the virtual displacement. The basic concept is to first locate the plastic hinges. Where there are several possible mechanisms, independent and combined, the combined mechanism may be the best first choice. The second step is to give the mechanism a virtual displacement. With this virtual displacement, the instant center is located. The student may want to refer back to his basic mechanics course for review of instant center. The instant center is the intersection of lines connecting the natural and plastic hinges.

The gable frame in Fig. 15.9 is selected to demonstrate the method. The frame has both vertical loads and a lateral load. The plastic section modulus (Z) is different for the columns than for the girder. Possible locations for plastic hinges are numbered. Since the most likely mechanism to form first in a gabled frame is the combined mechanism, the plastic hinges will be selected at points ③ and ⑥. Because of foundation conditions, it is decided to have the columns pinned at the bases. Therefore, natural hinges are at ① and ⑦.

The next step is to give the frame (now a mechanism) a virtual displacement. This displacement is shown in Fig. 15.10. The instant center (I) is

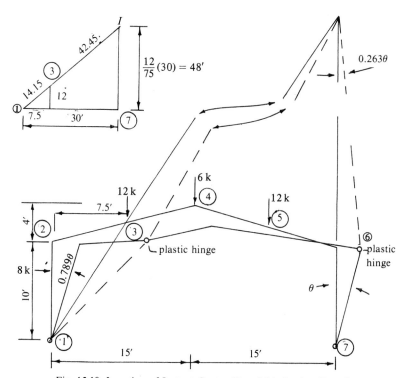

Fig. 15.10. Location of Instant Center (I) and Mechanism Rotation.

located by the intersection of two lines. One line drawn through the natural hinge at ① and the plastic hinge at ③. The second line is drawn through the natural hinge at ⑦ and the plastic hinge at ⑥. From geometry it is seen that I is located 48 ft directly above point ⑦. With the instant center located, all points on the frame can be considered as rotating about this point during the virtual displacement. It is well to remember, that the virtual displacement is a very small value so that chords of arcs can be taken as equal to the length of the arcs.

The next step is to select the rotation of one part of the mechanism as an arbitrary value θ. In Fig. 15.10 the rotation fo the right column about the natural hinge at ⑦ is given the rotation θ. If this is so, then the rotation of point ⑥ with respect to I is equal to $\frac{10}{38}(\theta)$. This then tells us that the entire frame rotates through an angle of $\frac{10}{38}(\theta) = 0.263\theta$ about the instant center.

If the rotation at the instant center I is 0.263θ, then the rotation at point ① can be calculated by considering the displacement of point ③ which rotates about both point ① and I. Rotation of the left column will be

$$\frac{(42.45)(0.263\theta)}{14.15} = 0.789\theta$$

With all the rotations known, the rotations at the plastic hinges and the displacements at the points of loading, in the direction of the loads, can be calculated.

The rotation at the plastic hinge ③ will be the rotation at ① plus the rotation at I. This will be

Hinge rotation @ ③ $= 0.789\theta + 0.263\theta = 1.052\theta$

Hinge rotation @ ⑥ $= \theta + 0.263\theta = 1.263\theta$

Horizontal displacement of 8 k load @ ② $= 0.789\theta\,(10) = 7.89\theta$ ft

Vertical displacement of load @ ③ $= 0.263\theta\,(22.5) = 5.92\theta$ ft

Vertical displacement of load @ ④ $= 0.263\theta\,(15) = 3.95\theta$ ft

Vertical displacement of load @ ⑤ $= 0.263\theta\,(7.5) = 1.97\theta$ ft

Now with all rotations and displacements calculated the internal work can be equated to the external work.

Total internal work $= \begin{cases} \sigma_y(1.62Z_1)(1.052\theta) = 1.70\sigma_y Z_1\theta \\ \sigma_y(Z_1)(1.263\theta) = 1.263\sigma_y Z_1\theta \\ \hspace{3.5cm} \Sigma = 2.963\sigma_y Z_1\theta \end{cases}$

Total external work $= \begin{cases} 8(7.89\theta) = 63.12\theta \text{ ft-k} \\ 12(5.92\theta) = 71.04\theta \text{ ft-k} \\ 6(3.95\theta) = 23.70\theta \text{ ft-k} \\ 12(1.97\theta) = \underline{23.64\theta} \text{ ft-k} \\ \hspace{2cm} \Sigma = 181.5\theta \text{ ft-k} \end{cases}$

$$2.936\sigma_y Z_1\theta = 181.5\theta \text{ ft-k}$$

$$M_{p_1} = \sigma_y Z_1 = 61.3 \text{ ft-k} = 735.6 \text{ in-k}$$

$$M_{p_2} = 1.62M_{p_1} = 99.3 \text{ ft-k}$$

The required plastic section modulus required for the columns if $\sigma_y = 36$ ksi is

$$Z_1 = \frac{735.6}{36} = 20.4 \text{ in}^3$$

$$Z_2 = 20.4(1.62) = 33.0 \text{ in}^3$$

The above required values of Z are based on bending only. The additional axial force in the sections may require a larger section. This subject is considered in Section 15.6.

It should be realized that there are other possible mechanisms. From a previous section, it was shown that an independent sway and beam mechanism is possible. These can be investigated separately as was previously done or the bending moments at the other possible plastic hinge locations can be determined. A check is made to see if the moments anywhere exceed the plastic hinge moment.

The first step in calculating the bending moments is to determine the reactions. First, the horizontal reaction at ⑦ is found.

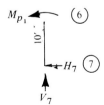

$$H_7 = \frac{M_{p_1}}{10} = 0.1 M_{p_1}$$

$$H_1 = 8 - 0.1 M_{p_1} = 8 - 0.1(61.3) = 1.87\,\text{k}$$

$$V_7 = \frac{30}{2} + \frac{8(10)}{30} = 17.67\,\text{k}$$

$$V_1 = 30 - 17.67 = 12.33\,\text{k}$$

moment at

$$2 = H_1(10) = (8 - 0.1 M_{p_1})10$$

$$= 80 - 61.3 = 18.7 < M_{p_1} = 61.3\,\text{ft-k}$$

Other possible hinge locations are at ④ and ⑤.

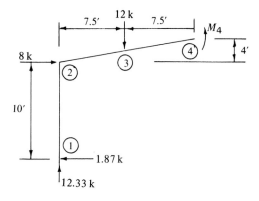

$$M_4 = 12.33(15) + 1.87(14) - 8(4) - 12(7.5) = 89.1 < M_{p_2} = 99.3\,\text{ft-k}$$

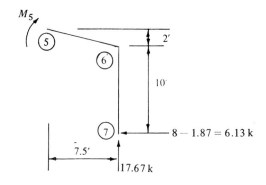

$$M_5 = 17.67(7.5) - 6.13(12) = 59 < M_{p_2} = 99.3 \text{ ft-k}$$

Since the bneding moments at ②, ④, and ⑤, are less than the plastic hinge moments at these locations, the mechanism selected is the correct one. It gives the lower bound of collapse load.

15.6 Effect of Axial Load on Plastic Hinge Development

Before any design procedures can be developed, the effect of axial load on the plastic hinge formation must be investigated. If an I-shaped section is considered to have both an axial load and a bending moment acting on the cross section, then the area of the compression block will be greater than the area of the tension block at the fully plastic condition, as shown in Fig. 15.11. The stress block in Fig. 15.11 (b) is due to the axial load P and the stress block in (c) is due to the bending moment M, which at the fully plastic condition will be represented by M_{pc} (the plastic hinge moment when the section is subjected to an axial load P).

The axial load P is then equal to the volume of the stress block (b) which is:

$$P = 2\sigma_y(y)t_w$$

and

$$M_{pc} = \sigma_y(Z - t_w y^2)$$

where Z is the plastic section modulus for the entire cross-section.
Solving for

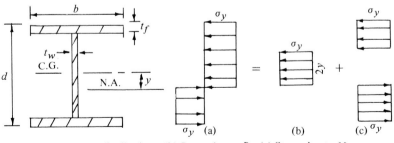

(a) Total stress distribution   (b) Stress due to P (c) Stress due to M

Fig. 15.11. Stress on I-shaped section at fully plastic condition, both axial load and bending moment.

$$y = \frac{P}{2\sigma_y t_w}$$

from the first equation and substituting into the second equation

$$M_{pc} = M_p - \frac{P^2}{4\sigma_y t_w}. \tag{15.2}$$

Equation 15.2 says that M_{pc} is equal to M_p less the amount

$$\frac{P}{4\sigma_y t_w}.$$

Equation 15.2 is valid only when the neutral axis is within the web. When the neutral axis is within the flange (large values of P)

$$P = \sigma_y[A - b(d - 2y)]$$

(A is the cross-sectional area of the beam).
   The value of M_{pc} is then

$$M_{pc} = \frac{\sigma y}{2}\left[d\left(A - \frac{P}{\sigma y} \right) - \frac{1}{2b}\left(A - \frac{P}{\sigma y} \right)^2 \right] \tag{15.3}$$

Equation 15.3 is valid when

$$P > \sigma y t_w (d - 2t_f)$$

For design purposes, A.I.S.C. recommends that M_{pc} be taken equal to M_p when

$$\frac{P}{P_y} \lesssim 0.15$$

where $P_y = A \sigma_y$.
   A.I.S.C has also recommended the use of the following simple equation for M_{pc} in place of Equations 15.2 and 15.3 when $\frac{P}{P_y} > 0.15$.

$$M_{pc} = 1.18 M_p \left(1 - \frac{P}{P_y} \right) \tag{15.4}$$

15.7 Plastic Design

It is now possible to consider the plastic design procedure. This will be done by several example problems. First, beams will be considered and then frames. In beams only M_p is involved while in frames M_{pc} may have to be used for the plastic hinge moment. The student is referred to Part 2 of the American Institute of Steel Construction *Manual of Steel Construction* for values of Z and M_p for standard rolled steel sections.

In the design process, the design loads are multiplied by load factors to arrive at loads to be used in the plastic design procedure. The values of load factors to be used are given in the building codes.

$$P_u = \text{Load factor } xP$$

EXAMPLE PROBLEM 15.3   Determine the minimum weight W steel section for a beam as shown, for a value of $P_u = 90$ k.

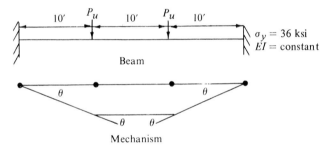

Beam

Mechanism

Internal work $= M_p(4\theta)$

External work $= P_u(10\theta) + P_u(10\theta) = 20P_u\theta$

$$M_p = 5P_u = 5(90) = 450 \text{ ft-k}$$

$$Z = \frac{450(12)}{36} = 150 \text{ in}^3$$

A W 24 $\times$ 61 section provides $Z = 152 \text{ in}^3$.

EXAMPLE PROBLEM 15.4   A two span continuous beam is to have a constant section of $Z = 100 \text{ in}^3$ and cover plates welded top and bottom to increase the Z over the support to 120 in^3. The ultimate load in the short span is to be 200 k and in the long span 150 k. Is the section adequate to carry these loads before collapse?

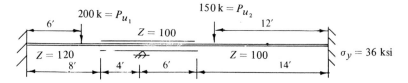

The value of P_u will be calculated for each beam span and compared with the given ultimate loads.

Left span:

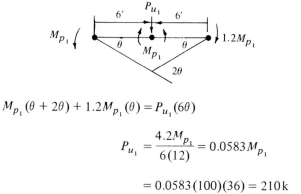

$$M_{p_1}(\theta + 2\theta) + 1.2M_{p_1}(\theta) = P_{u_1}(6\theta)$$

$$P_{u_1} = \frac{4.2M_{p_1}}{6(12)} = 0.0583M_{p_1}$$

$$= 0.0583(100)(36) = 210k$$

Since $P_{u_1} > 200$ the section appears to be adequate for the load in the left span. However, the analysis was based on a hinge forming over the interior support. If the cover plate is not long enough, the hinge can form a t the end of the cover plate in the smaller section. This can easily be checked by a moment diagram.

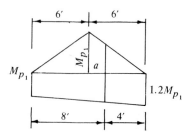

The net moment at the end of the cover plate (point a) is

$$M_a = M_{p_1}\left(\frac{4}{6}\right) - M_{p_1} - \frac{8}{12}(0.2M_{p_1})$$

$$= 2/3M_{p_1} - M_{p_1} - \frac{1.6}{12} = 0.467M_{p_1} < M_{p_1}$$

Therefore, since $M_a < M_{p_1}$ a plastic hinge will not form at the end of the cover plate. In fact, the length of the cover plate could be much less. The calculation for minimum cover plate length x is

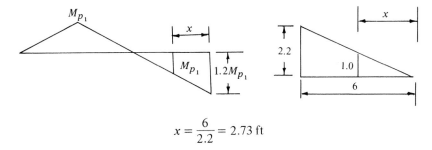

$$x = \frac{6}{2.2} = 2.73 \text{ ft}$$

The right span will now be checked for its value of P_u.

$$M_{p_1}(1.2\theta + 1.67\theta - 0.66\theta) = 0.75P_{u_1}(8\theta)$$

$$3.53M_{p_1}\theta = 6.0P_{u_1}\theta$$

$$P_{u_1} = 0.589M_{p_1} = 0.589(300) = 177\,\text{k}$$

Since the magnitude of loads on the two spans were set at the ratio of 1/0.75, then the limit of P_u would be 177 k. At this value of P_u, a mechanism would form in the right span but would not form in the left span, since the value of P_u for that span is 210 k.

A check of the right span would show that no plastic hinge will form at that end of the cover plate before the plastic hinge forms over the support.

EXAMPLE PROBLEM 15.5 Determine the ultimate load P_u for the frame shown. First ignore the effect of axial load on the plastic hinge moment and then check for this effect.

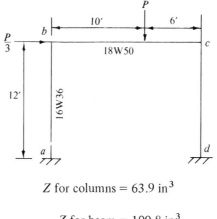

$$Z \text{ for columns} = 63.9 \text{ in}^3$$

$$Z \text{ for beam} = 100.8 \text{ in}^3$$

$$\sigma_y = 36 \text{ ksi}$$

$$M_{pcol} = 36 \times 6.39 = 2300 \text{ in-k} = 192 \text{ ft-k}$$

$$M_{pbeam} = 36 \times 100.8 = 3629 \text{ in-k} = 302 \text{ ft-k}$$

The frame will be checked for the ultimate load P_u for the two independent mechanisms plus the combined mechanism. The analysis for the beam mechanism is:

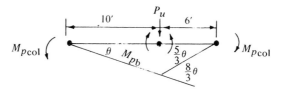

The hinges at the corners would form in the columns not the beams since $M_{pcol} < M_{pbeam}$.

$$P_u(10\theta) = M_{pcol}\left(\frac{8}{3}\theta\right) = M_{pb}\left(\frac{8}{3}\theta\right)$$

$$P_u = \frac{8}{30}(M_{pcol} + M_{pb})$$

$$P_u = \frac{8}{30}(192 + 302) = 132 \text{ k}$$

The analysis for the sway mechanism gives

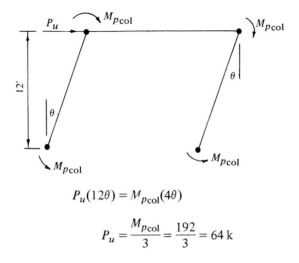

$$P_u(12\theta) = M_{pcol}(4\theta)$$

$$P_u = \frac{M_{pcol}}{3} = \frac{192}{3} = 64 \text{ k}$$

Since the applied lateral load is only $\frac{1}{3}$ of the vertical load, the beam mechanism would occur before the sway mechanism since $\frac{132}{3} < 64$.

The combined mechanism will now be investigated. The rotation at the base will be taken as θ.

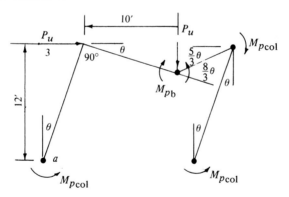

$$\frac{P_u}{3}(12\theta) + P_u(10\theta) = M_{pcol}\left(\theta + \theta + \frac{8}{3}\theta\right) + M_{pb}\left(\frac{8}{3}\theta\right)$$

$$P_u = \frac{M_{pcol}}{3} + \frac{4M_{pb}}{21}$$

$$= \frac{192}{3} + \frac{1208}{21} = 121.5 \text{ k}$$

This is the minimum value of P_u. The reactions and axial loads will now be calculated.

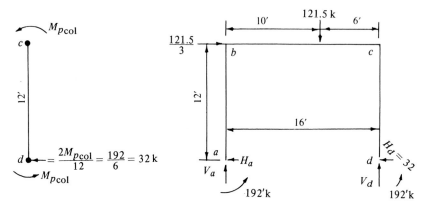

The horizontal reaction at a will be $8.5\,\text{k} = \frac{121.5}{3} - 32.0$.

$$V_d = \frac{121.5(10) + 40.5(12) - 2(192)}{16} = 82.3\,\text{k}$$

$$V_a = 121.5 - 82.3 = 39.2\,\text{k}$$

The maximum axial load in the columns will be 82.3 k and in the beams 32 k.

The values of M_{pc} will now be determined.

For columns

$$P_y = 10.6(36) = 382\,\text{k}; \qquad \frac{P}{P_y} = \frac{82.3}{382} = 0.22$$

For beams

$$P_y = 14.7(36) = 529\,\text{k}; \qquad \frac{P}{P_y} = \frac{32}{529} = 0.06$$

Since $\frac{P}{P_y}$ is less than 0.15 for the beam, then $M_{pc} = M_p$. However, the ratio of $\frac{P}{P_y}$ for the column is greater than 0.15; therefore, the value of M_{pcol} will have to be reduced.

$$y = \frac{P}{2\sigma_y t_w} = \frac{82.3}{2(36)(0.3)} = 3.81\,\text{in} < 7.5\ (\tfrac{1}{2}\ \text{web depth})$$

Since the neutral axis is within the web,

$$M_{pc} = M_p - \frac{P^2}{4\sigma_y t_w} = 2300 - \frac{(82.3)}{4(36)(0.3)}$$

$$= 2300 - 158 = 2142 \text{ in-k} = 178.5 \text{ ft-k}$$

It is seen, that M_{pc} is only slightly less than M_p for the column. A check will now be made to determine the reduction in P_u. The combined mechanism was controlling for the frame and the previous analysis gave

$$P_u = \frac{M_{pcol}}{3} + \frac{4M_{pb}}{21} = \frac{178.5}{3} + \frac{1208}{21} = 117 \text{ k}$$

Consideration of the reduction of the plastic hinge moment due to axial force in the column only reduced the ultimate load by 4.5 kips (about 4%).

There are other items that must be considered in the plastic design of structures. These are as follows.

(1) Overall buckling should not occur in any member before the ultimate load is reached. This will require adequate bracing.
(2) The sections chosen for structures should have restricted width to thickness ratios of all elements so that local buckling will not occur below the ultimate.
(3) The ultimate strength in shear should equal or exceed the ultimate load from flexure.
(4) Failure by fatigue should not be possible.
(5) Failure by brittle fracture of the steel should be prevented before the formation of a plastic hinge.
(6) All connections should be designed on the basis of ultimate strength so that the mechanisms will form before any connection would reach ultimate load.
(7) The deflection limitations at working loads should be met.

A discussion of all the above topics as well as complete design procedures for multi-story steel frames is contained in Ref. (3). Students interested in plastic design should study this reference.

In the design of structures, the working load is multiplied by a load factor to determine the ultimate load to be used in the plastic design method. The load factor would either be selected from the design code or determined by the engineer after a careful study of the particular conditions that relate to the specific structure.

References

1. Beedle, L. S., *Plastic Design of Steel Frames,* John Wiley & Sons, 1958.
2. Massonnet, C. E. and M. A. Save, *Plastic Analysis and Design,* Blaisdell Publishing Co., 1965.
3. American Institute of Steel Construction, *Plastic Design of Braced Multi-story Frames,* 1968.
4. Disque, R., *Applied Plastic Design in Steel,* D. VanNostrand Co., 1971.

Problems

Note: All Z values are in in^3.

1. Determine P_u for the suspended structure shown below. The beam is sufficiently large so that P_u is limited by the steel rods. $\sigma_y = 50$ ksi.

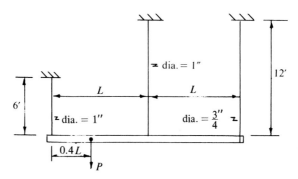

2. Determine M_p for the three steel beam cross-sections. All parts are welded together.

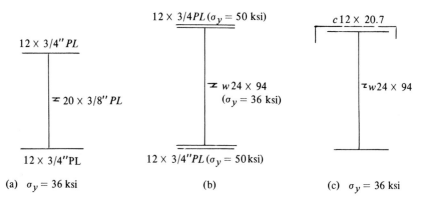

12 × 3/4PL ($\sigma_y = 50$ ksi)

c12 × 20.7

12 × 3/4" PL

$\pm$ w 24 × 94
($\sigma_y = 36$ ksi)

$\top$ w24 × 94

$\pm$ 20 × 3/8" PL

12 × 3/4"PL

12 × 3/4"PL ($\sigma_y = 50$ksi)

(a) $\sigma_y = 36$ ksi   (b)   (c) $\sigma_y = 36$ ksi

3. Determine w_u in terms of M_p and L for the continuous beam shown below.

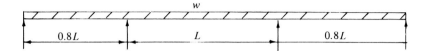

4. Determine P_u for the structure shown below and the minimum theoretical length of cover plate. $\sigma_y = 36$ ksi.

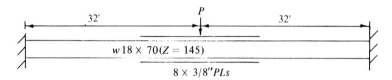

5. Determine P_u for the beams structure shown below in terms of M_p and L. Constant section.

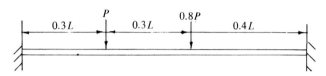

6. Determine P_u for the beam structure shown below. Determine the theoretical minimum length of cover plate. $\sigma_y = 36$ ksi.

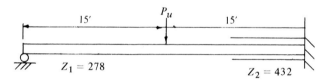

7. Determine the limit of w_u for the continuous beam shown below. $\sigma_y = 50$ ksi.

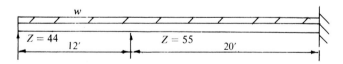

8. Determine P_u for the frame shown in Fig. 15.12. $\sigma_y = 36$ ksi. Ignore effects of axial load on M_p values.

9. Determine P_u for the frame shown in Fig. 15.13. $\sigma_y = 40$ ksi. Ignore effects of axial loads on M_p values.

10. Determine P_u for the frame shown in Fig. 15.14. Ignore effects of axial loads on M_p values. $\sigma_y = 36$ ksi

11. A W14 X 34 steel section is to be used in a plastic designed steel frame.

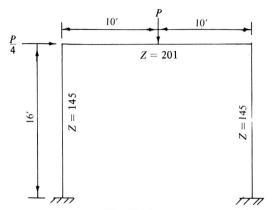

Fig. 15.12

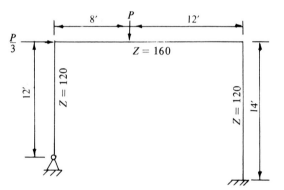

Fig. 15.13

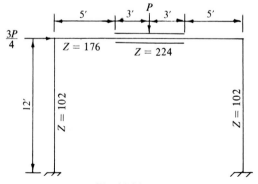

Fig. 15.14

The steel will be A36 with $\sigma_y = 36$ ksi. This section has the following properties. $A = 10$ in^2, $Z = 54.6$ in^3, $d = 14$ in, $b = 6.75$ in, Flange thickness $= 0.453$ in, $t_w = 0.287$ in. Determine the value of M_{pc} if the axial load is

(a) 50 k   (b) 100 k   (c) 200 k

Use the exact equations for M_{pc}.

12. Work problem 11 using the AISC approximate equation and compare results with those found using the more exact equations.

13. Work problem 8 but consider effects of axial load on M_p. AISC has suggested using the approximate equation

$$\frac{M_{pc}}{M_p} = 1.18 \left(1 - \frac{P}{P_y} \right)$$

The cross sectional area of the column is 24.7 in^2 and the area of the beam is 22.4 in^2.

14. Determine P_u for the frame shown below using instant center. Take M_{pc} for columns 220 ft-k and for the girder 300 ft-k.

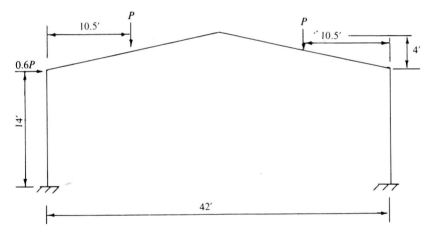

15. Find P_u for the frame in Prob. 14 with the bases pinned.

INFLUENCE LINES
AND ELASTIC CENTER METHOD

16.1 Introduction

The previous four chapters have presented theories and methods of analysis of statically indeterminate structures. The skilled structural engineer should be familiar with this subject matter. The method selected by the engineer for any specific problem will depend upon several factors; the size and complexity of the problem, the type of structure, and last but not least the availability and cost of computer software and hardware. The small engineering office, handling smaller projects of less complexity, will have procedures different from the large office with computer programs and facilities readily available. Future computer developments may change operational procedures.

Because the engineering student does not know what will be demanded of him, he should become familiar with as many methods and procedures as possible within his limit of time.

This chapter will present the subjects of influence diagrams for statically indeterminate structures, and the analysis of bents and arches by the elastic center method.

16.2 Influence Diagrams

The definition of influence line was covered in Chapters 4 and 5 and will not be repeated here. For statically indeterminate beam structures, influence

diagrams are necessary for shear and bending moment, and for truss structures, bar force influence diagrams are required. In continuous beams that
are subjected to moving loads or to loads that can be placed only on parts
of the structure, then influence diagrams are necessary to indicate the correct
placement of the loads in order to calculate maximum stresses. This eliminates
any time consuming "cut and try" procedure.

The procedure for developing influence diagrams is essentially that which
has already been shown in previous chapters. A unit load is moved across the
structure and the bending moment, shear, or axial force is calculated. As can
be realized, this objective can first require the calculation of the magnitude
of the redundants. As a result, it will be necessary to generate a rather large
volume of numerical work. The structural engineer should take advantage of
summetry wherever possible to reduce the amount of work.

A bridge consisting of three continuous spans will be used to illustrate the
above points. It is desired to develop the influence diagram for the bending
moment over the interior support for the three span continuous beam in
Fig. 16.1.

This beam has two redundants which can either be two of the reactions or
the two bending moments over the interior supports. In most of the methods
of indeterminate analysis, it would first be necessary to find the bending
moments in the beam at b and c before finding the reactions. The bending
moment in the beam at b can be determined quite readily by either the
three-moment equation or by moment distribution. The three-moment equation (Section 12.2 (b)) will be used for this example.

In order to plot the influence diagram for M_b, it appears necessary to place
a unit concentrated load at many locations from a to d on the beam and
then perform a separate analysis for each location of the unit load. In performing a solution, it will be necessary to calculate both of the redundant
bending moments M_b and M_c. With both of these bending moments determined, the unit load need only be placed at selected locations from a to
the midpoint of the middle span. A case in point is shown by referring to
Fig. 16.1. The bending moment at c with a unit load at location m is equal
to the bending moment at b with a unit load at location n. It should now be
clear that bending moments at b with unit loads to the right of the center
line of the structure can be determined from the bending moment at c with
the unit load at the corresponding locations to the left of the center line of
the structure.

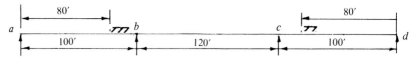

Fig. 16.1. Continuous beam (constant EI)

The work of developing the influence diagram will now proceed. A unit load is placed at various locations along the beam. The number of locations are such that an accruate plot can be made. Since the spans are an even multiple of twnety feet, the load is placed at twenty foot intervals along the beam. The three-moment equation is used as the tool in determining the bending moment at the interior support.

The three-moment equations for a unit load at a distance x from the end support is

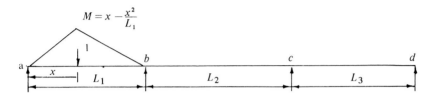

(1) $2M_b(L_1 + L_2) + M_c L_1$

$$= -\frac{6}{L_1}\left[\left(x - \frac{x^2}{L_1}\right)\left(\frac{x^2}{3}\right) + \left(x - \frac{x^2}{L_1}\right)\left(\frac{L_1 - x}{2}\right)\left(x + \frac{L_1 - x}{3}\right)\right]$$

(2) $M_b(L_2) + 2M_c(L_2 + L_3) = 0$ (no load in spans 2 and 3)

$$M_c = \frac{M_b(L_2)}{2(L_2 + L_3)}$$

Substituting the value of M_c obtained from equation (2) into the first equation and with $L_1 = 100$ ft $= L_3; L_2 = 120$ ft and $x = 20$ ft.

$$407.3 M_b = -1920$$

$$M_b = -4.71 \text{ ft-k}$$

$$M_c = +1.28$$

Redundant bending moments for other values of x at 20 ft. increments are:

@ $x = 40$ ft. $M_b = -8.25$ ft-k $M_c = +2.25$

@ $x = 60$ ft. $M_b = -9.43$ ft-k $M_c = +2.57$

@ $x = 80$ ft. $M_b = -7.09$ ft-k $M_c = +1.93$

It is now necessary to find M_b with the unit load at twenty foot intervals in the center span. The two three-moment equations are now

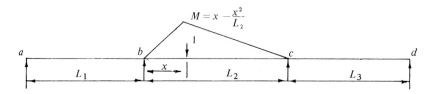

(1) $2M_b(L_1 + L_2) + M_c(L_2)$

$$= -\frac{6}{L_2}\left[\left(x - \frac{x^2}{L_2}\right)\left(\frac{x}{2}\right)\left(L_2 - \frac{2x}{3}\right) + \left(x - \frac{x^2}{L_2}\right)(L_2 - x)\left(\frac{L_2 - x}{3}\right)\right]$$

(2) $M_b(L_2) + 2M_c(L_2 + L_3)$

$$= -\frac{6}{L_2}\left[\left(x - \frac{x^2}{L_2}\right)\left(\frac{x^2}{3}\right) + \left(x - \frac{x^2}{L_2}\right)\left(\frac{L_2 - x}{2}\right)\left(x + \frac{L_2 - x}{3}\right)\right]$$

For

$x = 20$	$M_b = -7.44$ ft·k	$M_c = -3.28$ ft·k
$x = 40$	$M_b = -10.25$ ft·k	$M_c = -6.91$ ft·k
$x = 60$	$M_b = -9.64$ ft·k	$M_c = -9.64$ ft·k

A plot of these values will give the influence diagram as shown in Fig. 16.2. Other methods such as moment distribution could have been used to find the values of M_b for a unit load at twenty foot intervals along the beam. The influence diagram can be used to find the value of M_b for any load system just as was used in Chapters 4 and 5. When the load consists of a uniformly distributed load, the value of M_b is equal to the product of the area of the

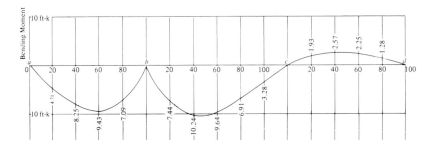

Fig. 16.2.   Influence line diagram for M_b.

influence diagram and the intensity of the load per unit length. This would require the determination of the area of the influence diagram by some means. Another way is to divide the uniform load into a number of equal equivalent concentrated loads (about 8 to 10 per span) and then to multiply each concentrated load by the respective ordinate to the influence diagram at the location of each concentrated load. The value of M_b is then the algebraic summation of these products. As an example, consider that the three span continuous beam of Fig. 16.1 is loaded with a uniformly distributed live load which is the HS-20 equivalent lane load 0.64 k/ft. of variable length. Instead of using the actual area of the influence diagram, the uniform load over the first two spans is divided into a concentrated load placed at the mid-point of each 20 ft. incremental length of the beam.

The actual influence diagram can be idealized as having straight lines between points at twenty foot spacing along the beam. The ordinate of the influence diagram to be used in conjunction with each concentrated load is the average of the influence line ordinates at each end of each twenty foot segment. A second method is to plot the influence line diagram (Fig. 16.2) and to obtain the ordinate for each twenty foot increment by reading from the influence line. The values from both methods are as follows:

Segment (Measured from point a)	Average Ordinate	Ordinate from Influence Line Diagram
0-20	2.35 ft-k/k	2.6
20-40	6.48	6.5
40-60	8.84	9.0
60-80	8.26	8.7
80-100	3.55	4.3
100-120	3.72	4.4
120-140	8.84	9.4
140-160	9.94	10.4
160-180	8.28	8.4
180-200	5.10	5.0
200-220	1.64	1.6
$\Sigma =$	67.00 ft-k/k	$\Sigma = 70.3$

The concentrated load for each 20 ft. length is $0.64(20) = 12.8$ k. Max. bending moment at $b = 12.8(67.0) = 858$ ft-k or $12.8(70.3) = 900$ ft-k.

The maximum bending moment at b obtained by considering the true area of the influence diagram is 886 ft-k. The approximate method gives an error of 3% by average ordinate or 1.5% by values from influence line diagram. If the distributed load had been divided into eight to ten concentrated loads per span, even less error would have been ovtained.

To obtain the bending moment for the entire HS-20 live load a moment due to a concentrated load of 18 kips in each of the two adjacent spans has to be added to the moment due to the uniform load (AASHTO Design Specifications). This additional moment is obtained by summing the two maximum ordinates as read from the influence line diagram (Fig. 16.2). The sum of the maximum ordinates is $9.5 + 10.5 = 20.0$.

Multiplying this sum by 18 k, the bending moment at point b due to the two concentrated loads is:

$$M_b = 18(20) = 360 \text{ ft-k}$$

Total M_b(HS-20 loading) $= 858 + 360 = 1218$ ft-k or

$$900 + 360 = 1260 \text{ ft-k.}$$

The previous procedure shows that a small increase in accuracy is obtained by drawing the influence line diagram and obtaining the ordinates to the influence line diagram by reading from the plotted curve. The line can be drawn to check the accuracy of the calculated values. A value not lying on the curve would indicate an erroneous value.

16.3 Relationship of Deflected Shape to Influence Line Diagrams

Professor Müller-Breslau (1851–1925) of Germany discovered that there was a relationship between the deflected shape of a structure and the influence line diagram. This relationship is; *the influence line for a reaction or internal stress is, to some scale, the elastic curve of the structure when deflected by an action similar to the reaction or stress.*

An example of this principle of Müller-Breslau is the case of determining the shape of the influence line for the interior reaction of a three-span continuous beam (Fig. 16.3). The shape of the influence diagram for R_b is produced by displacing the beam at the point over support b a unit distance. The resulting deflected shape of the beam is the influence diagram, to some scale, for R_b. The values of the ordinates could be determined by analytical

Fig. 16.3. Deflected Shape and Influence Line for Continuous Beam.

means or by model analysis. This principle is the basis for structural analysis by experimental models. Such model procedures, once commonly taught and used, have been largely replaced by mathematical models using the computer.

This principle of Müller-Breslau has a qualitative importance in structural design. It can be used in quickly sketching the influence line for any stress function for any structure, determinate or indeterminate. From the shape of the influence line diagram, the engineer can readily see the location of live load application to produce the maximum shear, bending moment, axial force, or reaction. The influence diagram of Fig. 16.3 shows positive values of influence in spans one and two and negative in span three. Therefore, the live load would be placed in the first two spans only to obtain maximum upward reaction of R_b.

The sketches in Fig. 16.4 show several other influence diagrams generated from the deflected curve of the structure. Figure 16.4 (a) shows the generation of the influence line for shear at a location just to the right of b, by cutting the beam at this location and then displacing the ends in a manner consistant with a shear distortion. This shows that for maximum shear just to the right of support b the first and second spans would be loaded but not

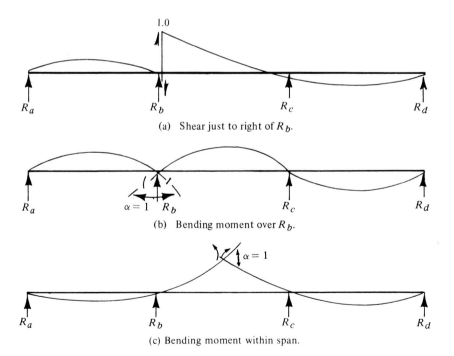

(a)   Shear just to right of R_b.

(b)   Bending moment over R_b.

(c) Bending moment within span.

Fig. 16.4.   Influence lines from deflected shape of continuous beam.

the third span since it is of negative sign. Sketches (b) and (c) show the generation of influence lines for bending moments in a continuous beam. To obtain maximum bending moment over the interior support, the first two spans but not the third should be loaded with the live load. Part (c) shows live loading in the middle span only for maximum bending moment in the middle span.

This same procedure can be used for building frames. Figure 16.5 (a) shows a building frame and the influence line for bending moment at an interior point in the beam of the middle-bay lower story. A unit rotation at

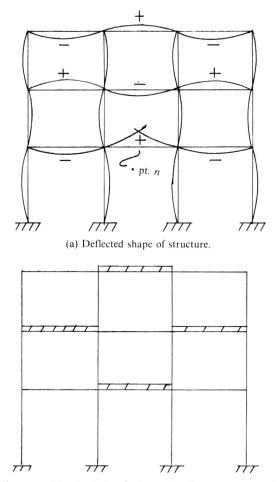

(a) Deflected shape of structure.

(b) Placement of live loads for maximum bending moment at point n.

Fig. 16.5. Influence lines for building frame.

this location will cause the deflected shape of the structure as shown. In sketching the deflected shape, the continuous elastic line of the bottom story beam is drawn first. The matching elastic lines of the columns are then drawn, keeping the angle of intersection of the beams and the columns at ninety degrees. The deflected shapes of the beams of the next two stories are then drawn. Right angle intersection of beams and columns must be maintained.

The deflected shape, which is the influence line to some scale, shows which bays and which floors should be loaded with the live load to obtain maximum bending moment at the location of the impressed distortion. Figure 16.5 (b) shows where the live load is placed as indicated by the shape of the influence diagram. This indicates a "checkerboard" plan of loading. Building codes require this loading pattern.

To obtain maximum bending moment or shears at other locations on the structural frame, rotation or shear distortions can be impressed on the frame as was done for the beam in Fig. 16.4. The resulting influence diagram, obtained from a sketch of the deflected shape, will show the loading pattern.

Quantitative values of the influence diagram could be obtained by a ratio of the deflection at any location to the impressed distortion. This is the procedure in model analysis such as the Beggs Deformator method.

16.4 Elastic Center Method

The elastic center method is a special formulation of the method of consistant displacements for frames and arches. It is only useable for single bent frames and arches. Although it can be applied to bents that have supports at different elevations (1), it is quite involved for this case and other methods may be preferable. It can be most favorably used for a bent with fixed supports and several independent loading conditions. The method has advantages for hand calculations of a fixed arch. The method can be easily programmed for a computer or for hand calculators that have programmable capacity.

A bent is shown in Fig. 16.6. Part (a) shows the real structure acted upon by any loading system P. Part (b) of this figure shows the primary structure and (c) the primary structure with the redundant reactions. The redundants are taken in the positive direction. A coordinate system is established as shown in (c). Figure 16.6 (d) shows the unit forces acting at the location and in the direction of the redundants.

The method of virtual work is used to develop the equations for displacement of support a. If the real structure has no support displacement, the general equation can be written

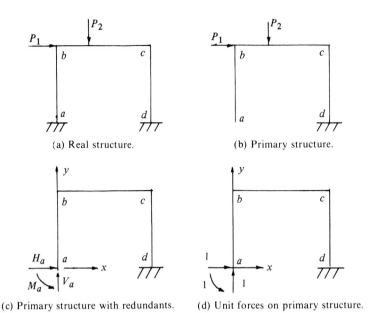

(a) Real structure.   (b) Primary structure.

(c) Primary structure with redundants.   (d) Unit forces on primary structure.

Fig. 16.6. Fixed Support Frame.

$$D = \int_0^S \frac{Mmds}{EI} + R\int_0^S \frac{m^2 ds}{EI} = 0 \qquad (16.1)$$

Since there are three possible D displacements and three reactions R, equation 16.1 will have to be written three times. In considering a sign convention for bending moments in the frame, it is immaterial what convention is established as long as there is consistancy in sign between the M and m moments. The convention adopted is that moments causing tension on the inside of the frame are positive.

The direction of M_a and the corresponding unit moment at a is shown as counterclockwise. This is purely arbitrary.

With the unit forces acting at point a of the frame, the m bending moments can be established in terms of the coordinate distances x and y. On this basis the m moment for the unit load at a in the x direction is

$$m_1 = -(1)y = -y \text{ (tension on outside)}$$

for unit load in the y direction

$$m_2 = +(1)x = +x \text{ (tension on inside)}$$

and for a unit moment at a in the counterclockwise direction

$$m_3 = -1 \text{ (tension on outside)}$$

These bending moments are consistant with the bending moment sign convention; tension on the inside is positive.

The three simultaneous equations of displacement (Eq. 16.1) can now be written.

$$\Delta_x = -\int \frac{My ds}{EI} + M_a \int \frac{y ds}{EI} - V_a \int \frac{xy ds}{EI} + H_a \int \frac{y^2 ds}{EI} = 0$$

In writing the above equation, it should be remembered that the m^2 term in Equation 16.1 consists of two bending moment conditions. One m value is due to a unit load in the direction of the displacement for which the equation is being written and the second value of m is for a unit load in the direction of the redundant reaction. The second and third displacement equations are

$$\int \frac{M_x ds}{EI} - M_a \int \frac{x ds}{EI} + V_a \int \frac{x^2 ds}{EI} + H_a \int \frac{xy ds}{EI} = 0$$

$$\theta_a = -\int \frac{M ds}{EI} + M_a \int \frac{ds}{EI} + V_a \int \frac{x ds}{EI} + H_a \int \frac{y ds}{EI} = 0$$

The three redundant reactions could be determined from the above displacement equations by evaluating all the integrals and solving for the R matrix. This is actually the flexibility method of Chapter 14. The integrals corresponding to the redundants are the flexibility coefficients. All this is not necessary if we change location of the origin of the coordinates. This new origin will be ta the elastic cneter of the bent. The elastic center is the center of gravity of the bent considering all the members have a width in the plane of the elevation view equal to $1/I$ of the member (E is considered constant for the entire frame). It is then considered that an infinitely rigid member is attached to the frame at a and extends from a to 0 (the elastic center) as shown in Fig. 16.7.

The redundant reactions are now V_0, H_0, and M_0. The three displacement equations are now (eliminating E):

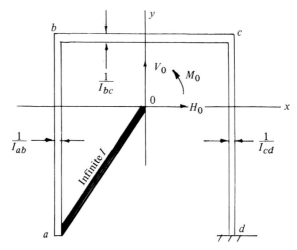

Figure 16.7. Frame with Rigid Arm $a0$.

$$\Delta_{0x} = -\int \frac{Myds}{I} + M_0 \int \frac{yds}{I} - V_0 \int \frac{xyds}{I} + H_0 \int \frac{y^2ds}{I} = 0$$

$$\Delta_{0y} = \int \frac{Mxds}{I} - M_0 \int \frac{xds}{I} + V_0 \int \frac{x^2ds}{I} + H_0 \int \frac{xyds}{I} = 0$$

$$\theta_0 = -\int \frac{Mds}{I} + M_0 \int \frac{ds}{I} + V_0 \int \frac{xds}{I} + H_0 \int \frac{yds}{I} = 0$$

It can be noted that all the flexibility coefficients corresponding to the redundants V_0, H_0, and M_0 correspond to the mathematical equations for section properties for a geometrical figure whose width is equal to $\frac{1}{I}$. If the reference axes are centroidal axes, then the products of inertia

$$\int xydA = \int \frac{xyds}{I} = 0$$

and also the first moments of the area

$$\int \frac{xds}{I} = \int \frac{yds}{I} = 0$$

The displacement equations can then be simplified by eliminating the zero terms

$$-\int \frac{Myds}{I} + H_0 \int \frac{y^2 ds}{I} = 0$$

$$\int \frac{Mxds}{I} + V_0 \int \frac{x^2 ds}{I} = 0$$

$$-\int \frac{Mds}{I} + M_0 \int \frac{ds}{I} = 0$$

and

$$H_0 = \frac{\int \dfrac{Myds}{I}}{\int \dfrac{y^2 ds}{I}} = \frac{\int \dfrac{Myds}{I}}{I_x}$$

$$V_0 = -\frac{\int \dfrac{Mxds}{I}}{\int \dfrac{x^2 ds}{I}} = -\frac{\int \dfrac{Myds}{I}}{I_y} \qquad (16.2)$$

$$M_0 = \frac{\int \dfrac{Mds}{I}}{\int \dfrac{ds}{I}} = \frac{\int \dfrac{Mds}{I}}{A_e}$$

The terms I_x, I_y, and A_e are the elastic moments of inertia and area, respectively.

Positive values of H_0, V_0, and M_0 will mean that their directions will be as shown in Fig. 16.7.

The redundant reactions (H_0, V_0, and M_0) can now be determined by the three independent equations (16.2). With H_0, V_0, and M_0 determined H_a, V_a, and M_a can be calculated from statics.

The application of this method can be applied to a frame as follows:

EXAMPLE PROBLEM 16.1   (a) Using the elastic center method, determine the reactions at a and d for the frame shown. (b) Change the support conditions to pin ends instead of fixed, and calculate the reactions.

(a) Fixed Base

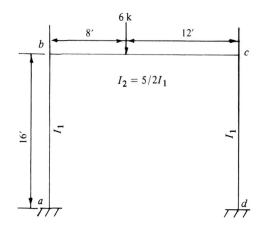

$$\bar{y} = \frac{2(16)(8) + \frac{2}{5}(20)(16)}{2(16) + \frac{2}{5}(20)} = 9.6 \text{ ft. (from axis thru } ad)$$

Seg.	ds/I	x	y	$\dfrac{x^2 ds}{I}$	$\dfrac{y^2 ds}{I}$	I_{0x}	I_{0y}
ab	16	-10	-1.6	1600	41	$\dfrac{16(16)^2}{12} = 341$	0
bc	$20(2/5) = 8$	0	$+6.4$	$--$	328	0	$\dfrac{2}{5}\left(\dfrac{20}{12}\right)(20)^2 = 267$
cd	16	$+10$	-1.6	1600	41	341	0
Σ	40			3200	210	682	267

$$I_x = 410 + 682 = 1092 \qquad I_y = 3200 + 267 = 3467$$

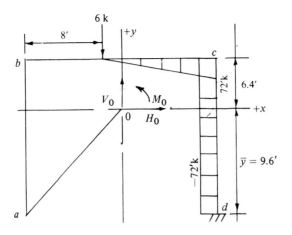

Seg.	$\dfrac{Mds}{I}$	x	y	$\dfrac{Mxds}{I}$	$\dfrac{Myds}{I}$
ab	0			0	0
bc	$\dfrac{2}{5}\dfrac{-72}{2}(12) = -172.8$	$+6$	$+6.4$	$-1{,}307$	-1106
cd	$(-72)(16) = -1152.0$	$+10$	-1.6	$-11{,}520$	$+1843$
Σ	-1324.8			$-12{,}557$	$+\ 737$

$$V_0 = -\frac{\Sigma\dfrac{Mxds}{I}}{I_y} = -\frac{-12{,}557}{3.467} = +3.62 \uparrow$$

$$H_0 = \frac{\Sigma\dfrac{Myds}{I}}{I_x} = \frac{737}{1092} = 0.675 \rightarrow$$

$$M_0 = \frac{\Sigma\dfrac{Mds}{I}}{A_c} = \frac{-1324.8}{40} = -33.1 \circlearrowright$$

$$V_a = 3.62$$

$$H_a = 0.675$$

$$M_a = 3.62(10) - 0.68(9.6) - 33.1 = -3.4 \circlearrowright$$

The minus (−) sign for M_a indicates that the bending moment applied at a for the column ab is in the clockwise direction.

(b) If the bottom of the columns were pinned, the location of the elastic center would be on the x axis passing through the pins. The reasoning behind this is that the pins have no stiffness so the equivalent I is zero. Then the value of $1/I$ (the elastic area) is infinite and the centroidal axis would then pass through the pins. However, this area would be infinitely wide but with very small height so that

$$\frac{y^2\,ds}{I}$$

for this area resulting from the pin would be very small and can be neglected. Since there is only one redundant, the only equation is

$$H_0 = \frac{\Sigma \dfrac{My\,ds}{I}}{I_x}$$

EXAMPLE PROBLEM 16.2   Rework Example Problem 16.1 but with frame hinged at base.

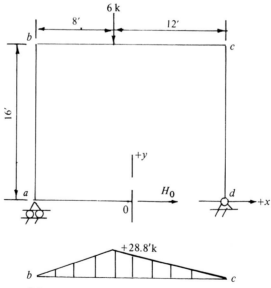

Primary structure with bending moment diagram.

Seg	$\dfrac{y^2 ds}{I}$	$\dfrac{Mds}{I}$	y	$\dfrac{Myds}{I}$
ab	$\dfrac{(16)^3}{3} = 1365$	0		0
bc	$\dfrac{2(20)(16)^2}{5} = 2048$	$\dfrac{+28.8(20)}{2}\left(\dfrac{2}{5}\right) = +115.2$	$+16$	1843
cd	$= 1365$			0
Σ	4778			1843

$$H_0 = \frac{\Sigma \dfrac{Myds}{I}}{\Sigma \dfrac{y^2 ds}{I}} = \frac{1843}{4778} = 0.386\,\text{k} \rightarrow$$

$$H_a = 3.86\,\text{k} \rightarrow \qquad V_a = 3.6\,\text{k} \uparrow$$

$$H_d = 3.86\,\text{k} \leftarrow \qquad V_d = 2.4\,\text{k} \uparrow$$

EXAMPLE PROBLEM 16.3   Using the elastic center method, determine the reactions at supports a and d of the frame shown below

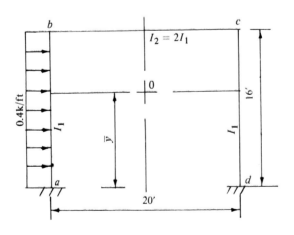

$$\bar{y} = \frac{2(16)(8) + \dfrac{20}{2}(16)}{2(16) + \dfrac{20}{2}} = 9.9\,\text{ft.}$$

Seg.	ds/I	x	y	$\dfrac{x^2 ds}{I}$	$\dfrac{y^2 ds}{I}$	I_{0x}	I_{0x}
ab	16	+10	−1.9	1600	57.8	341	- - -
bc	10	0	+6.1	- - -	372	- - -	333
cd	16	−10	−1.9	1600	57.8	341	- - -
z	42			3200	487.6	682	333

$$I_x = 488 + 682 = 1170 \qquad I_y = 3200 + 333 = 3533$$

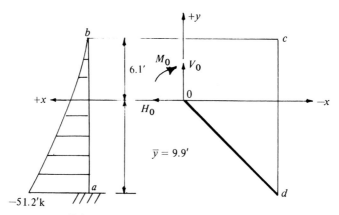

Primary structure with redundant reactions.

Seg.	$\dfrac{Mds}{I}$	x	y	$\dfrac{Mxds}{I}$	$\dfrac{Myds}{I}$
ab	$\dfrac{-51.6(16)}{3} = -273.1$	+10	−5.9	−2731	1611
bc	0			0	0
cd	0			0	0

$$V_0 = -\frac{\Sigma \dfrac{Mxds}{I}}{I_y} = -\frac{-2731}{3533} = 0.77 \uparrow$$

$$H_0 = \frac{\Sigma \dfrac{Myds}{I}}{I_x} = \frac{1611}{1170} = 1.38 \leftarrow$$

$$M_0 = \frac{\sum \dfrac{Mds}{I}}{A} = \frac{-273}{42} = -6.53\,\text{⌒}$$

$$V_d = 0.77\uparrow$$

$$H_d = 1.38\leftarrow$$

$$M_d = 1.38(9.9) - 0.77(10) + 6.53 = 12.5\,\text{⌒}$$

$$V_a = 0.77\downarrow$$

$$H_a = 6.4 - 1.38 = 502\leftarrow$$

$$M_a = -0.77(20) + 6.4(8) - 12.5 = 23.3\,\text{⌒}$$

16.5 Elastic Center Method Applied to Arches

The most likely application of the elastic center method is to a fixed end symmetrical arch (Fig. 16.8). The use of three simultaneous equations, as

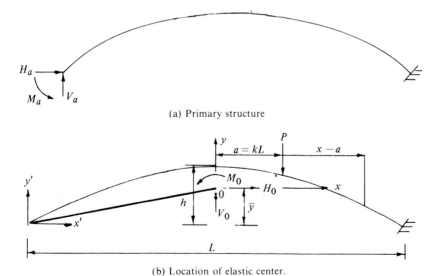

(a) Primary structure

(b) Location of elastic center.

Fig. 16.8. Fixed-end Parabolic Arch.

required in the general (flexibility) method or the stiffness method, is not necessary. The number of flexibility coefficients are also reduced. Once the elastic center is located, the values of the terms in Eqs. 16.1 are all that are required.

The fixed-end parabolic arch of Fig. 16.8 will be used to illustrate the method. The following relationships are considered valid.

$$ds = dx(\sec\alpha)$$

$$I = I_c(\sec\alpha)$$

The angle α is the angle the tangent to the arch axis forms with the horizontal, and I_c is the moment of inertia of the arch rib at the crown. From the above relationships

$$\frac{ds}{EI} = \frac{dx}{EI_c}.$$

If the elastic width (as seen in elevation) of the arch rib is $\frac{1}{EI}$, the elastic area of the arch rib in a ds length is

$$\frac{ds}{EI} = \frac{dx}{EI_c}.$$

The total elastic area is

$$A_e = 2\int_0^{L/2} \frac{dx}{EI_c} = \frac{L}{EI_c}.$$

The moment of this elastic area taken about a line through the springings (X-axis) is

$$\int_0^L y\,\frac{dx}{EI_c}.$$

For a parabola

$$y = 4h\left(\frac{x}{L} - \frac{x^2}{L^2}\right)$$

The moment of the elastic area about the x axis is

$$M_e = \frac{4h}{EI_c}\int_0^L\left(\frac{x}{L}-\frac{x^2}{L^2}\right)dx = \frac{2hL}{3EI_c}$$

The centroid of the elastic area from the x axis is

$$\bar{y} = \frac{M_e}{A_e} = \frac{2hL}{3EI_c}\frac{L}{EI_c} = \frac{2}{3}h.$$

With the elastic center known, x and y axes can be drawn through this center. Figure 16.7 (b) shows this geometrical figure and from this new set of axes, the values of y in terms of L, h, and x can be evaluated.

$$y = y' - \bar{y}$$

$$x = x' - \frac{L}{2}$$

$$y = 4h\left(\frac{x'}{L}-\frac{x'^2}{L^2}\right) - \frac{2}{3}h$$

substituting $x' = x + \frac{L}{2}$ in the above equation,

$$y = \frac{h}{3L^2}(L^2 - 12x^2)^2$$

The evaluation of the denominators for equations 16.1 are

$$I_x = \int_{-\frac{L}{2}}^{+\frac{L}{2}}\frac{(y)^2 dx}{EI_c} = \frac{h}{9L^4 EI_c}\int_{-\frac{L}{2}}^{+\frac{L}{2}}(L^2-12x^2)^2 dx = \frac{4h^2L}{45EI_c}$$

$$I_y = \int_{-\frac{L}{2}}^{+\frac{L}{2}}\frac{x^2 dx}{EI_c} = \frac{L}{12EI_c}$$

The numerators for these equations are next determined.

(a) From

$$x = a \qquad \text{to} \qquad x = \frac{L}{2}$$

$$M = -P(x-a)$$

$$\int_a^{\frac{L}{2}} \frac{Mdx}{EI_c} = \frac{1}{EI_c} \int_a^{\frac{L}{2}} -P(x-a)dx = -\frac{PL}{8EI_c}(1-2k)^2$$

$$\int_a^{\frac{L}{2}} \frac{Mxdx}{EI_c} = \frac{1}{EI_c} \int_a^{\frac{L}{2}} -P(x-a)xdx = -\frac{PL^3}{24EI_c}[(1-2k)^2(1+k)]$$

$$\int_a^{\frac{L}{2}} \frac{Mydx}{EI_c} = \frac{1}{EI_c} \int_a^{\frac{L}{2}} P(x-a)\left[\frac{h}{3L^2}(L^2-12x^2)\right]dx$$

$$= \frac{PhL^2}{48EI_c}(1-8k^2+16k^4)$$

There is bending moment in the primary structure only between $x = a$ to $x = \frac{L}{2}$.

Substituting the values just determined into the equations 16.1, the values for H_0, V_0, and M_0 are:

$$H_0 = \frac{\int \frac{Mydx}{EI_c}}{I_x} = \frac{PhL^2}{48EI_c}\left[\frac{(1-8k^2+16k^4)}{\frac{4}{45}\frac{h^2L}{EI_c}}\right] = \frac{16}{54}\frac{PL}{h}(1-8k^2+16k^4)$$

$$V_0 = -\frac{\int \frac{Mxdx}{EI_c}}{I_y} = \frac{PL}{24EI_c}\left[\frac{(1-2k)^2(1+k)}{\frac{L^3}{12EI_c}}\right] = \frac{P}{2}(1-2k)^2(1+k)$$

$$M_0 = \frac{\int \frac{Mdx}{EI_c}}{A_e} = \frac{PL}{3EI_c}\left[\frac{(1-2k)^2}{\frac{L}{EI_c}}\right] = \frac{PL}{8}(1-2k)^2$$

Table 16.1

1	2	3	4	5	6	7	8	9	10	11	12	13	14	15
Seg.	ΔS	I	$\frac{\Delta S}{I}$	X'	Y'	$\frac{Y'\Delta S}{I}$	X	Y	$\frac{X^2\Delta S}{I}$	$\frac{Y^2\Delta S}{I}$	M	$M\frac{\Delta S}{I}$	$MX\frac{\Delta S}{I}$	$MY\frac{\Delta S}{I}$
1 ↓ 20														
			$\Sigma 4$			$\Sigma 7$			$\Sigma 10$	$\Sigma 11$		$\Sigma 13$	$\Sigma 14$	$\Sigma 15$

The values of H_a, V_a, and M_a can be determined by simple statics when the values of H_0, V_0, and M_0 are calculated.

It may be that the fixed-end arch is of an unusual curve, I is not equal to $I_c \sec \alpha$, or the loading is very complex. If any one of these conditions occur, it will be difficult to perform the mathematical integration. A numerical integration of equation 16.1 can be done. If a numerical integration is to be performed, a systematic procedure is necessary. One half of the length of the arch is divided into at least ten segments with the length of each segment along the arch axis of equal value. Table 16.1 can be used to systematically calculate the values of H_0, V_0, and M_0.

Each column of the table is numbered. Column 2 is the length of each segment of arch axis length. Column 3 is the average moment of inertia of the cross section of the arch rib for each segment. The value of I at the mid-length of each segment is usually sufficiently accurate for design purposes.

The values x' and y' (columns 5 and 6) are the distances from the springing to the centroid of each length ΔS. The values x and y are measured from the elastic center. This measurement will have to be performed after the location of the elastic center is known. The numbers in all the other columns are calculated. Table 16.1 is valid only for a fixed-end arch with a y axis of symmetry. The elastic-center method can be used for an arch without an axis of symmetry but it is questionable if this is the best method for the non-symmetrical case.

The primary structure for a general case of arch with fixed-ends is shown in Fig. 16.9.

For a fixed ended arch, the following equations are valid if x and y are measured from the elastic center.

$$H_0 = \frac{\int \dfrac{Myds}{I}}{\int \dfrac{y^2 ds}{I}} \qquad V_0 = -\frac{\int \dfrac{Mxds}{I}}{\int \dfrac{x^2 ds}{I}} \qquad M_0 = \frac{\int \dfrac{Mds}{I}}{\int \dfrac{ds}{I}}$$

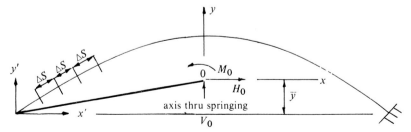

Fig. 16.9. Fixed-end Arch.

If the above integrations cannot be easily performed, then numerical integration can be used by dividing the arch into 20 or more segments, preferably of equal ΔS length. Table 16.1 can be used to determine the six quantities in the above equations with $\int$ replaced with Σ.

$$y_0 = \frac{\Sigma Y \frac{\Delta S}{I}}{\Sigma \frac{\Delta S}{I}} = \frac{\Sigma 7}{\Sigma 4} \qquad H_0 = \frac{\Sigma 15}{\Sigma 11} \qquad V_0 = -\frac{\Sigma 14}{\Sigma 10} \qquad M_0 = \frac{\Sigma 13}{\Sigma 4}$$

The values of X' and Y' are measured from springing and X and Y are measured from elastic center.

A similar procedure can be performed for an arch without an axis of symmetry. The procedure is more complicated and space does not permit this case. Reference (1) has a treatment of the unsymmetrical arch.

References

1. Carpenter, S. T., *Structural Mechanics*, Robert E. Krieger Publishing Co., Inc., 1971.

Problems

1. Plot the influence line diagram for shear just to the right of support b for the beam in Fig. 16.1.
2. Plot the influence line diagram for bending moment at the midspan point of the middle span of the beam in Fig. 16.1.
3. Using the Muller-Breslau principle, sketch the influence lines and determine which floors should be loaded with live load for the maximum value of:
    (a) Shear in beam just to right of column in second story.
    (b) Positive bending moment in the beam of first bay, second story. Use frame in Fig. 16.5.
4. Using the elastic center method, determine the reactions for the frame of Fig. 16.10.
5. Repeat Problem 5, but with column bases pinned.
6. Using the elastic center method, determine the reactions for the frame of Fig. 16.11. *Ans.* $H_a = 0.72$ k, $V_a = 0.40$ k, $M_a = 5.5$ ft-k.
7. Use the elastic center method to determine the reactions for the gabled frame of Fig. 16.12. Roof load is 4 k/ft of horizontal projection.
8. Using Table 16.1, determine the reactions for a parabolic arch of 80 ft. span and 16 ft. rise. Load is a concentrated gravity load of 12 k at quarter points of span. Assume $I = I_c \sec \alpha$. Remember

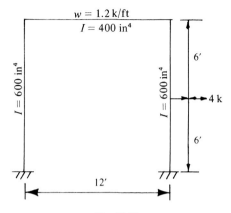

Fig. 16.11.

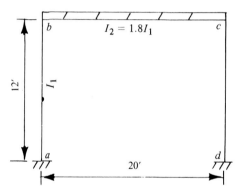

Fig. 16.10.

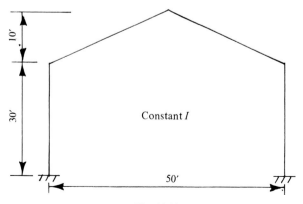

Fig. 16.12.

$$S = L\left[1 + \frac{8}{3}\left(\frac{h}{L}\right)^2 - \frac{32}{5}\left(\frac{h}{L}\right)^4\right]$$

9. A fixed-end arch is to be in the shape of a catenary (See Chapter 9) with span of 800 ft. and rise of 120 ft. The load is a gravity load of one kip per ft. length of arch I is constant. Determine reactions at base. Use Table 16.1.

STRUCTURAL ANALYSIS UNDER SI

17.1 Introduction to SI Units

In the not too distant future, the United States will be changing its units of measure to the metric system. To most people, this means using metres*, kilometres*, and kilograms instead of the present units for length, distance, and weight. To most people, the advantage appears to be using a multiple ten digit system instead of a system that has no common multiplier. To engineers who have had to detail structures in fractions of an inch, inches, and feet, this new system of decimal addition of lengths appears to be a welcome relief.

The change to the metric system is more than this to the engineer. The system to be adopted in the U.S.A. is that of a system agreed to in 1960 by the Eleventh General Conference on Weights and Measures. This system, called the International System of Units (abbreviation is SI in all languages) is, in addition to being a metre-kilogram-second system, an absolute system and not a gravitational system as the presently used English system. This then means that in SI the value of mass and force are not interchangable as in the English system.

---

*In SI the adopted spelling of the unit of length is metre (not meter as common in some countries).

17.2   Classes of SI Units

SI units are divided into three classes: base units, derived units, and supplementary units. The Eleventh General Conference selected the base units as the metre, kilogram, second, ampere, kelvin, mole, and candela. Structural engineers will likely be involved only with the first three named units. These three basic units represent the following quantities.

Basic Units

Quantity	Name of SI Unit	Symbol
Length	Metre	m
Mass	Kilogram	kg
Time	Second	s
Plane angle*	Radian	rad

*Angular measurement is termed a supplementary unit.

There are many derived units developed from applying the base units to plysical quantities. Those derived units that are most likely to be used by the structural engineer are listed in the following table.

Derived Units

Quantity	Name of Unit	Symbol
Area	square metre	m^2
Volume	cubic metre	m^3
Force	newton	N
Pressure, stress	pascal	$Pa(N/m^2)$
Moment, torque	newton-metre	$N \cdot m$
Work, energy	jjoule (newton-metre)	$J = N \cdot m$
Velocity-linear	metre/sec	m/s
Velocity-angular	radian/sec	rad/s
Acceleration-linear	metre/sec^2	m/s^2
Acceleration-angular	radian/sec^2	rad/s^2
Frequency	hertz=cycles/sec	Hz
Density	kilogram/cubic metre	kg/m^2
Power	watt=joule/sec	W

17.3   Mass, Force, and Pressure

It is noted that derived units are expressed algebraically in terms of base units or supplementary units. Some of their symbols have special names, but most are the mathematical signs of multiplication or division.

The derived units that are of special importance to structural engineers are the newton (unit of force) and pascal (unit of pressure or stress). As was stated previously, the SI is an absolute system instead of a gravitational system. This means that the unit of force is not interchangeable with mass as in the English system or the metric systems used in some countries. In SI, a unit of force is defined as that force which will impart to a one kilogram mass an acceleration of one metre per second per second. This unit force is termed a newton. The algebraic expression ofr this derived unit is $kg \cdot m/sec^2$. On the earth, the acceleration of gravity is 9.81 metres/sec^2 (32.2 ft/sec^2). Therefore, a one kilogram mass on the earth will have a force exerted on it that will give it an acceleration of 9.81 m/sec^2. This gravitational force would then be equal to 9.81 newtons. Therefore, if a one kilogram mass is hanging by a chord in the earth's gravitational field, the force in the chord (SI) is 9.81 N. If this same one kilogram mass were transported to the moon, its mass would remain the same (meaning of absolute system). However, if this one kilogram mass was again hung from a chord, the force in the chord would be less since the gravitational field is less on the moon. The force in the chord would be 1.6 N. The converting of mass to force will be an additional step in the analysis of structures that will be required in SI that is not now performed in the English system or in some metric systems.

The procedure for structural analysis in SI will be to convert all weights (in kilograms) into force in newtons by multiplying the kilograms by 9.81 (on earth) to obtain newtons. The analysis will then proceed using newtons, newton-metre (moments), and newtons per square metre or pascals.

17.4   Multiplying Factors

If forces are given in units of newtons, the number will be quite large for most civil engineering type structures. It is recommended that under SI the numerical values used in computations be between 0.1 and 1,000. Therefore, prefixes will be necessary where numbers get below 0.1 or above 1,000. A list of approved prefixes are given in the following table. When structures are analyzed for masses that are in units of thousands of kilograms, then it is quite likely that the engineer will want to work with kilonewtons (kN) or meganewtons (MN)'

Recommended Multiples

Factor by Which Unit is Multiplied	Prefix Name	Symbol
10^{12}	tera	T
10^{9}	giga	G
10^{6}	mega	M
10^{3}	kilo	k
10^{-3}	milli	m
10^{-6}	micro	μ
10^{-9}	nano	n
10^{-12}	pica	p

There are multiples in power of ten between 10^3 and 10^{-3} but their use is not recommended. Dimensions of cross-sections will normally be given in millimeters (m $\times$ 10^{-3}) and length dimensions in metres. In applying the prefixes to the basic units, no hyphen is used. The hyphen is used between the basic unit words (newton-metre).

There is some preference to using powers of ten instead of prefix symbols in numerical calculations. This can avoid possible errors. One possible error is confusing the letter m which is used as a symbol for both metre and the prefix milli.

It has been common practice in European countries that use the "metric system" to use the unit of "tonne" or "ton" for 1,000 kilograms. This is not recognized under SI, but it could become standard usage since it is common in Europe.

17.5 Conversion between the SI and Gravitational English Systems

There will undoubtedly be a long transistional period between the time of first use of the SI and full adoption. Even after complete adoption, there will be conversion of previous data. Engineers will have to carry double numbers in their memory for a period of time. Extensive conversion tables are readily available in many publications along with details of SI (1) (2). For handy reference and comparison of units between the two systems, the following table gives conversion factors that will be commonly used in structural engineering.

Conversion Factors

English to SI	SI to English

Length
1 in = 25.40 mm	1 mm = 0.0394 in
1 ft = 0.3048 m	1 m = 3.281 ft
1 in = 0.0254 m	1 m = 39.37 in
1 chain = 20.117 m	1 m = 0.0497 chains
1 mile = 1.609 km	1 km = 0.622 mile

Area
1 in^2 = 645.2 mm^2	1 cm^2 = 0.155 in^2
1 ft^2 = 0.0929 m^2	1 m^2 = 10.76 ft^2
1 yd^2 = 0.835 m^2	1 m^2 = 1.196 yd^2

Volume
1 in^3 = 16.387 mm^3	1 cm^3 = 0.0610 in^3
1 ft^3 = 0.0283 m^3	1 m^3 = 35.31 ft^3
1 yd^3 = 0.765 m^3	1 m^3 = 1.564 yd^3

Moment of Inertia
$$1 \text{ in}^4 = 41.62 \text{ cm}^4 = 416{,}200 \text{ mm}^4$$
$$= 0.4162 \times 10^{-6} \text{ m}^4$$

Mass
1 lb = 0.454 kg	1 kg = 2.205 lbs
1 ton (2000 lbs) = 907.2 Kg	1 Mg = 1,102 t (2000 lbs)
1 tonne (metric) = 1.102 ton (2000 lbs.)	

Force
1 lb = 4.448 N	1 N = 0.2248 lb
1 Kip = 4.448 kN	

Stress
1 psi = 6.895 kPa (kN/m^2)	1 MPa = 145.0 psi
1 ksi = 6.895 MN/m^2	

Pressure
1 psf = 0.0479 kN/m^2 or kilopascal (kPa)	
1 ksf = 0.479 MN/m^2	1 N/m^2 = Pa = 20.89 psf
1 ton force/ft^2 = 95.76 kPa	

Bending Moment or Torque
1 ft-lb = 1.356 N·m	1 N·m = 0.738 ft-lb

Work or Energy
1 ft-lb = 1.356 J	1 J = 0.738 ft-lb

Power
1 hp = 745.6 W	1 kW = 1.341 hp

The following are a few common comparisons from English units to SI units.

(a) Modulus of elasticity of steel:
29.6×10^6 psi $= 204.1 \times 10^9$ Pa
29.6×10^3 ksi $= 204.1$ GN/m^2 $= 204.1$ kN/mm^2

Using E as 29×10^3 ksi, the SI equivalent is 200 kN/mm^2 or if pascals are preferred, 200 GPa. It is quite likely that structural engineers will round off E to the 200 value.

(b) Allowable tensile stress of steel:
20,000 psi $= 137.9$ MPa or MN/m^2 or MN/m^2 $= 137.9$ N/mm^2
(c) Ultimate strength of concrete:
3,000 psi $= 20.7$ n/mm^2

It is quite likely that in SI, the allowable stresses of 20,000 psi and 3,000 psi, now so commonly used, will become 138 N/mm^2 and 21 N/mm^2, respectively. In August, 1974, a joint committee of AISC and AISI selected *mm* as cross-sectional dimensions for all steel products. Since cross-sectional areas will be in mm^2, it is quite logical that structural engineers will use stress as N/mm^2 although kilopascal or megapascal is generally recommended.

In the past, engineers have converted, when necessary, from the English stress units of pounds per square inch to the metric units of kilograms per square centimeter. The old metric unit of kilograms force does not exist in SI. This will be a new learning point to structural engineers.

17.6   Example in SI

1.   A beam shown below is required to support the load as shown. Draw the shear and bending moment diagrams. All lengths are in metres and loads in megagrams.

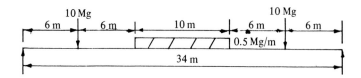

The first step is to change the load (mass) into newtons (force). The length-force diagram is then as follows (1 kilogram mass converts to 9.81 newtons).

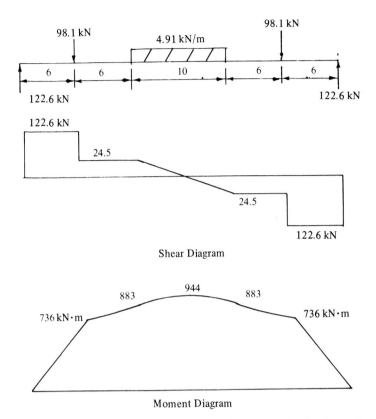

Shear Diagram

Moment Diagram

The units used in the force diagram were kilonewtons, in the shear diagram, kilonewtons, and in the moment diagram, kilonewton metres.

2. A bridge truss is shown with the dimensions in metres. Draw the influence diagram for member $U_2 U_3$ and then determine the maximum force in this member for a moving vehicle of weight and dimension as shown.

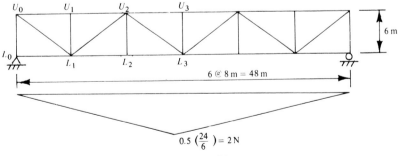

Influence Line Diagram

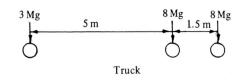

Truck

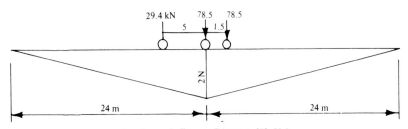

Truck on Influence Diagram $(U_3 U_4)$
(Note mass in Mg has been changed to force in kN)

The vehicle has been placed in the position for maximum force in member $U_3 U_4$.

$$F_{U_3 U_4} = 78.5 \left(2 + 2 \times \frac{22.5}{24} \right) + 29.4 \left(2 \times \frac{19}{24} \right)$$

$$= 304.2 + 46.6 = 350.8 \text{ kN}$$

3. Determine the deflection in the cantilever beam shown if the beam is 160 mm wide and 200 mm in depth. It is an aluminum beam with $E = 73$ kN/mm^2. Weight of beam is 100 kg/m and concentrated load is 400 kg.

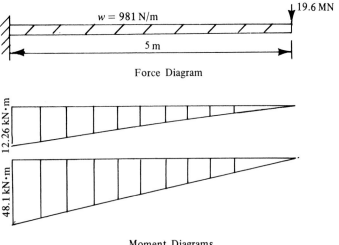

Force Diagram

Moment Diagrams

Using moment-area method, the deflection at the end of the cantilever is equal to the moment of the M/EI diagram taken about the cantilever end. This calculation is as follows.

$$EI\Delta = 12.26\left(\frac{5}{3}\right)\left(\frac{15}{4}\right) + 48.1\left(\frac{5}{2}\right)\left(\frac{10}{3}\right) = 477.5 \text{ kN} \cdot \text{m}^3$$

$$E = 73 \text{ kN/mm}^2$$

$$I = \frac{160(200)^3}{12} = 106.7 \times 10^6 \text{mm}^4$$

$$\Delta = \frac{477.5 \times 10^9}{(73)(106.7 \times 10^6)} = 61 \text{ mm}$$

$$= \frac{\text{kN} \cdot \text{m}^3}{\dfrac{\text{kN}}{\text{mm}^2}(\text{mm}^4)} = \frac{\text{m}^2}{\text{mm}^2} \times 10^9 = \text{mm}$$

It is observed that the above problem had a mixture of metres and milli-metres. The numerator was changed from metres^3 to millimetres^3, so that the answer would be in millimetres.

4. Determine the deflection at the end of the cantilever truss if the mass M is 2×10^3 kg. The area of all members is 300 mm^2. The members are all steel with $E = 200$ kN/mm^2.

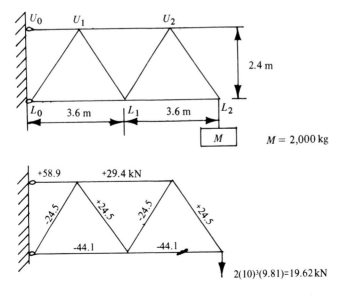

The deflection is required at the point where the force of 19.62 kN is applied. Therefore, by virtual work, the next step would be to apply a unit force at the same location as the 19.62 kN is applied. The analysis for the unit force would give bar forces the same as above divided by 19.62.

Bar	S (kN)	L (m)	$S^2 L$
$U_0 U_1$	+58.9	1.8	6245
$U_1 U_2$	+29.4	3.6	3112
$L_0 L_1$	−44.1	3.6	7001
$L_1 L_2$	−14.7	3.6	778
$L_0 U_1$	−24.5	3.0	1801
$U_1 L_1$	+24.5	3.0	1801
$L_1 U_2$	−24.5	3.0	1801
$U_2 L_2$	+24.5	3.0	1801
			24,340 (kN)² ·m

$$\Delta_{L_2} = \frac{\Sigma SLu}{\Sigma AE} = \frac{24{,}340 \times 10^3}{19.62(3 \times 10^2)(200)} = 20.7 \text{ mm}$$

$$= \frac{(kN)^2 \cdot m (10^3)}{kN(mm^2)\left(\dfrac{kN}{mm^2}\right)} = mm$$

The example problems show the advantage of working with units that are multiples of ten. It also shows that it is necessary to change mass to newtons force before analysis can proceed. Since one kg of mass produces a force on earth of 9.81 newtons, problems with masses in whole numbers will result in newtons in odd numbers. This will be some irritation, but mostly in academic circles. In real world problems, weights do not usually have magnitudes that result in round numbers.

Newtons are such a small quantity that the forces on civil engineering type structures will most likely be in kilonewtons or meganewtons and will have to be symbolized by two letters (kN or MN) instead of one (k = kips) in the English system.

There is one additional advantage to SI. It will force structural engineers (I hope) from using the term stress when the correct terminology for the force in a bar of a truss is *force*.

References

1. LeMaraic, A. L., and J. P. Ciaramella, *The Complete Metric System with the International System of Units,* Abbey Books, New York, 1973.

2. Leffel, R. E., "Civil Engineering Calculations Using SI Units," Engineering Issues, ASCE, Jan. 1976.

APPENDIX A

INFLUENCE LINES FOR CONTINUOUS BEAMS ON ELASTIC SUPPORTS

W = stiffness factor = $\dfrac{6EIK}{L^3}$

E = modulus of elasticity of the beam.

I = moment of inertia of the beam.

K = elastic constant of the support in deformation per unit load.

L = span of beam between supports—all spans are equal.

r = distance from unit load to the right reaction. This distance is given as a proportion of the span length in the charts on the following pages.

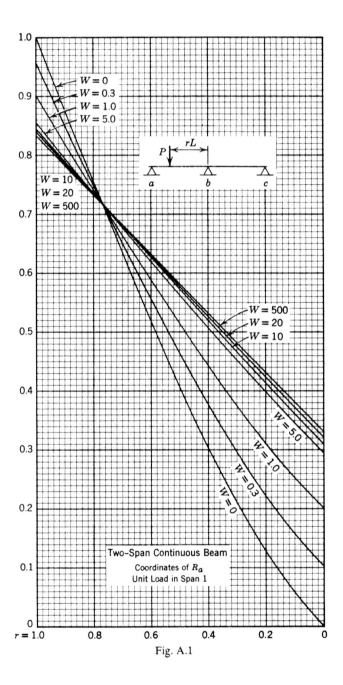

Fig. A.1

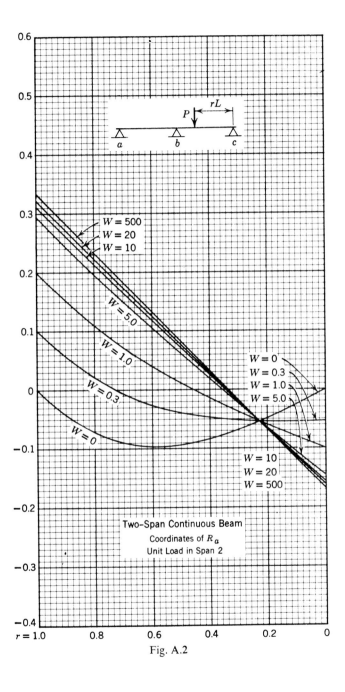

Fig. A.2

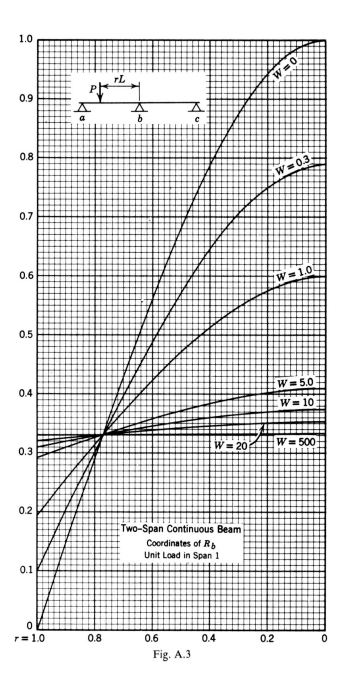

Fig. A.3

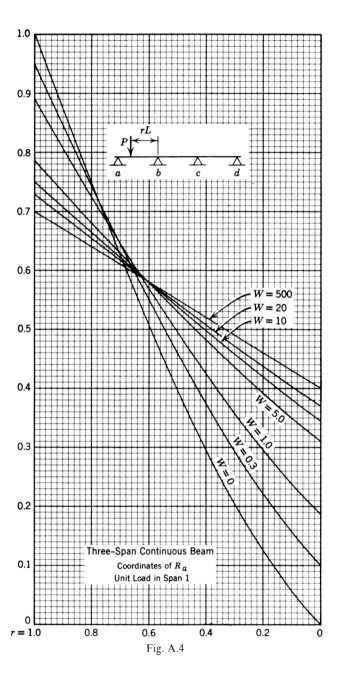

Fig. A.4

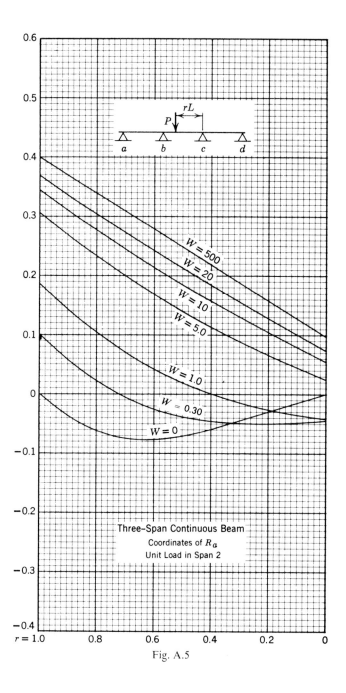

Fig. A.5

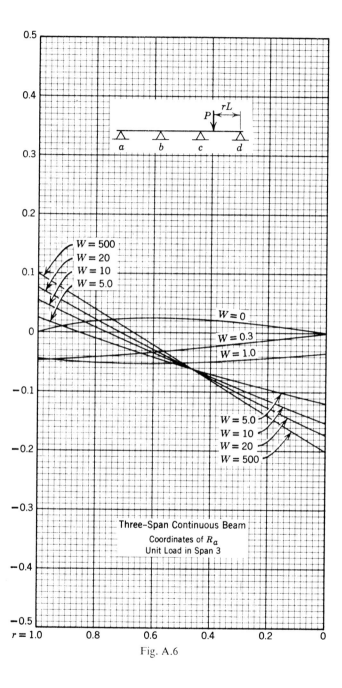

Fig. A.6

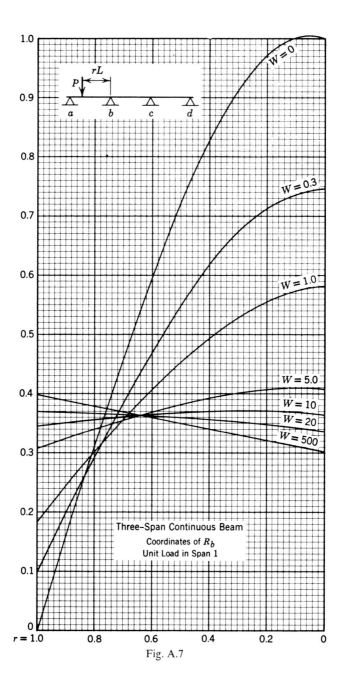

Fig. A.7

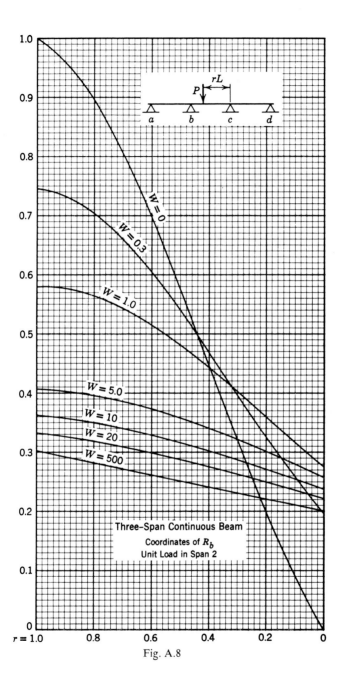

Fig. A.8

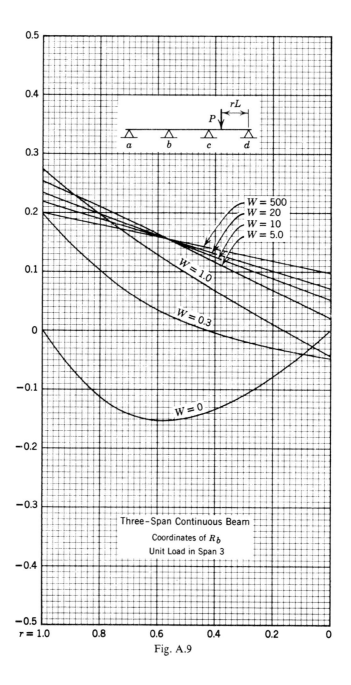

Fig. A.9

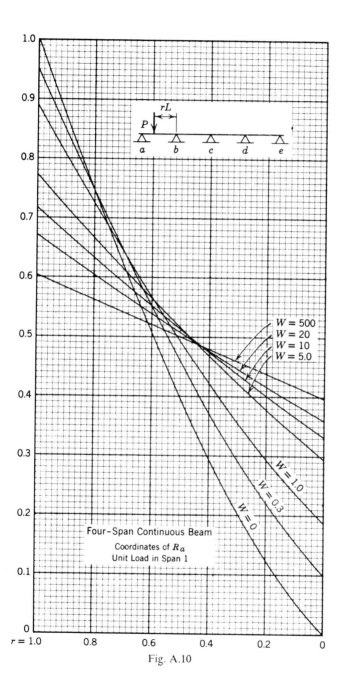

Fig. A.10

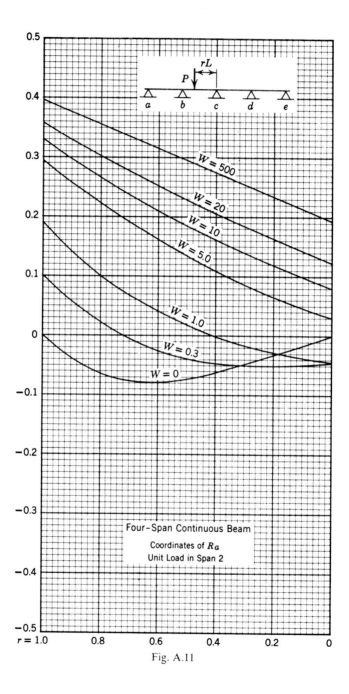

Fig. A.11

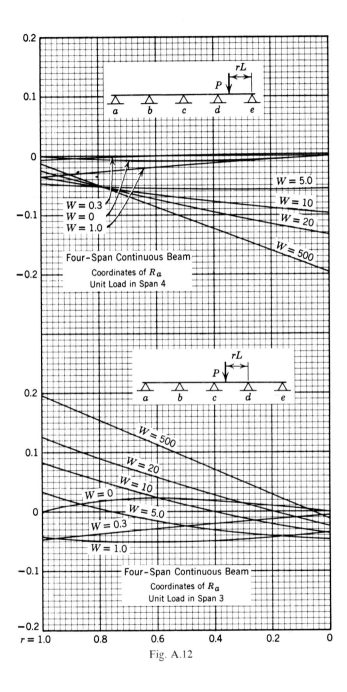

Fig. A.12

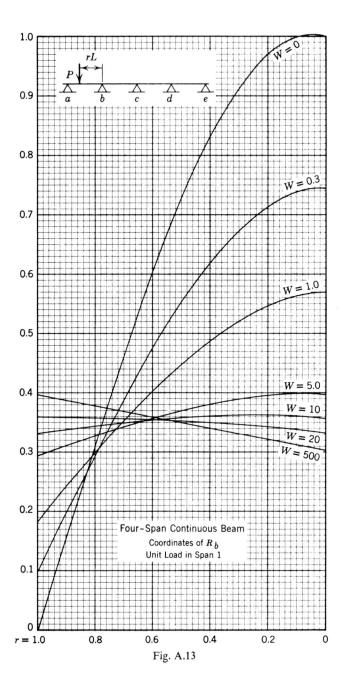

Fig. A.13

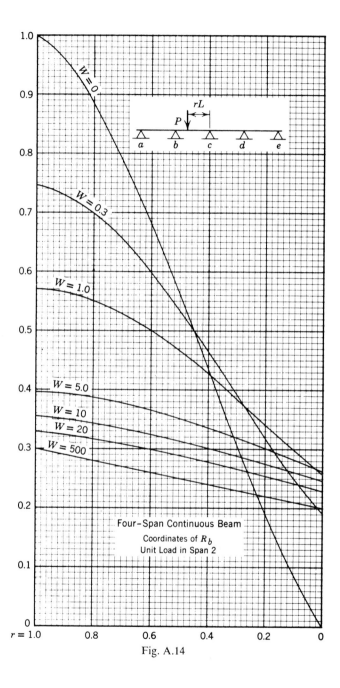

Fig. A.14

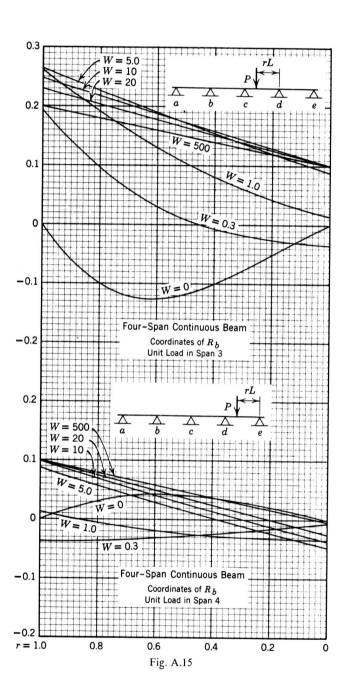

Fig. A.15

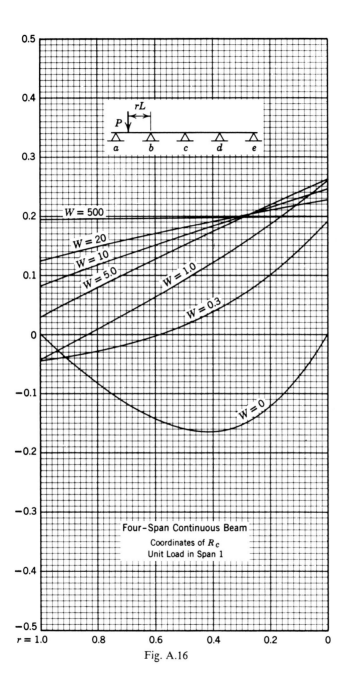

Fig. A.16

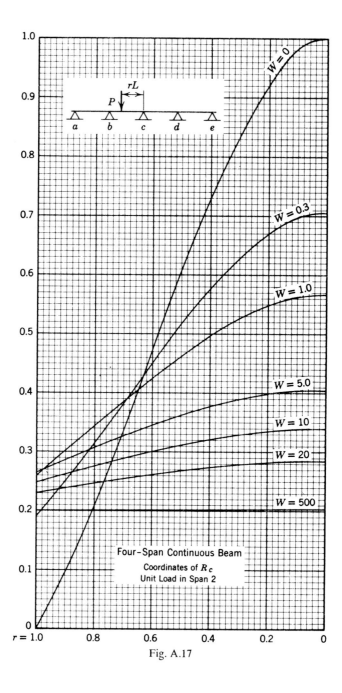

Fig. A.17

APPENDIX B[1]
BRIDGE STRAND

For Main, Wind, and Suspender Cables on Suspension Bridges, and other Tension Members in Straight-Line Pull with Ends Socketed.

For Guys for Television, Radio, and Aerial Tramway Towers, Derricks, Stacks, and Fixed-Length Booms on Large Shovels and Dragline Excavators.

Fig. B.1

Multiple-wire bridge strand is furnished in varying numbers of wires (from 19 to 161) and layers of wires depending on the diameter, and in lengths up to approximately 5000 ft. Because of its higher unit strength, this bridge strand can be used with smaller diameters than multiple-strand bridge rope, and its use for suspension-bridge cables usually results in less costly structures.

Bridge strand of this type can be prestretched in any length that can be made or shipped and can be furnished with the heavier bethanized zinc coatings.

[1] This Appendix through courtesy Bethlehem Steel Company.

Diam, inches	Construction	Weight per foot Approx, pound	Metallic Area, square inches	Min Breaking Strength, tons
$\frac{1}{2}$	1 × 19	0.52	0.1499	15
$\frac{9}{16}$	1 × 19	0.63	0.1816	19
$\frac{5}{8}$	1 × 19	0.80	0.2302	24
$\frac{11}{16}$	1 × 19	0.95	0.2733	29
$\frac{3}{4}$	1 × 19	1.14	0.3283	34
$\frac{13}{16}$	1 × 19	1.39	0.3978	40
$\frac{7}{8}$	1 × 19	1.54	0.4428	46
$\frac{15}{16}$	1 × 31	1.92	0.5496	54
1	1 × 31	2.14	0.6138	61
$1\frac{1}{16}$	1 × 31	2.40	0.6883	69
$1\frac{1}{8}$	1 × 37	2.70	0.7760	78
$1\frac{3}{16}$	1 × 37	3.04	0.8733	86
$1\frac{1}{4}$	1 × 43	3.45	0.9898	96
$1\frac{5}{16}$	1 × 43	3.76	1.0785	106
$1\frac{3}{8}$	1 × 51	3.98	1.1422	116
$1\frac{7}{16}$	1 × 51	4.33	1.2435	126
$1\frac{1}{2}$	1 × 59	4.73	1.3571	138
$1\frac{9}{16}$	1 × 59	5.17	1.4840	150

Diam, inches	Construction	Weight per foot Approx, pound	Metallic Area, square inches	Min Breaking Strength, tons
$1\frac{5}{8}$	1 × 67	5.58	1.6006	162
$1\frac{11}{16}$	1 × 67	6.08	1.7447	176
$1\frac{3}{4}$	1 × 77	6.29	1.8039	188
$1\frac{13}{16}$	1 × 77	6.85	1.9652	202
$1\frac{7}{8}$	1 × 87	7.37	2.1134	216
$1\frac{15}{16}$	1 × 87	7.88	2.2612	230
2	1 × 97	8.40	2.4118	245
$2\frac{1}{16}$	1 × 97	8.98	2.5775	261
$2\frac{1}{8}$	1 × 109	9.32	2.6751	277
$2\frac{3}{16}$	1 × 109	9.94	2.8535	293
$2\frac{1}{4}$	1 × 121	10.45	2.9979	310
$2\frac{5}{16}$	1 × 121	11.15	3.1985	327
$2\frac{3}{8}$	1 × 133	11.69	3.3548	344
$2\frac{7}{16}$	1 × 133	12.47	3.5791	360
$2\frac{1}{2}$	1 × 147	12.78	3.6672	376
$2\frac{9}{16}$	1 × 147	13.63	3.9112	392
$2\frac{5}{8}$	1 × 147	14.47	4.1340	417
$2\frac{11}{16}$	1 × 161	15.16	4.3340	432
$2\frac{3}{4}$	1 × 161	15.88	4.5380	452

Minimum moduli of elasticity of the above strands, when prestretched, are as follows:

$\frac{1}{2}$-in. to $2\frac{1}{16}$-in. diam, 24,000,000 psi; $2\frac{1}{8}$-in. to $2\frac{9}{16}$-in. diam, 23,500,000 psi; $2\frac{5}{8}$-in. to $2\frac{3}{4}$-in. diam, 23,000,000 psi.

INDEX